Numerical Modeling of Seismic Responses from Fractured Reservoirs by the Grid-characteristic Method

Vladimir Leviant

Igor Kvasov

Igor Petrov

Geophysical Developments Series No. 17

Raul Cova, managing editor
Naum Y. Marmalevsky, scientific editor

SOCIETY OF EXPLORATION
GEOPHYSICISTS

ISBN 978-0-931830-41-9 (Series)
ISBN 978-1-56080-366-9 (Volume)

Library of Congress Control Number: 2019945420

Society of Exploration Geophysicists
8801 S. Yale Ave., Ste. 500
Tulsa, OK 74137-3575

Published 2019
Printed in the United States of America

Contents

About the authors

Vladimir Leviant graduated from Moscow Petroleum Institute in 1954 as an exploration geophysicist. From 1954 to 1960, he worked as a seismic explorationist at the Volgograd Geophysical Co. and from 1960 to 1970 as chief geophysicist of the Experimentation–Methodology Expedition where he developed a methodology for mapping deep-seated horizons in the Lower Volga region. He began a PhD study at the Moscow Geological Prospecting Institute in 1961 and received his PhD in 1967. From 1970 to 2009, Leviant was an employee of Central Geophysical Expedition (Moscow) holding the positions of chief geophysicist/geologist (1973–1996) and focusing on improving the efficiency and depth of seismic data processing. He was the main author of a number of industry-specific guidelines (2002, 2006, and 2010). In 1996–1997 Leviant was an international expert in India at the invitation of ONGC. Since 2010, Leviant has been a geophysical consultant at the Computer Science Department of Moscow Institute of Physics and Technology (MIPT) where he and the Department's team simulate seismic responses from fractured reservoirs by using models with parameters highly close to the actual ones and applying the grid-characteristic method (GCM). The goal is to identify the properties of waves from fractured reservoirs for the latter's direct detection. Leviant is the winner of the USSR State Prize in 1987 and was awarded the Vernadsky Medal from the Russian Academy of Natural Sciences.

Igor Kvasov graduated from Moscow Institute of Physics and Technology (MIPT), Russia, in 2008 (physics and applied mathematics). Having successfully completed his thesis on numerical simulation of wave processes in heterogeneous solid elastic bodies, Kvasov received his PhD in physics and computer science from MIPT in 2011. During 2006–2015, he served as an associate professor at the MIPT Computer Science and Computational Mathematics Department. Currently, he works as an engineering manager at Google in California. As part of his research in the numerical simulation of wave processes, Kvasov has made significant improvements in and broadened the abilities of the grid-characteristic method (GCM), including the ability to handle hundreds of thousands of fractures, which corresponds to the conditions of real fractured reservoirs, and the ability to model such 3D fractured prospects using unstructured grids. His research on the properties of seismic responses from fractured reservoirs helps solve the problem of their direct detection.

Igor Petrov graduated from Moscow Institute of Physics and Technology (MIPT) in 1976 and was an assistant professor at the MIPT Computational Mathematics Department. He received a PhD in 1983 and an ScD in physics and mathematics in 1991, and was promoted to a professor position at the same department. From 2002 to 2015, Petrov was the head of the MIPT Computer Science Department, and since 2015 has been heading the MIPT Computer Science and Computational Mathematics Department. Petrov's scientific activities include mathematical modeling, numerical methods, computing on supercomputers, continuum mechanics problems, and geophysics. He is one of the main developers of the grid-characteristic method, numerical simulation/modeling, as well as the author of GCM adaptation to seismic exploration needs. Based on his solutions to a number of solid body deformation problems, he has investigated topical problems of various industry branches. In 2011, Petrov was elected an associate member of the Russian Academy of Sciences in the Division of Mathematical Sciences. Petrov was awarded a Medal of the Order for Services to the Fatherland, Class II, for his scientific accomplishments.

Foreword

Fracturing, a fundamental and significant characteristic of rocks, has been studied extensively and its investigation has provided plentiful information about the subsurface and properties of hydrocarbon reservoirs. Increasingly in recent years, dense fractured rocks have become the target of hydrocarbon prospecting and exploration. The task of mapping and parameterizing fractured porous reservoirs is a significant challenge for traditional methods of seismic exploration, for which solutions need to be found. This challenge has driven the development and application of new methods for investigating the physical properties and dimensions of real fractures, particularly crack openings measuring only millimeters or microns. To create the databases necessary to facilitate reservoir exploration and development, solutions to forward problems for fractured media are required. This can be achieved only through mathematical modeling and numerical simulation. The authors have developed mechanical-mathematical models for analyzing fractured media that will be of particular interest to scientists researching continuous medium mechanics and computing methods.

The proposed method for solving problems of seismic exploration for such models is based on a system of equations in continuous medium mechanics (continuum mechanics) and numerical solution of partial differential equations, and is able to account fully for the properties of fractured media incorporated in the models. This monograph appears to be the first in the Russian geoscience literature (perhaps not only Russian) to apply a numerical solution to the forward problems of seismic exploration concerned with the detection and analysis of fracturing.

Creation of this mathematical modeling/simulation method for 3D wavefields in complex media should not be limited only to detecting fracture zones; it can map the structure and physical properties of geologic medium in its various forms. With the increased capabilities of numerical modeling and simulation provided by supercomputers, we expect that the method will become more widely applied to many other difficult areas of seismic exploration and development, and seismology in general.

This monograph explores a new direction of geophysical investigations, the study of fractured media and their representative numerical models, through the creative interaction of mathematicians and geophysicists. The goal is to resolve the core problem of exploration geophysics – the discovery of oil and gas reservoirs.

New formal methods of computational mathematics and data interpretation always are needed in the geosciences, but human intelligence, experience, and creativity come first. This monograph is addressed to those in the community of geoscientists and mathematicians who have inquisitive minds and are not constrained by traditional views.

Prof. A.V. Nikolaev,
Associate Member, Russian Academy of Sciences

Authors' note

This monograph appears to be the first sizeable work to address the properties of seismic signatures of fractured media by describing and analyzing the various seismic wave responses from both the discrete fractured media and the surrounding rocks. It also proposes a methodology for the direct detection of fractured reservoirs, based on these analyses.

Because a complete analytical description of fractured media is unavailable at this time, and effective models generally simplify the problem, we believe the numerical simulation/modeling of wavefield propagation is the best possible tool for resolving the problem. This approach could be implemented only with the advent and application of the grid-characteristic method (GCM) and the infinitely thin fracture (ITF) model, which have facilitated discrete specification of any number of fractures and the use of realistic fracture parameters.

To investigate the many kinds of fractured media involved in the major types of relevant seismic-geologic sections, the strategy for employing the numerical simulation/modeling approach must be based on the principle, "from the particular to the general." Realizing this strategy, we have relied upon successively accumulating examples of wave responses (surface-recorded seismograms) for a set of models with either common physical conditions or a common objective. Analyzing information this way contributes to the identification of the common properties of the wave responses for these conditions, including those previously unknown. Some of the properties revealed are of a very general and fundamental nature (e.g., the absence of polarity reversal of the compressional and converted diffractions from fluid-saturated, near-vertical fractures on the horizontal component of seismic reflection data).

We are convinced that our work will be the start of a trend for the direct detection and parameterization of fractured geologic features through the investigation of scattered waves.

We have used the ideas advanced by Kozlov (2006) regarding the four-level gradation of fractured features: micro-, meso-, macro-, and megafractures. Certainly, understanding of the structure and character of real fracturing has broadened and changed during the past decade; these new findings should be incorporated into the numerical models of fractured media, and the changes revealed should be assessed for their impact on previous conclusions. All of this may suggest directions for further investigations.

Many other aspects of the problem have been left untouched or only partly investigated to date. Therefore, our work may be regarded as only the beginning of a series of possible new investigations focused on one of the most topical problems in present-day seismic exploration for oil and gas – the detection and study of fractured reservoirs.

The principal chapters devoted to seismic responses from macrofracture systems and megafractures (or fracture corridors) have been largely rewritten with due regard for the capability of modern seismic long CMP observation/acquisition systems. Newly discovered properties of scattered compressional and converted wave responses underlie the proposed methods of detecting these fractured features.

Acknowledgments

The authors express their sincere gratitude to the scientific editor of this work, Naum Y. Marmalevskyi, chief geophysicist, Tesseral Technologies Inc. and TetraSeis Inc., for consultations, useful discussions, and recommendations on all sections of the monograph, as well as to Vladimir Grechka, senior technical consultant, Marathon Oil Company, for substantial support for preparing the monograph for publication under the auspices of SEG and for final editing of the work.

The authors are grateful to Prof. Alexey Stovas, Norwegian University of Science and Technology (NTNU), for valuable advice and support of the idea of producing an English version of our monograph. The authors also are grateful to Alexandre G. Averbukh, ScD (Tech.), for useful comments and advice on individual issues of the investigation. Special note should be made of the highly professional technical translation by Yuriy Tyapkin, ScD (Tech.). The authors deeply appreciate the thorough copyediting of the book, including the English language editing by Susan Stamm, SEG books manager.

Assistance and support by the EAGE Russia & CIS Office (Mikhail Novikov, director, and Irina Orlova, publications coordinator) is gratefully acknowledged for publishing the Russian version and their efforts to make the English language version a reality.

The authors thank Maxim Muratov, Vlad Miryakha, and Vasiliy Golubev, contributors to individual sections of research in this book, as well as Anton Orlov for graphics preparation according to SEG standards.

Special thanks go to Zora I. Zharn for invaluable assistance in preparing the work for publication in Russia, and to Alexey P. Bolbasov for active involvement in preparing the English language version of the monograph.

This work in Russia was conducted with the financial support of the Russian Foundation for Basic Research (RFBR), project no. 19-11-00023.

Introduction

As the domain of exploration for oil and gas expands, tight rocks such as carbonates and deep-seated brittle sandstones are targeted increasingly. Generally, productive beds of tight rocks are pervaded by systems of near-vertical, fluid-saturated, uniformly oriented fractures of different sizes. These largely determine the filtration and capacity characteristics of reservoirs and provide the basis for building filtration models for hydrocarbon fields as well as validation of field development regimes. Defining the positions, concentrations, and other characteristics of fractures, therefore, is critically important.

Seismic characterization of fracture systems in hydrocarbon reservoirs.—Fracturing in productive sedimentary rocks has been the subject of numerous studies in the last two decades. Numerical analysis of wave processes that occur in fractured-cavernous media has become one of principal tools in this research.

Typically, media that contain such fractures have been considered in the framework of effective media theories (e.g., Hudson, 1991; Coates and Schoenberg, 1995; Grechka and Kachanov, 2006; and Kozlov, 2006). Based on this approach, many investigations have been undertaken by F. Hilterman, L. Thomsen, A. Rüger, I. Tsvankin, E. Kozlov, V. Grechka, and others. Thomsen's effective parameters resulting from these studies are used extensively to describe fractured media. In particular, Thomsen's parameters have been used to develop the processing and interpretation methods of wide-azimuth seismic data, leading to assessments of anisotropic parameters of seismic reflections and subsequently to the determination of characteristics of fractures field (Helbig and Thomsen, 2005).

It has been asserted that determining the anisotropy of kinematic and dynamic characteristics of reflections through fractured rocks can help assess the fracture-induced anisotropy of the media and the characteristics of fracturing. This approach uses indirect estimates of the fracture parameters, derived from the properties of waves generated by other subsurface features (e.g., deeper reflectors). Based on this approach, technology has been developed for determining azimuthal anisotropy of the medium, which has been applied widely in the last decade. Then the obtained data has been used to determine the parameters of fracturing and fracture systems.

Although wide-azimuth 3D surveys have been used extensively in detailed seismic exploration, the adequacy of the inferred subsurface fractures to the real image in productive intervals has not always been confirmed by subsequent assessment of the fractured reservoir productivity. In most cases, the details of the imaged fractures also have been poor, because the reflection characteristics to assess azimuthal anisotropy were obtained using multi-km receiver arrays.

In recent years, a new approach to detecting and characterizing fractured reservoir zones has been under development, based on the use of the scattered component (diffractions) of the wavefield (Shlenkin et al., 2000; Kozlov et al., 2004; Kuznetsov et al., 2004; Pozdniakov, 2005; and Kremlev et al., 2008). The techniques used by these authors have enabled the construction of maps showing the energy fields of scattered waves (SWEF) and the plotting of SWEF anomalies on seismic sections.

However, the use of anomalies as an indicator of fractured reservoir zones has faced a problem of "sorting" such anomalies, because amplitude anomalies in the energy field may be related to other factors. The inherent ambiguity of geologic interpretation of these anomalies is the main weakness of this line of research. It appears that the measurement of amplitude and the use only of vertical component data of the scattered waves are not sufficient for the purpose.

So, we may conclude that the problem of reliably detecting and mapping fracture systems in a hydrocarbon reservoir from seismic data still remains unresolved.

Substantiation of the objectives and goals of the study.—The preceding discussion has underlined the importance and relevance of finding a solution to the problem of detecting fracture zones and defining their parameters.

Using the direct seismic responses of fractures and fracture zones in productive formations to obtain information about their locations and characteristics is an essentially new and promising direction of investigation.

The database required to develop a method for mapping fracture systems in productive formations needs to include detailed characteristics of wave responses from single fractures and the major types of fracture systems. This requires the solution of forward problems of seismic modeling for the types of fractures typical for hydrocarbon reservoirs in tight rocks.

Therefore, the purpose of our work has been to (1) investigate the feasibility of using direct responses from fractures and fracture accumulations for mapping fracture development zones and (2) generate the database needed to underpin the development of methods for their detection and estimate their key parameters (orientation, concentration, etc.). There are practically no analytical solutions that would enable the respective wavefields to be computed, while physical modeling is not feasible for a study on such a scale. Only mathematical modeling is capable of addressing the problem. To achieve this goal, we need to determine the character of wave responses from the different types of fractures and sets of fractures; that is, we need to solve the forward problem of seismic modeling for fractured media by applying numerical simulations of wavefields.

Substantiation of the choice of the grid-characteristic method as the mathematical modeling tool.—We considered that it was important to perform the task by applying a method based as much as possible on fundamental physics and the laws of mechanics for continuous media, using a minimum of empirical coefficients and dependencies.

We also set the following requirements for the modeling process:

- the ability to specify discretely each fracture in a fractured medium;
- the ability to specify explicitly the boundary conditions on the surface of each fracture;
- the use of the most realistic linear size-to-thickness (opening) ratio for the fractures;
- the use of the most complete information of the wave response, particularly by recording not only the vertical *Z*-component of data but also the horizontal *X*- and *Y*-components of data; and
- high computation efficiency.

In the last decade, a numerical modeling method based on a finite-difference algorithm proposed by Coates and Schoenberg (1995) has been used extensively to address these problems. In their method, the discrete specification of fractures in the models is coupled with a limitation on fracture thickness, as determined by the mesh cell size. A method of local downscaling of the grid, proposed by Lisitsa et al. (2013), allows a minimum mesh cell side size of 0.1 m, which is still two orders of magnitude larger than the fracture opening on the order of 1 mm. This method is unrealistic for studying wave responses from beds with thousands of discretely positioned fractures with a 1 mm opening.

The grid-characteristic method for solving direct problems of seismic exploration is developed and discussed by Petrov and Kholodov (1984), Petrov et al. (1989), Petrov et al. (1990), and Ivanov et al. (1999). This method allows the full characteristic properties of dynamic systems of hyperbolic type to be taken into consideration; only by using these hyperbolic systems can we create the most accurate algorithms to define wave propagation across the boundaries of the integration region, as well as on the model interfaces and on the fracture surfaces.

Subsequent to these earlier studies, the method has been much improved, allowing it to be applied effectively to several seismic exploration tasks.

- Provision was made for monotonous numerical solution to prevent spurious oscillations in schemes in which the order of magnitude of approximation exceeds one. Methods were developed for achieving the fifth order of accuracy and hierarchical mesh downscaling.
- It became possible to use a non-structured, triangular, differential grid, which enabled the inclusion of inhomogeneities such as voids, cavities, and planar fractures. Explicit setting of boundary conditions on the interfaces and fracture surfaces allowed us to:

    a) relax restrictions on the position and orientation of inhomogeneities;
    b) account for heterogeneities with practically zero volume but finite surface area (fractures); and
    c) enhance the computational accuracy for micro-heterogeneities (vugs) of dimensions comparable with one grid voxel.

- In recent years, the grid-characteristic method was extended to 3D numerical modeling, thanks to advances in computing technology.

Part One

Application of the Grid-characteristic Method to Seismic Modeling

Problem Statement

Basic requirements for the numerical modeling/simulation method to ensure its effective use

The development of methods to detect and study fractured reservoirs is one of most topical issues in contemporary hydrocarbon exploration. Even so, the lack of analytical solutions for computing wavefields – the wave responses from fractured media – has hindered the development of methods that would honor the specific characteristics of these responses. This deficiency has prompted research to seek solutions in terms of effective models of fractured media. The results of this approach, however, have not always been able to satisfy the needs of the oil and gas industry.

Under these circumstances, the solution to the problem, as we see it, lies in numerical modeling and simulation of seismic responses from fractured media, and in particular from fractured reservoirs with realistic properties. The geophysical literature appears to lack sufficient information regarding the numerical modeling/simulation of waves generated by fractured beds containing tens of thousands of discrete fractures, whose fracture linear size-to-opening ratios lie between three and five orders of magnitude. These are the conditions for real geological fractured reservoirs, but for which currently there is no adequate numerical modeling method.

In our opinion, a successful method for resolving the problem should satisfy the following requirements:

- be based as much as possible on fundamental physics and the laws of mechanics of continuous media, and with only a minimum involvement of empirical coefficients and dependencies;
- use the most effective algorithms of numerical methods that account for the specific characteristics of wave processes in fractured media;
- enable the development of a mathematically accurate model for fractured media, in which we can:
  a) specify each fracture in the fractured medium without substantially limiting the number of fractures,
  b) explicitly specify the boundary conditions (including those relating to the type of fracture infill) on the surface of each fracture,
  c) apply as closely as possible the real ratio of a fracture linear size-to-thickness (opening) which typically is three to five orders of magnitude (1000 to 100,000 times), with real-fracture thicknesses ranging from one millimeter to a few decimals of a millimeter, and
  d) ensure highly accurate computations and avoid spurious oscillations;
- enable a high computing speed that allows the numerical modeling of wavefields from real sections of fractured reservoirs to be completed within a reasonable time.

Applicability of existing numerical methods

We have considered and compared the adequacy of the primary approaches available with the above criteria in terms of both computational efficiency and modeling methods.

Finite difference method

The class of finite difference methods is quite broad. It includes the simplest classical schemes (Lax, Lax-Wendroff, and Courant-Isaacson-Rees schemes) and more complex high-order schemes.

On a structured grid with a relatively uniform distribution of cell sizes, these methods are relatively easy to implement and flexible enough for complex integration domains to be applied. Moreover, they provide acceptable accuracy and computational efficiency.

In seismic studies of fractured media, finite-difference-grid algorithms have been introduced by Coates and Schoenberg (1995). In the last decade, the finite-difference method has been utilized extensively in studies related to the fracturing problems (for example, Wu et al., 2005; Fang et al., 2010; Kostin et al., 2011; and Lisitsa et al., 2013).

Typically, the fracture models in the finite-difference schemes are specified as a set of grid cells inside a fracture. This requires substantial grid-downscaling, which incurs enormous computations and makes numerical modeling of fractured media containing multitudes of fractures impractical. Considerable efforts have been applied to devising approaches to overcome this limitation, particularly the use of local space-time downscaling of the grid (Lisitsa et al., 2012). To date, however, these efforts have failed to resolve the problem.

Another method for specifying fractures, the linear-slip model (LSM), requires the boundary and contact conditions of fractures to be defined. Even so, in the method of Coates and Schoenberg (1995), all fracture cells have parameters of an averaged equivalent medium with anisotropic properties.

In terms of modeling wave responses from fractured reservoirs, limitations of the finite-difference method are as follows:

- Although schemes designed to cancel spurious oscillations are numerous, these methods lead, in the case of large numbers of fractures, to inaccurate solutions because they do not honor the characteristic properties of the initial system of equations (wave propagation along the characteristic surfaces), and also because of accumulated spurious oscillations;
- With the large number of fractures, modeling them using grid cells inside the fracture of only a few fractions of a millimeter thick by some meters high and long requires an extraordinary level of grid downscaling. This entails unacceptably large computational expenses, even when the entire process is optimized.

Finite element method

The finite element method incorporates an entire class of possible algorithms. The method has been described by Newmark (1959), Hughes (1987), Bathe (1996), and Zienkiewcz and Taylor (2000).

The method is highly generalized, is extensively utilized in various areas of physics, and allows a wide class of problems to be modeled. The primary merit of the method is its fast computation speed, so it is attractive for solving seismic modeling problems, particularly problems of seismic wave propagation in models containing fractures (Frehner et al., 2008; and Frehner and Schmalholz, 2010).

If inhomogeneities (cavities and fractures) are very numerous, the method may, however, yield less accurate results. This may be due to accumulated errors because of failure to account for the characteristic properties of the initial system of equations (wave propagation along characteristic surfaces), and also because of accumulated spurious oscillations. This drawback imposes significant limitations on high-precision modeling of wave propagation in media containing tens of thousands of fractures.

The mathematical description of wave processes on the fractures requires that additional conditions be introduced, such as linear slip, which is the basic hypothesis in Thomsen (1986), Hsu and Schoenberg (1993), and Hudson et al. (2001).

The fractures are modeled by specifying a set of grid cells, so that the grids must be downscaled considerably (considering how thin the fractures actually may be). Although numerical modeling of wave responses from single fractures or from tens of fractures is possible, the modeling/simulation of wavefields from reservoirs with tens of thousands of fractures is an unrealistic objective.

The discontinuous Galerkin method

The discontinuous Galerkin method has a number of desirable features (Hesthaven and Warburton, 2008), the most important of which is its higher-order space-time convergence due to its explicit consideration of wave-like processes. The discontinuous Galerkin method enables the original system of equations to be solved with an arbitrary order of approximation in space, which in practice is limited by floating-point arithmetic precision only.

Fractures can be specified either in a finite opening fracture model (with a series of grid cells inside the fracture) or an infinitely thin fracture (ITF) model. The high accuracy of the method has been verified by comparing its modeling results against those produced for an identical model where fractures were modeled explicitly, including grid construction inside the fracture (Wu et al., 2005). This comparison has shown a good correspondence between the two approaches.

The Galerkin method is applicable for handling large volumes of inhomogeneities (fractures) and ensures high computation accuracy. Unfortunately, the method requires substantially greater computing resources for the same models than the other methods discussed.

Spectral methods and pseudo-spectral methods

Spectral methods enable us to consider the system of equations of elastic deformation in the Fourier frequency domain, where the partial derivatives are expressed in algebraic form. Because of that (as opposed to numerical discretization of partial derivatives by other numerical methods), spectral methods can provide very effective and accurate solutions (Vershinin and Sabitov, 2011).

At the same time, however, the spectral method has significant limitations associated with the geometry of the integration domain. Specifically, the use of the method for complex areas with dipping and curved boundaries of layers is extremely difficult. In this case, the use of pseudo-spectral methods is justified, in which the direct and inverse Fourier transforms are applied only to approximate partial derivatives in the frequency domain (Kosloff and Baysal, 1982; Druskin and Knizhnerman, 1988; Priolo et al., 1994; and Seriani and Priolo, 1994). Despite these proposed improvements, the applicability of the method to our problem is still limited because the explicit identification of a large number of inhomogeneities appears to be impossible in practice.

So, we have to conclude that none of the above methods of numerical modeling/simulation meets all of the requirements for solving the stated problem. The same conclusion can be drawn from an overview of the existing numerical methods used to investigate fracturing (Hall and Wang, 2012).

Features of the grid-characteristic method which demonstrate its effectiveness

The features of the grid-characteristic method relevant to the problem are as follows:

1) With the grid-characteristic method, the computational algorithm is able to account for the characteristic properties of the dynamic system of partial differential equations of the hyperbolic type; the dynamic system of linear elasticity is one of these. The method explicitly uses a property of the hyperbolic system to transport the solution along the characteristics (Magomedov and Kholodov, 2017). This enables an improved account of the physics of waves, in particular, wave motion and the propagation of elastic disturbances along characteristic surfaces. Because this approach incorporates a full account of the characteristic properties of nonstationary equations of the elasticity theory (Magomedov and Kholodov, 2017), unlike numerical methods that do not account for these properties, the most appropriate computational algorithms can be devised to account for the boundaries of integration domains, on interfaces between media, contact boundaries, and fractures.

   For this approach, the numerical solutions are typically monotonous, which excludes spurious oscillations and ensures high-order computational accuracy. This, in turn, ensures greater computational accuracy for the seismic wave response when the number of fractures inside the integration domain is large, as is typical for fractured reservoirs. The efficiency of this method is comparable to that of the finite-difference grid method.
2) We adapt the computation grid to the boundaries of inhomogeneities and explicitly specify the contact conditions on the inhomogeneities (fractures). This will allow us to
   a) lift the constraints on the positions and orientations of the inhomogeneities,
   b) take into consideration inhomogeneities with practically zero volume but finite surface area (i.e., fractures), and
   c) enhance the computational accuracy for inhomogeneities with sizes comparable to a single grid cell.
3) Specifying a fractured medium in the model as a multitude of infinitely thin fractures (ITF) in the integration domain along with explicitly specifying boundary conditions on each fracture surface enable us to compute the wavefields, the responses from each fracture, with a minimum of computational resources. Previously, we have proven the validity of using ITF models for fluid-saturated mesofractures and macrofractures, the principal types of fractures in oil and gas reservoirs, in a dedicated study (Leviant et al., 2015).

Chapter 1

Numerical Solution of Equations in Linear Elasticity Theory

The numerical solution of linear equations in the theory of elasticity is of great value in addressing questions of solid mechanics, despite the fact that real materials satisfy these equations only when external loads are relatively small and short. The numerical study of complex rheological models (elastoplastic, elastoviscoplastic, and other media) by considering the different physical processes can be reduced to some modification of the elastic solution (Kukudzhanov, 2004).

In this chapter we will discuss schemes for applying the grid-characteristic method to solving elasticity equations in the case of an arbitrary choice of the basis in space, which depends on the local shape of a difference grid. To represent these schemes, we propose an explicit form that has computational advantages over the standard formulation (Petrov and Kholodov, 1984). Particular attention is paid to the solution on the surface of a domain of integration with different types of boundary conditions, as well as the case of contact between two elastic media.

1.1. Mathematical model

Let us formulate the equations that, according to Eringen (1974), govern the state of an infinitesimal volume element of a continuous linear elastic medium. The first equation is the local equation of motion

$$\rho\dot{\vec{v}} = \boldsymbol{\nabla}\cdot\boldsymbol{\sigma}^{\mathrm{T}}, \tag{1.1}$$

where ρ is material density, $\vec{v}$ is particle velocity, $\boldsymbol{\sigma}$ is the Cauchy stress tensor, the dot over $\vec{v}$ indicates partial derivative with respect to time, T is tensor transpose operator, the dot between the gradient and stress tensor is their product, and the gradient is denoted by

$$\boldsymbol{\nabla} = \begin{Vmatrix} \nabla_1 \\ \nabla_2 \\ \nabla_3 \end{Vmatrix} = \begin{Vmatrix} \dfrac{\partial}{\partial \hat{x}_1} \\ \dfrac{\partial}{\partial \hat{x}_2} \\ \dfrac{\partial}{\partial \hat{x}_3} \end{Vmatrix}. \tag{1.2}$$

By virtue of the law of pairing of shear stresses (Kondaurov and Fortov, 2002), $\boldsymbol{\sigma}$ is a symmetric tensor, and we can omit its transpose in equation 1.1. The components of the dot product of the gradient vector with this tensor are given by the equality

$$\left(\boldsymbol{\nabla}\cdot\boldsymbol{\sigma}^{\mathrm{T}}\right)^k = \sum_j \nabla_j \left(\boldsymbol{\sigma}^{\mathrm{T}}\right)^{kj} = \sum_j \nabla_j \sigma^{jk}, \tag{1.3}$$

where ∇_j is the scalar differential operator. In the following discussion, it is assumed that the indices $(i, j, k, l, \ldots)$ progress through the values 1, 2, and 3.

Let us introduce the symmetric strain tensor,

$$\mathbf{e} = \frac{1}{2}\left(\boldsymbol{\nabla}\otimes\vec{u} + (\boldsymbol{\nabla}\otimes\vec{u})^{\mathrm{T}}\right), \tag{1.4}$$

where $\vec{u}$ is the displacement field ($\vec{x} = \vec{X} + \vec{u}$), $\vec{X}$ is the original (unstressed) position vector of a given point, and $\vec{x}$ is the position vector of the same point in deformed state, and $\otimes$ denotes the tensor product: $(\vec{a}\otimes\vec{b})^{ij} = a^i b^j$. Furthermore, $\left(\nabla\otimes\vec{u} + (\nabla\otimes\vec{u})^{\mathrm{T}}\right)^{ij} = \nabla_i u^j + \nabla_j u^i$. In the following, the displacement field is no longer considered, and the symbol $\vec{u}$ is used in a different sense.

Linear elasticity of a material implies the following strain-stress relationship (Hooke's law) in isotropic media,

$$\boldsymbol{\sigma} = \lambda(\mathbf{e} \div \mathbf{I})\mathbf{I} + 2\mu\mathbf{e}, \tag{1.5}$$

where λ and μ are the Lamé parameters, which determine the elastic properties of the material, $\mathbf{I}$ is the unit tensor, and $\div$ represents the double dot product

$$\mathbf{a} \div \mathbf{b} = \sum_{i,j} a^{ij} b^{ij}. \tag{1.6}$$

To obtain a closed system of equations in terms of stresses and particle velocities, we can differentiate the strain-stress relationship only with respect to time:

$$\begin{aligned}\sigma^{ij} &= \lambda\left(\sum_{k,l} e^{kl}\delta^{kl}\right)\delta^{ij} + 2\mu e^{ij} = \lambda\delta^{ij}\sum_{k} e^{kk} + 2\mu e^{ij} \\ &= \lambda\delta^{ij}\left(\sum_{k}\frac{1}{2}\left(\nabla_k u^k + \nabla_k u^k\right)\right) + 2\mu\frac{1}{2}\left(\nabla_i u^j + \nabla_j u^i\right) \\ &= \lambda\delta^{ij}\sum_{k}\nabla_k u^k + 2\mu\frac{1}{2}\left(\nabla_i u^j + \nabla_j u^i\right)\end{aligned} \tag{1.7}$$

where δ^{ij} is Kronecker delta.

Using the change of variables $\dot{\vec{u}} = \vec{v}$, we obtain

$$\begin{aligned}\dot{\sigma}^{ij} &= \lambda\delta^{ij}\left(\sum_{k}\nabla_k \dot{u}^k\right) + 2\mu\frac{1}{2}\left(\nabla_i \dot{u}^j + \nabla_j \dot{u}^i\right) \\ &= \lambda\delta^{ij}\left(\sum_{k}\nabla_k v^k\right) + 2\mu\frac{1}{2}\left(\nabla_i v^j + \nabla_j v^i\right) \\ &= \lambda(\nabla\cdot\vec{v})\delta^{ij} + \mu\left(\nabla\otimes\vec{v} + (\nabla\otimes\vec{v})^{\mathrm{T}}\right)^{ij}.\end{aligned} \tag{1.8}$$

Thus, the final system of equations is

$$\begin{cases}\rho\dot{\vec{v}} = \nabla\cdot\boldsymbol{\sigma} \\ \dot{\boldsymbol{\sigma}} = \lambda(\nabla\cdot\vec{v})\mathbf{I} + \mu\left(\nabla\otimes\vec{v} + (\nabla\otimes\vec{v})^{\mathrm{T}}\right).\end{cases} \tag{1.9}$$

This system of equations governs the dynamic of a solid and will be used throughout this text. The first equation is Newton's second law, and the second equation is Hooke's law. Next, this system of equations will be solved numerically.

1.2. Selection of a coordinate system

In the numerical integration of differential equations is that, in a software implementation of this process, it is convenient to store the state of a medium (in this case the components of velocity and stress) in a fixed Cartesian coordinate system $(\hat{x}_1, \hat{x}_2, \hat{x}_3)$. At the same time, derivatives are approximately computed in directions $\hat{\xi}_1, \hat{\xi}_2$, and $\hat{\xi}_3$, the selection of which depends on the local configuration of grid nodes.

Let us derive an expression for the components of the gradient ∇ in the new coordinate system. By definition, the derivative in the direction $\hat{\xi}_j$ is

$$\frac{\partial\hat{x}_i}{\partial\hat{\xi}_j} = \left(\hat{\xi}_j\cdot\nabla\right)\hat{x}_i. \tag{1.10}$$

Now introduce a^k either as the projection of the vector $\vec{a}$ onto the axis x_k or as its kth component in the original Cartesian coordinate system, and let σ^{kl} represent the components of the second-rank tensor $\boldsymbol{\sigma}$ in the original Cartesian coordinate system.

Let us represent $\frac{\partial\hat{x}_i}{\partial\hat{\xi}_j} = \left(\hat{\xi}_j\cdot\nabla\right)\hat{x}_i$ as the sum

$$\begin{aligned}\frac{\partial\hat{x}_i}{\partial\hat{\xi}_j} &= \left(\hat{\xi}_j\cdot\nabla\right)\hat{x}_i = \sum_k\left(\hat{\xi}_j\right)^k\frac{\partial\hat{x}_i}{\partial\hat{x}_k} \\ &= \sum_k\left(\hat{\xi}_j\right)^k\delta^{ik} = \left(\hat{\xi}_j\right)^i,\end{aligned} \tag{1.11}$$

which can be written in matrix notation as

$$\begin{aligned}\frac{\partial\left(\hat{x}_1,\hat{x}_2,\hat{x}_3\right)}{\partial\left(\hat{\xi}_1,\hat{\xi}_2,\hat{\xi}_3\right)} &= \begin{bmatrix}\left(\hat{\xi}_1\right)^1 & \left(\hat{\xi}_2\right)^1 & \left(\hat{\xi}_3\right)^1 \\ \left(\hat{\xi}_1\right)^2 & \left(\hat{\xi}_2\right)^2 & \left(\hat{\xi}_3\right)^2 \\ \left(\hat{\xi}_1\right)^3 & \left(\hat{\xi}_2\right)^3 & \left(\hat{\xi}_3\right)^3\end{bmatrix} \\ &= \left[\hat{\xi}_1 \quad \hat{\xi}_2 \quad \hat{\xi}_3\right],\end{aligned} \tag{1.12}$$

where

$$\frac{\partial\left(a^1,a^2,a^3\right)}{\partial\left(b^1,b^2,b^3\right)} = \begin{bmatrix}\frac{\partial a^1}{\partial b^1} & \frac{\partial a^1}{\partial b^2} & \frac{\partial a^1}{\partial b^3} \\ \frac{\partial a^2}{\partial b^1} & \frac{\partial a^2}{\partial b^2} & \frac{\partial a^2}{\partial b^3} \\ \frac{\partial a^3}{\partial b^1} & \frac{\partial a^3}{\partial b^2} & \frac{\partial a^3}{\partial b^3}\end{bmatrix}. \tag{1.13}$$

Next, we introduce the notation

$$\begin{bmatrix} \vec{q}_1^{\,\mathrm{T}} \\ \vec{q}_2^{\,\mathrm{T}} \\ \vec{q}_3^{\,\mathrm{T}} \end{bmatrix} = \begin{bmatrix} (q_1)^1 & (q_1)^2 & (q_1)^3 \\ (q_2)^1 & (q_2)^2 & (q_2)^3 \\ (q_3)^1 & (q_3)^2 & (q_3)^3 \end{bmatrix} = \frac{\partial(\hat{\xi}_1,\hat{\xi}_2,\hat{\xi}_3)}{\partial(\hat{x}_1,\hat{x}_2,\hat{x}_3)} = \left(\frac{\partial(\hat{x}_1,\hat{x}_2,\hat{x}_3)}{\partial(\hat{\xi}_1,\hat{\xi}_2,\hat{\xi}_3)}\right)^{-1} \quad (1.14)$$

where $\vec{q}_i \cdot \hat{\xi}_j = \delta^{ij}$.

Thus, $(\vec{q}_1,\vec{q}_2,\vec{q}_3)$ is the basis biorthogonal relative to $(\hat{\xi}_1,\hat{\xi}_2,\hat{\xi}_3)$

$$\vec{q}_1 = \frac{\hat{\xi}_2 \times \hat{\xi}_3}{(\hat{\xi}_1,\hat{\xi}_2,\hat{\xi}_3)}, \quad \vec{q}_2 = \frac{\hat{\xi}_3 \times \hat{\xi}_1}{(\hat{\xi}_1,\hat{\xi}_2,\hat{\xi}_3)}, \quad \text{and}$$

$$\vec{q}_3 = \frac{\hat{\xi}_1 \times \hat{\xi}_2}{(\hat{\xi}_1,\hat{\xi}_2,\hat{\xi}_3)}. \quad (1.15)$$

where the numerators are cross (vector) products, and the denominators are mixed (scalar triple) products.

To simplify the subsequent formulae significantly, let us associate each of the vectors $\vec{q}_j$ with an orthonormal basis $(\hat{n}_{j,0},\hat{n}_{j,1},\hat{n}_{j,2})$, in which $\hat{n}_{j,0}$ is collinear to $\vec{q}_j$, i.e.,

$$\hat{n}_{j,0} = \frac{\vec{q}_j}{\|\vec{q}_j\|} \quad \text{and} \quad l_j = \|\vec{q}_j\|, \quad (1.16)$$

whereas the other two unit vectors, $\hat{n}_{j,1}$ and $\hat{n}_{j,2}$, are chosen arbitrarily. In the following discussion, we omit the first index j and, in $\hat{n}_{j,0}$, the second index 0, if their absence is obvious from the context.

Equations (1.9) are written in a form independent of the choice of a particular coordinate system. It is only necessary to specify the operators which should be substituted for the gradient components. In the Cartesian coordinate system,

$$\nabla = \{\nabla_1,\nabla_2,\nabla_3\} = \frac{\partial}{\partial \hat{x}}, \quad (1.17)$$

where $\nabla_i = \dfrac{\partial}{\partial \hat{x}_i}$.

Otherwise, we apply the change of variables formula:

$$\nabla_i = \sum_j \frac{\partial \hat{\xi}_j}{\partial \hat{x}_i}\frac{\partial}{\partial \hat{\xi}_j} \quad (1.18)$$

and $\nabla = \{\nabla_1,\nabla_2,\nabla_3\}$.

Given this formula, equations (1.9) can be rewritten in any coordinate system.

1.3. Notation generalization for differential equations

To analyze a system of equations in computational mathematics and construct different schemes, it is customary to represent the system in the following canonical form:

$$\dot{\vec{u}} + \sum_j \left(\mathbf{A}_j \cdot \frac{\partial \vec{u}}{\partial \hat{\xi}_j}\right) = 0, \quad (1.19)$$

where the vector $\vec{u}$ comprises all of the variables of the system. In our case of equations (1.9), the vector is

$$\vec{u} = \{\vec{v},\boldsymbol{\sigma}\}^{\mathrm{T}} = \{v^1,v^2,v^3,\sigma^{11},\sigma^{22},\sigma^{33},\sigma^{23},\sigma^{13},\sigma^{12}\}^{\mathrm{T}}. \quad (1.20)$$

To obtain the desired expression for the matrices $\mathbf{A}_j$, we represent the gradient operator from equation (1.18) as the sum of three operators, each of which contains only the derivatives in one direction, $\hat{\xi}_j$:

$$\nabla_i = \sum_j \frac{\partial \hat{\xi}_j}{\partial \hat{x}_i}\frac{\partial}{\partial \hat{\xi}_j} = \sum_j l_j(\hat{n}_j)^i \frac{\partial}{\partial \hat{\xi}_j}. \quad (1.21)$$

The initial system,

$$\begin{cases} \dot{\vec{v}} = \dfrac{1}{\rho}\nabla\cdot\boldsymbol{\sigma} \\ \dot{\boldsymbol{\sigma}} = \lambda(\nabla\cdot\vec{v})\mathbf{I} + \mu\left(\nabla\otimes\vec{v} + (\nabla\otimes\vec{v})^{\mathrm{T}}\right), \end{cases} \quad (1.22)$$

can be written coordinate-wise as

$$\begin{cases} \dot{v}^m = \dfrac{1}{\rho}\displaystyle\sum_i \nabla_i\sigma^{im} \\ \dot{\sigma}^{mn} = \lambda\left(\displaystyle\sum_i \nabla_i v^i\right)\delta^{mn} + \mu\left(\nabla_m v^n + \nabla_n v^m\right). \end{cases} \quad (1.23)$$

Substituting equation (1.21) into this system yields

$$\begin{cases} \dot{v}^m = \dfrac{1}{\rho}\displaystyle\sum_i\sum_j l_j(\hat{n}_j)^i \frac{\partial \sigma^{im}}{\partial \hat{\xi}_j} \\ \dot{\sigma}^{mn} = \lambda\left(\displaystyle\sum_i\sum_j l_j(\hat{n}_j)^i \frac{\partial v^i}{\partial \hat{\xi}_j}\right)\delta^{mn} \\ \quad + \mu\left(\displaystyle\sum_j l_j(\hat{n}_j)^m \frac{\partial v^n}{\partial \hat{\xi}_j} + \sum_j l_j(\hat{n}_j)^n \frac{\partial v^m}{\partial \hat{\xi}_j}\right) \end{cases} \quad (1.24)$$

Changing the order of summation produces

$$\begin{cases} \dot{v}^m = \sum_j l_j \frac{1}{\rho}\left(\frac{\partial \boldsymbol{\sigma}}{\partial \hat{\xi}_j} \cdot \hat{n}_j\right)^m \\ \dot{\sigma}^{mn} = \sum_j l_j \left(\lambda\left(\hat{n}_j \cdot \frac{\partial \vec{v}}{\partial \hat{\xi}_j}\right)\delta^{mn} \right. \\ \left. \qquad + \mu\left((\hat{n}_j)^m \frac{\partial v^n}{\partial \hat{\xi}_j} + (\hat{n}_j)^n \frac{\partial v^m}{\partial \hat{\xi}_j} \right)\right) \end{cases} . \tag{1.25}$$

Now, we can return to the matrix notation,

$$\begin{cases} \dot{\vec{v}} = \sum_j l_j \frac{1}{\rho}\left(\frac{\partial \boldsymbol{\sigma}}{\partial \hat{\xi}_j} \cdot \hat{n}_j\right) \\ \dot{\boldsymbol{\sigma}} = \sum_j l_j \left(\lambda\left(\hat{n}_j \cdot \frac{\partial \vec{v}}{\partial \hat{\xi}_j}\right)\mathbf{I} \right. \\ \left. \qquad + \mu\left(\hat{n}_j \otimes \frac{\partial \vec{v}}{\partial \hat{\xi}_j} + \frac{\partial \vec{v}}{\partial \hat{\xi}_j} \otimes \hat{n}_j \right)\right) \end{cases} , \tag{1.26}$$

and finally obtain

$$\mathbf{A}_j \begin{bmatrix} \vec{v} \\ \boldsymbol{\sigma} \end{bmatrix} = -l_j \begin{bmatrix} \frac{1}{\rho}(\boldsymbol{\sigma} \cdot \hat{n}_j) \\ \lambda(\hat{n}_j \cdot \vec{v})\mathbf{I} + \mu(\hat{n}_j \otimes \vec{v} + \vec{v} \otimes \hat{n}_j) \end{bmatrix}. \tag{1.27}$$

We are motivated to develop explicit and the simplest difference schemes that utilize computer resources as efficiently as possible and optimize the computer code in terms of computation speed. To this end, we must strive to increase both the code accuracy, by reducing the round-off error, and the code stability, by avoiding situations that might interrupt the calculation, such as inverting a singular matrix.

Difference schemes contain the matrix $\mathbf{A}_j$ itself as well as various functions of it. To be able to calculate explicitly the differences, avoiding such procedures as matrix inversion and approximate calculation of eigenpairs, the following analysis of the properties of $\mathbf{A}_j$ is conducted.

1.4. Spectral study of the system

The purpose of this section is to determine all of the eigenvalues c and eigenvectors $\vec{u} = \{\vec{v}, \boldsymbol{\sigma}\}^{\mathrm{T}}$ of the matrices $\mathbf{A}_j$:

$$\mathbf{A}\vec{u} = c\vec{u} \tag{1.28}$$

(note that the subscript j is omitted until the end of this section). The eigenvalues must have a dimension of velocity so that this expression is consistent in terms of dimension.

If we assume that matrix $\mathbf{A}$ has a complete set of eigenvalues $\{c_j\}$ and eigenvectors $\{\vec{u}_j\}$, which will be shown next, we can introduce the diagonal matrix that comprises all of the eigenvalues of matrix $\mathbf{A}$,

$$\boldsymbol{\Lambda} = \mathrm{diag}\{c_j\} = \begin{bmatrix} c_1 & 0 & \cdots & 0 \\ 0 & c_2 & \cdots & 0 \\ \vdots & \vdots & \ddots & \vdots \\ 0 & 0 & \cdots & c_9 \end{bmatrix}, \tag{1.29}$$

and the matrix

$$\boldsymbol{\Omega}^{-1} = \{\vec{u}_i\} = \begin{bmatrix} (\vec{u}_1)_1 & (\vec{u}_2)_1 & \cdots & (\vec{u}_9)_1 \\ (\vec{u}_1)_2 & (\vec{u}_2)_2 & \cdots & (\vec{u}_9)_2 \\ \vdots & \vdots & \ddots & \vdots \\ (\vec{u}_1)_9 & (\vec{u}_2)_9 & \cdots & (\vec{u}_9)_9 \end{bmatrix}, \tag{1.30}$$

in which the columns are the eigenvectors of $\mathbf{A}$ and appear in the same order as the corresponding eigenvalues in $\boldsymbol{\Lambda}$. Here superscript "–1" denotes the matrix inversion operator; we use the inverted matrix in equation (1.30) for convenience which will become clear later. Because the complete set of eigenvectors form a linearly independent basis, we may define the nonsingular matrix formed by them as the inverse of some other nonsingular matrix. The condition that each column of $\boldsymbol{\Omega}^{-1}$ is an eigenvector of $\mathbf{A}$ can be written as

$$\mathbf{A}\boldsymbol{\Omega}^{-1} = \boldsymbol{\Omega}^{-1}\boldsymbol{\Lambda}. \tag{1.31}$$

Right-multiplication of this equation by $\boldsymbol{\Omega}$ yields:

$$\mathbf{A}\boldsymbol{\Omega}^{-1}\boldsymbol{\Omega} = \boldsymbol{\Omega}^{-1}\boldsymbol{\Lambda}\boldsymbol{\Omega} \quad \text{and} \quad \mathbf{A} = \boldsymbol{\Omega}^{-1}\boldsymbol{\Lambda}\boldsymbol{\Omega}. \tag{1.32}$$

Left-multiplication of this equation with $\boldsymbol{\Omega}$ yields:

$$\boldsymbol{\Omega}\mathbf{A} = \boldsymbol{\Omega}\boldsymbol{\Omega}^{-1}\boldsymbol{\Lambda}\boldsymbol{\Omega} \quad \text{and} \quad \boldsymbol{\Omega}\mathbf{A} = \boldsymbol{\Lambda}\boldsymbol{\Omega}. \tag{1.33}$$

It is clear from these equations that each row of $\boldsymbol{\Omega}$ is an eigenrow of $\mathbf{A}$, with $\vec{q}^{\mathrm{T}}$ being the eigenrow of $\mathbf{A}$ associated with the eigenvalue Q when the equality $\vec{q}^{\mathrm{T}}\mathbf{A} = \vec{q}^{\mathrm{T}}Q$ holds true. Being transposed, the eigenrows turn into the eigenvectors of $\mathbf{A}^{\mathrm{T}}$. These eigenvectors can be found as the solution of the adjoint problem with respect to equation (1.28):

$$\mathbf{A}^{\mathrm{T}}\vec{u} = c\vec{u}. \tag{1.34}$$

It follows from this mathematics that the adjoint problem has the same set of eigenvalues. Indeed, if

$\mathbf{A}\vec{u} = c\vec{u}$, then $\vec{u}$ is a row of $\mathbf{\Omega}$, and therefore $\mathbf{A}^{\mathrm{T}}\vec{u} = c\vec{u}$, and the eigenvectors can be determined to be coordinated:

$$\mathbf{\Omega}^{-1}\mathbf{\Omega} = \mathbf{I}. \tag{1.35}$$

It also follows that different elements of $\mathbf{\Omega}$ have dimensions of the reciprocal of velocity and the reciprocal of stress, which after left-multiplication by $\mathbf{\Omega}^{-1}$ yields dimensionless tensor $\mathbf{I}$.

1.4.1. Forward problem

Next, determining eigenvalues and eigenvectors of matrix $\mathbf{A}$ will be addressed. Returning from the generalized presentation of equation (1.28) via substituting equation (1.27) for the variables $\vec{v}$ and $\boldsymbol{\sigma}$, we obtain the system of constitutive equations:

$$-l\begin{bmatrix} \frac{1}{\rho}(\boldsymbol{\sigma}\cdot\hat{n}) \\ \lambda(\hat{n}\cdot\vec{v})\mathbf{I} + \mu(\hat{n}\otimes\vec{v} + \vec{v}\otimes\hat{n}) \end{bmatrix} = \mathbf{A}\begin{bmatrix}\vec{v}\\ \boldsymbol{\sigma}\end{bmatrix} = c\begin{bmatrix}\vec{v}\\ \boldsymbol{\sigma}\end{bmatrix}, \tag{1.36}$$

or

$$\begin{cases} -l\frac{1}{\rho}(\boldsymbol{\sigma}\cdot\hat{n}) = c\vec{v} \\ -l\big(\lambda(\hat{n}\cdot\vec{v})\mathbf{I} + \mu(\hat{n}\otimes\vec{v} + \vec{v}\otimes\hat{n})\big) = c\boldsymbol{\sigma} \end{cases}. \tag{1.37}$$

As a result, we have

$$\begin{cases} (\boldsymbol{\sigma}\cdot\hat{n}) = -l^{-1}c\rho\vec{v} \\ \lambda(\hat{n}\cdot\vec{v})\mathbf{I} + \mu(\hat{n}\otimes\vec{v} + \vec{v}\otimes\hat{n}) = -l^{-1}c\boldsymbol{\sigma} \end{cases}. \tag{1.38}$$

Let us prove the identity

$$(\hat{n}\otimes\vec{v} + \vec{v}\otimes\hat{n})\cdot\hat{n} = (\hat{n}\cdot\vec{v})\hat{n} + \vec{v}. \tag{1.39}$$

Indeed,

$$\begin{aligned} \big((\hat{n}\otimes\vec{v} + \vec{v}\otimes\hat{n})\cdot\hat{n}\big)^k &= \sum_i n^i(\hat{n}\otimes\vec{v} + \vec{v}\otimes\hat{n})^{ik} \\ &= \sum_i n^i\left(n^i v^k + v^i n^k\right) \\ &= n^k\sum_i n^i v^i + v^k\sum_i n^i n^i \\ &= n^k\sum_i n^i v^i + v^k\|n\|^2 \\ &= n^k\sum_i n^i v^i + v^k = \big((\hat{n}\cdot\vec{v})\hat{n} + \vec{v}\big)^k \end{aligned} \tag{1.40}$$

Substituting one equation into the other and recognizing the identity equation (1.39) yields

$$\boldsymbol{\sigma} = -lc^{-1}\big(\lambda(\hat{n}\cdot\vec{v})\mathbf{I} + \mu(\hat{n}\otimes\vec{v} + \vec{v}\otimes\hat{n})\big), \tag{1.41}$$

$$-l^{-1}c\rho\vec{v} = \boldsymbol{\sigma}\cdot\hat{n} = \Big(-lc^{-1}\big(\lambda(\hat{n}\cdot\vec{v})\mathbf{I} + \mu(\hat{n}\otimes\vec{v} + \vec{v}\otimes\hat{n})\big)\Big)\cdot\hat{n}, \tag{1.42}$$

and

$$\begin{aligned} l^{-2}c^2\rho\vec{v} &= \big(\lambda(\hat{n}\cdot\vec{v})\mathbf{I} + \mu(\hat{n}\otimes\vec{v} + \vec{v}\otimes\hat{n})\big)\cdot\hat{n} \\ &= \lambda(\hat{n}\cdot\vec{v})\hat{n} + \mu\cdot(\hat{n}\otimes\vec{v} + \vec{v}\otimes\hat{n})\cdot\hat{n} \\ &= \lambda(\hat{n}\cdot\vec{v})\hat{n} + \mu(\hat{n}\cdot\vec{v})\hat{n} + \mu\vec{v}, \end{aligned} \tag{1.43}$$

and we arrive at

$$(\lambda + \mu)(\hat{n}\cdot\vec{v})\hat{n} = \left(l^{-2}c^2\rho - \mu\right)\vec{v}. \tag{1.44}$$

This equation has the following three qualitatively different groups of solutions.

1) If vectors $\vec{v}$ and $\hat{n}$ are collinear and $\vec{v} \neq 0$, then

$$(\hat{n}\cdot\vec{v}) = \|\hat{n}\|\|\vec{v}\| = \|\vec{v}\| = v \text{ and } \vec{v} = v\hat{n}. \tag{1.45}$$

Substituting equation (1.45) into equation (1.44)

$$c^2 = l^2\frac{\lambda + 2\mu}{\rho}, \tag{1.46}$$

which follows from

$$(\lambda + \mu)v\hat{n} = \left(l^{-2}c^2\rho - \mu\right)v\hat{n} \tag{1.47}$$

and

$$\lambda + \mu = l^{-2}c^2\rho - \mu, \tag{1.48}$$

and introducing

$$c_1 = \sqrt{\frac{\lambda + 2\mu}{\rho}}, \tag{1.49}$$

which represents the velocity of pressure waves in a purely isotropic elastic material, we obtain the eigenpair $c = \pm lc_1$ and $\vec{v} = \alpha\hat{n}$. Considering equation

(1.41), then

$$\begin{aligned}\sigma^{ik} &= \left(-lc^{-1}\left(\lambda(\hat{n}\cdot\vec{v})\mathbf{I} + \mu(\hat{n}\otimes\vec{v} + \vec{v}\otimes\hat{n})\right)\right)^{ik} \\ &= -lc^{-1}\left(\lambda\alpha\delta^{ik} + \mu\alpha\left(n^i n^k + n^k n^i\right)\right) \\ &= -lc^{-1}\left(\lambda\alpha\delta^{ik} + 2\mu\alpha n^i n^k\right) \\ &= -lc^{-1}\alpha\left(\lambda\delta^{ik} + 2\mu(\hat{n}\otimes\hat{n})^{ik}\right)\end{aligned} \tag{1.50}$$

and

$$\boldsymbol{\sigma} = \mp\frac{\alpha}{c_1}\left(\lambda\mathbf{I} + 2\mu\hat{n}\otimes\hat{n}\right), \tag{1.51}$$

where α is an arbitrary nonzero multiplier with a dimension of velocity.

2) If vectors $\vec{v}$ and $\hat{n}$ are orthogonal and $\vec{v} \neq 0$, then the left side of equation (1.44) vanishes:

$$0 = \left(l^{-2}c^2\rho - \mu\right)\vec{v}. \tag{1.52}$$

This is valid only when the right side vanishes too, i.e.,

$$l^{-2}c^2\rho - \mu = 0, \tag{1.53}$$

where $c^2 = \dfrac{\mu l^2}{\rho}$.

After introducing the notation

$$c_2 = \sqrt{\frac{\mu}{\rho}}, \tag{1.54}$$

which represents the velocity of shear waves in isotropic elastic media, we can describe the set of eigenvalues and eigenvectors:

$$c = \pm lc_2, \quad \hat{n}\cdot\vec{v} = 0, \tag{1.55}$$

and

$$\begin{aligned}\boldsymbol{\sigma} &= -lc^{-1}\left(\lambda(\hat{n}\cdot\vec{v})\mathbf{I} + \mu(\hat{n}\otimes\vec{v} + \vec{v}\otimes\hat{n})\right) \\ &= -lc^{-1}\mu(\hat{n}\otimes\vec{v} + \vec{v}\otimes\hat{n}) \\ &= \mp\frac{\mu}{c_2}(\hat{n}\otimes\vec{v} + \vec{v}\otimes\hat{n}).\end{aligned} \tag{1.56}$$

Then, there are two linearly independent solutions to the equation,

$$\hat{n}\cdot\vec{v} = \sum_i n^i v^i = 0, \tag{1.57}$$

which we prove as follows.

Inasmuch as $||\hat{n}|| = 1$, there exists a nonzero component n^i, e.g., n^1. Then

$$v^1 + \frac{n^2}{n^1}v^2 + \frac{n^3}{n^1}v^3 = 0 \tag{1.58}$$

and

$$\vec{v} = \left\{-\frac{n^2}{n^1}\alpha - \frac{n^3}{n^1}\beta, \alpha, \beta\right\}, \tag{1.59}$$

where $\alpha, \beta \in \mathbb{R}$.

Therefore, there are only a total of four linearly independent eigenvectors $\{\vec{v}, \boldsymbol{\sigma}\}$ in this set and only two eigenvalues of the opposite sign.

3) If the eigenvalue is zero and the components of the eigenvector related to velocity are zero, we obtain, recalling equation (1.39), the following equation for the remaining components of the eigenvector:

$$c = 0, \quad \vec{v} = 0, \quad \text{and} \quad \boldsymbol{\sigma}\cdot\hat{n} = -l^{-1}c\rho\vec{v} = 0. \tag{1.60}$$

Then, there are three linearly independent solutions to the system

$$(\boldsymbol{\sigma}\cdot\hat{n})^k = \sum_i n^i\sigma^{ik} = 0, \quad k = 1, 2, \text{and } 3. \tag{1.61}$$

Indeed, because $||\hat{n}|| = 1$, there exists a nonzero component n^i, e.g., n^1. Then for $\forall k = 1,2,3$,

$$\sigma^{1k} + \frac{n^2}{n^1}\sigma^{2k} + \frac{n^3}{n^1}\sigma^{3k} = 0 \tag{1.62}$$

is valid, and

$$\left\{\sigma^{1k}, \sigma^{2k}, \sigma^{3k}\right\} = \left\{-\frac{n^2}{n^1}\alpha_k - \frac{n^3}{n^1}\beta_k, \alpha_k, \beta_k\right\}, \tag{1.63}$$

where $\alpha, \beta \in \mathbb{R}$.

Because tensor $\boldsymbol{\sigma}$ is symmetric, i.e., $\sigma^{ij} = \sigma^{ji}$, which yields three additional equations, three linearly independent solutions remain and, therefore, three linearly independent eigenvectors are associated with zero eigenvalue.

In items 1, 2, and 3, we produce two, four, and three eigenvectors, respectively. Thus, a total of nine linearly independent eigenvectors along with their eigenvalues are determined.

1.4.2. Adjoint problem

In this section we will focus on determining eigenvalues and eigenvectors of matrix $\mathbf{A}^{\mathrm{T}}$. The transpose of matrix $\mathbf{A}$ from equation (1.27) is derived from the equality

$$\vec{u}_1^{\mathrm{T}}\mathbf{A}^{\mathrm{T}}\vec{u} = \vec{u}^{\mathrm{T}}\mathbf{A}\vec{u}_1, \tag{1.64}$$

where $\vec{u}$ and $\vec{u}_1$ are arbitrary vectors.

Expanding the right side of equation (1.64) and rearranging its factors yields

$$\begin{aligned}
\begin{bmatrix}\vec{v}\\ \boldsymbol{\sigma}\end{bmatrix}^{\mathrm{T}} \mathbf{A} \begin{bmatrix}\vec{v}_1\\ \boldsymbol{\sigma}_1\end{bmatrix} &= \begin{bmatrix}\vec{v}\\ \boldsymbol{\sigma}\end{bmatrix}^{\mathrm{T}}\left(-l\begin{bmatrix}\frac{1}{\rho}(\boldsymbol{\sigma}_1\cdot\hat{n})\\ \lambda(\hat{n}\cdot\vec{v}_1)\mathbf{I}+\mu(\hat{n}\otimes\vec{v}_1+\vec{v}_1\otimes\hat{n})\end{bmatrix}\right)\\
&= -l\begin{bmatrix}\vec{v}\\ \boldsymbol{\sigma}\end{bmatrix}^{\mathrm{T}}\cdot\begin{bmatrix}\frac{1}{\rho}(\boldsymbol{\sigma}_1\cdot\hat{n})\\ \lambda(\hat{n}\cdot\vec{v}_1)\mathbf{I}+\mu(\hat{n}\otimes\vec{v}_1+\vec{v}_1\otimes\hat{n})\end{bmatrix}\\
&= -l\left(\frac{1}{\rho}\vec{v}\cdot(\boldsymbol{\sigma}_1\cdot\hat{n})+\lambda(\hat{n}\cdot\vec{v}_1)(\boldsymbol{\sigma}\div\mathbf{I})\right.\\
&\qquad\left.+\,\mu\big(\boldsymbol{\sigma}\div(\hat{n}\otimes\vec{v}_1+\vec{v}_1\otimes\hat{n})\big)\right)\\
&= -l\left(\frac{1}{\rho}\sum_i v^i\sum_k(\boldsymbol{\sigma}_1)^{ik}n^k+\lambda(\hat{n}\cdot\vec{v}_1)(\boldsymbol{\sigma}\div\mathbf{I})\right.\\
&\qquad\left.+\,\mu\sum_{i,k}\sigma^{ik}\left(n^i(\vec{v}_1)^k+(\vec{v}_1)^i n^k\right)\right)\\
&= -l\left(\lambda(\hat{n}\cdot\vec{v}_1)(\boldsymbol{\sigma}\div\mathbf{I})+\frac{1}{\rho}\sum_{i,k}v^i(\boldsymbol{\sigma}_1)^{ik}n^k\right.\\
&\qquad\left.+\,\mu\sum_{i,k}\sigma^{ik}n^i(\vec{v}_1)^k+\mu\sum_{i,k}\sigma^{ik}(\vec{v}_1)^i n^k\right)\\
&= -l\left(\lambda(\hat{n}\cdot\vec{v}_1)(\boldsymbol{\sigma}\div\mathbf{I})+\frac{1}{\rho}\sum_{i,k}v^i(\boldsymbol{\sigma}_1)^{ik}n^k\right.\\
&\qquad\left.+\,2\mu\sum_{i,k}\sigma^{ik}(\vec{v}_1)^i n^k\right),
\end{aligned}\tag{1.65}$$

where the components of tensors and vectors are denoted by superscripts.

On the other hand,

$$\begin{aligned}
&\begin{bmatrix}\vec{v}_1\\ \boldsymbol{\sigma}_1\end{bmatrix}^{\mathrm{T}}\left(-l\begin{bmatrix}\lambda(\boldsymbol{\sigma}\div\mathbf{I})\hat{n}+2\mu(\boldsymbol{\sigma}\cdot\hat{n})\\ \frac{1}{2\rho}(\vec{v}\otimes\hat{n}+\hat{n}\otimes\vec{v})\end{bmatrix}\right)\\
&= -l\begin{bmatrix}\vec{v}_1\\ \boldsymbol{\sigma}_1\end{bmatrix}^{\mathrm{T}}\begin{bmatrix}\lambda(\boldsymbol{\sigma}\div\mathbf{I})\hat{n}+2\mu(\boldsymbol{\sigma}\cdot\hat{n})\\ \frac{1}{2\rho}(\vec{v}\otimes\hat{n}+\hat{n}\otimes\vec{v})\end{bmatrix}\\
&= -l\left(\lambda(\boldsymbol{\sigma}\div\mathbf{I})(\vec{v}_1\cdot\hat{n})+2\mu\big(\vec{v}_1\cdot(\boldsymbol{\sigma}\cdot\hat{n})\big)\right.\\
&\qquad\left.+\,\frac{1}{2\rho}\big(\boldsymbol{\sigma}_1\div(\vec{v}\otimes\hat{n}+\hat{n}\otimes\vec{v})\big)\right)\\
&= -l\left(\lambda(\boldsymbol{\sigma}\div\mathbf{I})(\vec{v}_1\cdot\hat{n})+2\mu\sum_i(\vec{v}_1)^i\sum_k\sigma^{ik}n^k\right.\\
&\qquad\left.+\,\frac{1}{2\rho}\sum_{i,k}(\boldsymbol{\sigma}_1)^{ik}\left(v^i n^k+n^i v^k\right)\right)\\
&= -l\left(\lambda(\boldsymbol{\sigma}\div\mathbf{I})(\vec{v}_1\cdot\hat{n})+2\mu\sum_{i,k}\sigma^{ik}(\vec{v}_1)^i n^k\right.\\
&\qquad\left.+\,\frac{1}{2\rho}\sum_{i,k}(\boldsymbol{\sigma}_1)^{ik}v^i n^k+\frac{1}{2\rho}\sum_{i,k}(\boldsymbol{\sigma}_1)^{ik}n^i v^k\right)\\
&= -l\left(\lambda(\boldsymbol{\sigma}\div\mathbf{I})(\vec{v}_1\cdot\hat{n})+2\mu\sum_{i,k}\boldsymbol{\sigma}^{ik}(\vec{v}_1)^i n^k\right.\\
&\qquad\left.+\,\frac{1}{2\rho}\sum_{i,k}(\boldsymbol{\sigma}_1)^{ik}v^i n^k+\frac{1}{2\rho}\sum_{i,k}(\boldsymbol{\sigma}_1)^{ik}n^k v^i\right)\\
&= -l\left(\lambda(\boldsymbol{\sigma}\div\mathbf{I})(\vec{v}_1\cdot\hat{n})+2\mu\sum_{i,k}\sigma^{ik}(\vec{v}_1)^i n^k\right.\\
&\qquad\left.+\,\frac{1}{\rho}\sum_{i,k}(\boldsymbol{\sigma}_1)^{ik}v^i n^k\right).
\end{aligned}\tag{1.66}$$

Consequently,

$$\mathbf{A}^{\mathrm{T}}\begin{bmatrix}\vec{v}\\ \boldsymbol{\sigma}\end{bmatrix} = -l\begin{bmatrix}\lambda(\boldsymbol{\sigma}\div\mathbf{I})\hat{n}+2\mu(\boldsymbol{\sigma}\cdot\hat{n})\\ \frac{1}{2\rho}(\vec{v}\otimes\hat{n}+\hat{n}\otimes\vec{v})\end{bmatrix}. \tag{1.67}$$

As mentioned following equation (1.34), $\vec{v}$ and $\boldsymbol{\sigma}$ have a dimension of the reciprocal of velocity and the reciprocal of stress, respectively.

Reasoning similarly, we arrive at three groups of eigenvalues and eigenvectors,

$$c\begin{bmatrix}\vec{v}\\ \boldsymbol{\sigma}\end{bmatrix} = \mathbf{A}^{\mathrm{T}}\begin{bmatrix}\vec{v}\\ \boldsymbol{\sigma}\end{bmatrix} \tag{1.68}$$

and, substituting the expression determined for $\mathbf{A}^{\mathrm{T}}$,

$$c\begin{bmatrix}\vec{v}\\ \boldsymbol{\sigma}\end{bmatrix} = -l\begin{bmatrix}\lambda(\boldsymbol{\sigma} \div \mathbf{I})\hat{n} + 2\mu(\boldsymbol{\sigma}\cdot\hat{n})\\ \dfrac{1}{2\rho}(\vec{v}\otimes\hat{n} + \hat{n}\otimes\vec{v})\end{bmatrix} \tag{1.69}$$

and

$$\begin{cases} c\vec{v} = -l\big(\lambda(\boldsymbol{\sigma} \div \mathbf{I})\hat{n} + 2\mu(\boldsymbol{\sigma}\cdot\hat{n})\big)\\ c\boldsymbol{\sigma} = -l\dfrac{1}{2\rho}\ (\vec{v}\otimes\hat{n} + \hat{n}\otimes\vec{v})\end{cases}. \tag{1.70}$$

Let us substitute the second equation into the first. Then,

$$\boldsymbol{\sigma} = -\frac{l}{2\rho c}(\vec{v}\otimes\hat{n} + \hat{n}\otimes\vec{v}), \tag{1.71}$$

$$\begin{aligned} c\vec{v} &= -l\left(\lambda(\boldsymbol{\sigma}\div\mathbf{I})\hat{n} + 2\mu c^{-1}\left(-l\frac{1}{2\rho}(\vec{v}\otimes\hat{n} + \hat{n}\otimes\vec{v})\cdot\hat{n}\right)\right)\\ &= -l\left(\lambda\left(\left(-\frac{l}{2\rho c}(\vec{v}\otimes\hat{n} + \hat{n}\otimes\vec{v})\right)\div\mathbf{I}\right)\hat{n}\right.\\ &\qquad \left.+ 2\mu c^{-1}\left(-l\frac{1}{2\rho}(\vec{v}\otimes\hat{n} + \hat{n}\otimes\vec{v})\cdot\hat{n}\right)\right)\\ &= \left(\frac{l^2\lambda}{2\rho c}\sum_{i,k}(v^i n^k + v^k n^i)\delta^{ik}\right)\hat{n} + \frac{l^2\mu}{\rho c}((\hat{n}\cdot\vec{v})\hat{n} + \vec{v})\\ &= \left(\frac{l^2\lambda}{\rho c}\sum_i v^i n^i\right)\hat{n} + \frac{l^2\mu}{\rho c}((\hat{n}\cdot\vec{v})\hat{n} + \vec{v})\\ &= \frac{l^2(\lambda+\mu)}{\rho c}(\hat{n}\cdot\vec{v})\hat{n} + \frac{l^2\mu}{\rho c}\vec{v}, \end{aligned} \tag{1.72}$$

and

$$\frac{\rho c^2}{l^2}\vec{v} = (\lambda+\mu)(\hat{n}\cdot\vec{v})\hat{n} + \mu\vec{v}. \tag{1.73}$$

As a result, we have

$$(\rho c^2 l^{-2} - \mu)\vec{v} = (\lambda+\mu)(\hat{n}\cdot\vec{v})\hat{n}. \tag{1.74}$$

This equation also has three groups of qualitatively different solutions.

1) If vectors $\vec{v}$ and $\hat{n}$ are collinear and $\vec{v} \neq 0$, then $(\hat{n}\cdot\vec{v}) = \|\hat{n}\|\cdot\|\vec{v}\| = \|\vec{v}\| = v$ and $\vec{v} = v\cdot\hat{n}$. Substituting into equation (1.74)

$$c^2 = l^2\frac{\lambda+2\mu}{\rho}, \tag{1.75}$$

which follows from

$$(\lambda+\mu)v\hat{n} = (l^{-2}c^2\rho - \mu)v\hat{n} \tag{1.76}$$

and

$$\lambda+\mu = l^{-2}c^2\rho - \mu, \tag{1.77}$$

and recalling equation (1.49), we produce the eigenpair $c = \pm lc_1$ and $\vec{v} = \alpha\hat{n}$. Considering equation (1.71), we also obtain

$$\begin{aligned}\sigma^{ik} &= -\frac{l}{2\rho c}(v^i n^k + n^i v^k) = -\frac{l\alpha}{2\rho c}(n^i n^k + n^i n^k)\\ &= \mp\frac{\alpha}{\rho c_1}(\hat{n}\otimes\hat{n})^{ik}\end{aligned} \tag{1.78}$$

and

$$\boldsymbol{\sigma} = \mp\frac{\alpha}{\rho c_1}(\hat{n}\otimes\hat{n}), \tag{1.79}$$

where α is an arbitrary nonzero factor with a dimension of velocity.

2) If vectors $\vec{v}$ and $\hat{n}$ are orthogonal and $\vec{v} \neq 0$, then the left side of equation (1.74) vanishes,

$$0 = (l^{-2}c^2\rho - \mu)\vec{v}. \tag{1.80}$$

This is valid only when the right side vanishes too,

$$l^{-2}c^2\rho - \mu = 0, \tag{1.81}$$

where $c^2 = \dfrac{\mu l^2}{\rho}$.

Recalling equation (1.54), we can describe the set of eigenvalues and eigenvectors

$$c = \pm lc_2, \quad \hat{n}\cdot\vec{v} = 0, \tag{1.82}$$

and

$$\boldsymbol{\sigma} = -\frac{l}{2\rho c}(\vec{v} \otimes \hat{n} + \hat{n} \otimes \vec{v})$$
$$= \mp\frac{1}{2\rho c_2}(\vec{v} \otimes \hat{n} + \hat{n} \otimes \vec{v}). \quad (1.83)$$

There are two linearly independent solutions to the equation

$$\hat{n} \cdot \vec{v} = \sum_i n^i v^i = 0. \quad (1.84)$$

Indeed, because $\|\hat{n}\| = 1$, there exists a nonzero n^i, e.g., n^1. Then,

$$v^1 + \frac{n^2}{n^1}v^2 + \frac{n^3}{n^1}v^3 = 0 \quad (1.85)$$

and

$$\vec{v} = \left\{-\frac{n^2}{n^1}\alpha - \frac{n^3}{n^1}\beta, \alpha, \beta\right\}, \quad (1.86)$$

where $\alpha, \beta \in \mathbb{R}$.

Therefore, a total of four linearly independent eigenvectors $\{\vec{v}, \boldsymbol{\sigma}\}$ are in this set and only two eigenvalues of the opposite sign.

3) If the eigenvalue is zero and the components of the eigenvector related to velocity are zero, we obtain, recalling equation (1.70), the following equation for the remaining components of the eigenvector,

$$c = 0, \quad \vec{v} = 0, \quad \text{and}$$
$$0 = c\vec{v} = -l\big(\lambda(\boldsymbol{\sigma} \div \mathbf{I})\hat{n} + 2\mu(\boldsymbol{\sigma} \cdot \hat{n})\big), \quad (1.87)$$

i.e., for $\boldsymbol{\sigma}$,

$$0 = \lambda(\boldsymbol{\sigma} \div \mathbf{I})\hat{n} + 2\mu(\boldsymbol{\sigma} \cdot \hat{n}), \quad (1.88)$$

$$\lambda\Big(\sum_i \sigma^{ii}\Big)n^k + 2\mu(\boldsymbol{\sigma} \cdot \hat{n})^k = 0, \quad (1.89)$$

and

$$\lambda n^k \sum_i \sigma^{ii} + 2\mu \sum_i \sigma^{ik} n^i = 0. \quad (1.90)$$

There are three linearly independent solutions to the system

$$\lambda n^k \sum_i \sigma^{ii} + 2\mu \sum_i \sigma^{ik} n^i = 0, \quad k = 1, 2, \text{and } 3. \quad (1.91)$$

To prove this, let us represent the system as

$$\begin{cases} \lambda n^1\left(\sigma^{11} + \sigma^{22} + \sigma^{33}\right) + 2\mu\left(n^1\sigma^{11} + n^2\sigma^{12} + n^3\sigma^{13}\right) = 0 \\ \lambda n^2\left(\sigma^{11} + \sigma^{22} + \sigma^{33}\right) + 2\mu\left(n^1\sigma^{12} + n^2\sigma^{22} + n^3\sigma^{23}\right) = 0 \\ \lambda n^3\left(\sigma^{11} + \sigma^{22} + \sigma^{33}\right) + 2\mu\left(n^1\sigma^{13} + n^2\sigma^{23} + n^3\sigma^{33}\right) = 0. \end{cases} \quad (1.92)$$

A) If $n^2 = n^3 = 0$ and $n^1 = 1$, then

$$\begin{cases} \lambda\left(\sigma^{11} + \sigma^{22} + \sigma^{33}\right) + 2\mu\sigma^{11} = 0 \\ 2\mu\sigma^{12} = 0 \\ 2\mu\sigma^{13} = 0 \end{cases}, \quad (1.93)$$

or

$$\begin{bmatrix} \lambda + 2\mu & \lambda & \lambda & 0 & 0 & 0 \\ 0 & 0 & 0 & 0 & 0 & 2\mu \\ 0 & 0 & 0 & 0 & 2\mu & 0 \end{bmatrix} \begin{bmatrix} \sigma^{11} \\ \sigma^{22} \\ \sigma^{33} \\ \sigma^{23} \\ \sigma^{13} \\ \sigma^{12} \end{bmatrix} = \mathbf{B} \begin{bmatrix} \sigma^{11} \\ \sigma^{22} \\ \sigma^{33} \\ \sigma^{23} \\ \sigma^{13} \\ \sigma^{12} \end{bmatrix} = 0. \quad (1.94)$$

The rank of matrix **B** is three, which is why there are three linearly independent solutions.

B) If $n^1 \neq 0, n^2 \neq 0$, and $n^3 = 0$, then

$$\begin{cases} \lambda n^1\sigma^{11} + \lambda n^1\sigma^{22} + \lambda n^1\sigma^{33} + 2\mu n^1\sigma^{11} + 2\mu n^2\sigma^{12} = 0 \\ \lambda n^2\sigma^{11} + \lambda n^2\sigma^{22} + \lambda n^2\sigma^{33} + 2\mu n^1\sigma^{12} + 2\mu n^2\sigma^{22} = 0 \\ 2\mu n^1\sigma^{13} + 2\mu n^2\sigma^{23} = 0 \end{cases} \quad (1.95)$$

or

$$\begin{bmatrix} (\lambda + 2\mu)n^1 & \lambda n^1 & \lambda n^1 & 0 & 0 & 2\mu n^2 \\ \lambda n^2 & (\lambda + 2\mu)n^2 & \lambda n^2 & 0 & 0 & 2\mu n^1 \\ 0 & 0 & 0 & 2\mu n^2 & 2\mu n^1 & 0 \end{bmatrix}$$
$$\times \begin{bmatrix} \sigma^{11} \\ \sigma^{22} \\ \sigma^{33} \\ \sigma^{23} \\ \sigma^{13} \\ \sigma^{12} \end{bmatrix} = \mathbf{B} \begin{bmatrix} \sigma^{11} \\ \sigma^{22} \\ \sigma^{33} \\ \sigma^{23} \\ \sigma^{13} \\ \sigma^{12} \end{bmatrix} = 0. \quad (1.96)$$

Here, the third row is linearly independent from the other two.

Suppose that the first row is linearly dependent on the second. Then there exists some α, such that

$$\begin{cases} \lambda n^1 + 2\mu n^1 = \alpha\lambda n^2 \\ \lambda n^1 = \alpha\left(\lambda n^2 + 2\mu n^2\right) \\ \lambda n^1 = \alpha\lambda n^2 \\ 2\mu n^2 = \alpha 2\mu n^1 \end{cases}. \tag{1.97}$$

Where $n^1 = \alpha n^2 = \alpha^2 n^1$ and $\lambda + 2\mu = \lambda$, the result is impossible at a nonzero value of parameter μ. Therefore, in this case, the rank of matrix **B** is also three and there are three linearly independent solutions.

C) If $n^1 \neq 0, n^2 \neq 0$, and $n^3 \neq 0$, then

$$\begin{bmatrix} (\lambda+2\mu)n^1 & \lambda n^1 & \lambda n^1 & 0 & 2\mu n^3 & 2\mu n^2 \\ \lambda n^2 & (\lambda+2\mu)n^2 & \lambda n^2 & 2\mu n^3 & 0 & 2\mu n^1 \\ \lambda n^3 & \lambda n^3 & (\lambda+2\mu)n^3 & 2\mu n^2 & 2\mu n^1 & 0 \end{bmatrix} \times \begin{bmatrix} \sigma^{11} \\ \sigma^{22} \\ \sigma^{33} \\ \sigma^{23} \\ \sigma^{13} \\ \sigma^{12} \end{bmatrix} = \mathbf{B} \begin{bmatrix} \sigma^{11} \\ \sigma^{22} \\ \sigma^{33} \\ \sigma^{23} \\ \sigma^{13} \\ \sigma^{12} \end{bmatrix} = 0. \tag{1.98}$$

To prove the independence of the rows of matrix **B**, consider the matrix

$$2\mu\mathbf{C} = \begin{bmatrix} 0 & 2\mu n^3 & 2\mu n^2 \\ 2\mu n^3 & 0 & 2\mu n^1 \\ 2\mu n^2 & 2\mu n^1 & 0 \end{bmatrix} \tag{1.99}$$

or

$$\mathbf{C} = \begin{bmatrix} 0 & n^3 & n^2 \\ n^3 & 0 & n^1 \\ n^2 & n^1 & 0 \end{bmatrix}. \tag{1.100}$$

From which we have

$$\begin{bmatrix} 0 & n^3 & n^2 \\ n^3 & 0 & n^1 \\ n^2 & n^1 & 0 \end{bmatrix} \leftrightarrow \begin{bmatrix} n^2 & n^1 & 0 \\ n^3 & 0 & n^1 \\ 0 & n^3 & n^2 \end{bmatrix} \leftrightarrow \begin{bmatrix} 1 & \frac{n^1}{n^2} & 0 \\ 1 & 0 & \frac{n^1}{n^3} \\ 0 & n^3 & n^2 \end{bmatrix} \leftrightarrow \begin{bmatrix} 1 & \frac{n^1}{n^2} & 0 \\ 0 & -\frac{n^1}{n^2} & \frac{n^1}{n^3} \\ 0 & n^3 & n^2 \end{bmatrix}$$

$$\leftrightarrow \begin{bmatrix} \frac{n^2}{n^1} & 1 & 0 \\ 0 & -1 & \frac{n^2}{n^3} \\ 0 & 1 & \frac{n^2}{n^3} \end{bmatrix} \leftrightarrow \begin{bmatrix} \frac{n^2}{n^1} & 0 & \frac{n^2}{n^3} \\ 0 & 1 & -\frac{n^2}{n^3} \\ 0 & 0 & 2\frac{n^2}{n^3} \end{bmatrix} \leftrightarrow \begin{bmatrix} 1 & 0 & \frac{n^1}{n^3} \\ 0 & 1 & -\frac{n^2}{n^3} \\ 0 & 0 & 1 \end{bmatrix} \leftrightarrow \begin{bmatrix} 1 & 0 & 0 \\ 0 & 1 & 0 \\ 0 & 0 & 1 \end{bmatrix}. \tag{1.101}$$

Thus here, too, three independent solutions exist:

$$c = \pm l c_1, \quad \vec{v} = \alpha\hat{n}, \quad \boldsymbol{\sigma} = \mp\frac{\alpha}{c_1\rho}\hat{n}\otimes\hat{n}, \tag{1.102}$$

$$c = \pm l c_2, \quad \hat{n}\cdot\vec{v} = 0, \qquad \boldsymbol{\sigma} = \mp\frac{1}{2c_2\rho}\left(\hat{n}\otimes\vec{v} + \vec{v}\otimes\hat{n}\right), \tag{1.103}$$

and

$$c = 0, \quad \vec{v} = 0, \quad \lambda(\boldsymbol{\sigma} \div \mathbf{I})\hat{n} + 2\mu(\boldsymbol{\sigma}\cdot\hat{n}) = 0. \tag{1.104}$$

Thus, we have determined all nine linearly independent eigenvectors: two eigenvectors in group one, four eigenvectors in group two, and three eigenvectors in group three.

1.4.3. Important notations and relations

This subsection will introduce additional notations and the formulation of important identities and relationships between the introduced vectors and matrices, which will be referenced extensively in the subsequent text.

We define the symmetric matrices $\mathbf{N}_{ij} = \mathbf{N}_{ji}$ as

$$\mathbf{N}_{ij} = \frac{1}{2}\left(\hat{n}_i\otimes\hat{n}_j + \hat{n}_j\otimes\hat{n}_i\right), \quad i,j = 0, 1, \text{and } 2. \tag{1.105}$$

For matrices $\mathbf{N}_{ij}$, the indices i and j of the components δ^{ij} of unit tensor **I** are shifted by one, i.e., the numbering begins at zero.

The matrices $\mathbf{N}_{ij}$ are orthogonal to each other with respect to the operation of the double dot product:

$$\mathbf{N}_{ij} \div \mathbf{N}_{kl} = \frac{1}{2}\left(1+\delta^{ij}\right)\delta^{ik}\delta^{jl} \quad \text{for } \forall i \le j, \quad k \le l. \quad (1.106)$$

The action of $\mathbf{N}_{ij}$ on vector $\hat{n}_k$ for $\forall i, j, k$ produces

$$\mathbf{N}_{ij} \cdot \hat{n}_k = \frac{\hat{n}_i}{2}\left(\hat{n}_j \cdot \hat{n}_k\right) + \frac{\hat{n}_j}{2}\left(\hat{n}_i \cdot \hat{n}_k\right). \quad (1.107)$$

In particular,

$$\mathbf{N}_{ij} \cdot \hat{n} = \mathbf{N}_{ij} \cdot \hat{n}_0 = \frac{\hat{n}_i}{2}\left(\hat{n}_j \cdot \hat{n}_0\right) + \frac{\hat{n}_j}{2}\left(\hat{n}_i \cdot \hat{n}_0\right) = 0 \quad \text{for } i, j = 1, 2. \quad (1.108)$$

Another important property of matrices $\mathbf{N}_{ij}$ is the result of their double dot product with an arbitrary symmetric matrix $\boldsymbol{\sigma}$,

$$\forall i, j\text{: } \mathbf{N}_{ij} \div \boldsymbol{\sigma} = \hat{n}_i^{\mathrm{T}} \cdot \boldsymbol{\sigma} \cdot \hat{n}_j, \quad (1.109)$$

which, for the unit matrix, is

$$\forall i, j\text{: } \mathbf{N}_{ij} \div \mathbf{I} = \hat{n}_i^{\mathrm{T}} \cdot \mathbf{I} \cdot \hat{n}_j = \hat{n}_i^{\mathrm{T}} \cdot \hat{n}_j = \left(\hat{n}_i \cdot \hat{n}_j\right) = \delta^{ij}. \quad (1.110)$$

Also,

$$\mathbf{N}_{00} + \mathbf{N}_{11} + \mathbf{N}_{22} = \mathbf{I}. \quad (1.111)$$

The decomposition of an arbitrary vector $\vec{v}$ with respect to an orthonormal basis is

$$\vec{v} = \left(\hat{n} \cdot \vec{v}\right)\hat{n} + \left(\hat{n}_1 \cdot \vec{v}\right)\hat{n}_1 + \left(\hat{n}_2 \cdot \vec{v}\right)\hat{n}_2. \quad (1.112)$$

In particular,

$$\frac{1}{2}\left(\hat{n} \otimes \vec{v} + \vec{v} \otimes \hat{n}\right) = \left(\hat{n} \cdot \vec{v}\right)\mathbf{N}_{00} + \left(\hat{n}_1 \cdot \vec{v}\right)\mathbf{N}_{01} + \left(\hat{n}_2 \cdot \vec{v}\right)\mathbf{N}_{02}. \quad (1.113)$$

Let us prove that the system of matrixes $\mathbf{N}_{00}$, $\mathbf{N}_{01}$, $\mathbf{N}_{02}$, $\mathbf{N}_{12}$, $\mathbf{N}_{11} - \mathbf{N}_{22}$, and $\mathbf{N}_{11} + \mathbf{N}_{22}$ also is orthogonal with respect to a double dot product, but it is incompletely normalized $\left(\|\mathbf{N}\| = \sqrt{\mathbf{N} \div \mathbf{N}}\right)$:

$$\mathbf{N}_{00} \div \left(\mathbf{N}_{11} \pm \mathbf{N}_{22}\right) = \frac{1}{2}\left(1 + \delta^{00}\right)\delta^{01}\delta^{01} \pm \frac{1}{2}\left(1 + \delta^{00}\right)\delta^{02}\delta^{02} = 0, \quad (1.114)$$

$$\mathbf{N}_{01} \div \left(\mathbf{N}_{11} \pm \mathbf{N}_{22}\right) = \frac{1}{2}\left(1 + \delta^{01}\right)\delta^{01}\delta^{11} \pm \frac{1}{2}\left(1 + \delta^{01}\right)\delta^{02}\delta^{12} = 0, \quad (1.115)$$

$$\mathbf{N}_{02} \div \left(\mathbf{N}_{11} \pm \mathbf{N}_{22}\right) = \frac{1}{2}\left(1 + \delta^{02}\right)\delta^{01}\delta^{21} \pm \frac{1}{2}\left(1 + \delta^{01}\right)\delta^{02}\delta^{22} = 0, \quad (1.116)$$

$$\mathbf{N}_{12} \div \left(\mathbf{N}_{11} \pm \mathbf{N}_{22}\right) = \frac{1}{2}\left(1 + \delta^{12}\right)\delta^{11}\delta^{21} \pm \frac{1}{2}\left(1 + \delta^{12}\right)\delta^{12}\delta^{22} = 0, \quad (1.117)$$

$$\begin{aligned}\left(\mathbf{N}_{11} \pm \mathbf{N}_{22}\right) \div \left(\mathbf{N}_{11} \pm \mathbf{N}_{22}\right) &= \left(\mathbf{N}_{11} \div \mathbf{N}_{11}\right) + \left(\mathbf{N}_{22} \div \mathbf{N}_{22}\right) \pm \left(\mathbf{N}_{22} \div \mathbf{N}_{11}\right) \pm \left(\mathbf{N}_{11} \div \mathbf{N}_{22}\right) \\ &= \frac{1}{2}\left(1 + \delta^{11}\right)\delta^{11}\delta^{11} + \frac{1}{2}\left(1 + \delta^{22}\right)\delta^{22}\delta^{22} \\ &\quad \pm \frac{1}{2}\left(1 + \delta^{22}\right)\delta^{21}\delta^{21} \pm \frac{1}{2}\left(1 + \delta^{11}\right)\delta^{12}\delta^{12} = 2,\end{aligned} \quad (1.118)$$

and

$$\begin{aligned}\left(\mathbf{N}_{11} + \mathbf{N}_{22}\right) \div \left(\mathbf{N}_{11} - \mathbf{N}_{22}\right) &= \left(\mathbf{N}_{11} \div \mathbf{N}_{11}\right) - \left(\mathbf{N}_{22} \div \mathbf{N}_{22}\right) + \left(\mathbf{N}_{22} \div \mathbf{N}_{11}\right) - \left(\mathbf{N}_{11} \div \mathbf{N}_{22}\right) \\ &= \frac{1}{2}\left(1+\delta^{11}\right)\delta^{11}\delta^{11} - \frac{1}{2}\left(1+\delta^{22}\right)\delta^{22}\delta^{22} + \frac{1}{2}\left(1+\delta^{22}\right)\delta^{21}\delta^{21} \\ &\quad - \frac{1}{2}\left(1 + \delta^{11}\right)\delta^{12}\delta^{12} = 0.\end{aligned} \quad (1.119)$$

Consequently, an arbitrary symmetric matrix $\boldsymbol{\sigma}$ can be expanded as

$$\begin{aligned}\boldsymbol{\sigma} = {} & (\mathbf{N}_{00} \div \boldsymbol{\sigma})\mathbf{N}_{00} + 2(\mathbf{N}_{01} \div \boldsymbol{\sigma})\mathbf{N}_{01} + 2(\mathbf{N}_{02} \div \boldsymbol{\sigma})\mathbf{N}_{02} \\ & + 2(\mathbf{N}_{12} \div \boldsymbol{\sigma})\mathbf{N}_{12} + \frac{1}{2}\left((\mathbf{N}_{11} - \mathbf{N}_{22}) \div \boldsymbol{\sigma}\right)(\mathbf{N}_{11} - \mathbf{N}_{22}) \\ & + \frac{1}{2}\left((\mathbf{N}_{11} + \mathbf{N}_{22}) \div \boldsymbol{\sigma}\right)(\mathbf{N}_{11} + \mathbf{N}_{22}).\end{aligned} \quad (1.120)$$

Applying equation (1.112) to the vector $\boldsymbol{\sigma} \cdot \hat{n}$ yields

$$\boldsymbol{\sigma} \cdot \hat{n} = \left(\hat{n} \cdot \boldsymbol{\sigma} \cdot \hat{n}\right)\hat{n} + \left(\hat{n}_1 \cdot \boldsymbol{\sigma} \cdot \hat{n}\right)\hat{n}_1 + \left(\hat{n}_2 \cdot \boldsymbol{\sigma} \cdot \hat{n}\right)\hat{n}_2$$

and

$$= \left(\mathbf{N}_{00} \div \boldsymbol{\sigma}\right)\hat{n} + \left(\mathbf{N}_{01} \div \boldsymbol{\sigma}\right)\hat{n}_1 + \left(\mathbf{N}_{02} \div \boldsymbol{\sigma}\right)\hat{n}_2, \tag{1.121}$$

and applying equation (1.113) to the vector $\boldsymbol{\sigma} \cdot \hat{n}$ yields

$$\frac{1}{2}\left(\hat{n} \otimes (\boldsymbol{\sigma} \cdot \hat{n}) + (\boldsymbol{\sigma} \cdot \hat{n}) \otimes \hat{n}\right) = \left(\hat{n} \cdot (\boldsymbol{\sigma} \cdot \hat{n})\right)\mathbf{N}_{00} + \left(\hat{n}_1 \cdot (\boldsymbol{\sigma} \cdot \hat{n})\right)\mathbf{N}_{01} + \left(\hat{n}_2 \cdot (\boldsymbol{\sigma} \cdot \hat{n})\right)\mathbf{N}_{02}$$

and

$$= (\boldsymbol{\sigma} \div \mathbf{N}_{00})\mathbf{N}_{00} + (\boldsymbol{\sigma} \div \mathbf{N}_{01})\mathbf{N}_{01} + (\boldsymbol{\sigma} \div \mathbf{N}_{02})\mathbf{N}_{02}. \tag{1.122}$$

1.4.4. Normalization of nonzero eigenvalues

For many practical applications, it is convenient to have eigenvectors normalized. The previous equations for eigenvectors have some arbitrary nonzero coefficients, helping facilitate their normalization. This section will use a similar approach for nonzero eigenvalues (groups one and two) and the following section for zero eigenvalues (groups of three).

First, we define the meaning of normalization. The dot product of eigenvectors $\{\vec{v}_d, \boldsymbol{\sigma}_d\}$ of the forward problem, which constitute the columns of $\boldsymbol{\Omega}^{-1}$, and eigenvectors $\{\vec{v}_c, \boldsymbol{\sigma}_c\}$ of the adjoint problem, which constitute the rows of $\boldsymbol{\Omega}$, must add to the unity for the corresponding pairs of vectors, i.e.,

$$\vec{v}_d \cdot \vec{v}_c + \boldsymbol{\sigma}_d \div \boldsymbol{\sigma}_c = 1, \tag{1.123}$$

and zero otherwise.

Let us substitute into this equation the eigenvector values which have been determined, so that the normalization condition is satisfied. For the first group of eigenvectors,

$$c = \pm lc_1, \quad \vec{v}_d = \alpha\hat{n}, \quad \boldsymbol{\sigma}_d = \mp\frac{\alpha}{c_1}\left(\lambda\mathbf{I} + 2\mu(\hat{n} \otimes \hat{n})\right),$$

$$\vec{v}_c = \hat{n}, \quad \text{and} \quad \boldsymbol{\sigma}_c = \mp\frac{1}{c_1\rho}(\hat{n} \otimes \hat{n}), \tag{1.124}$$

we have

$$\begin{aligned} 1 &= \vec{v}_d \cdot \vec{v}_c + \boldsymbol{\sigma}_d \div \boldsymbol{\sigma}_c = \alpha(\hat{n} \cdot \hat{n}) + \left(\mp\frac{\alpha}{c_1}\left(\lambda\mathbf{I} + 2\mu(\hat{n} \otimes \hat{n})\right)\right) \div \left(\mp\frac{1}{c_1\rho}(\hat{n} \otimes \hat{n})\right) \\ &= \alpha + \frac{\alpha}{c_1^2\rho}\left(\lambda\mathbf{I} + 2\mu(\hat{n} \otimes \hat{n})\right) \div (\hat{n} \otimes \hat{n}) \\ &= \alpha\left(1 + \frac{\lambda}{c_1^2\rho}(\mathbf{N}_{00} \div \mathbf{I}) + \frac{2\mu}{c_1^2\rho}(\mathbf{N}_{00} \div \mathbf{N}_{00})\right). \end{aligned} \tag{1.125}$$

Using equation (1.110) yields

$$\begin{aligned} 1 &= \alpha\left(1 + \frac{\lambda}{c_1^2\rho}(\mathbf{N}_{00} \div \mathbf{I}) + \frac{2\mu}{c_1^2\rho}(\mathbf{N}_{00} \div \mathbf{N}_{00})\right) \\ &= \alpha\left(1 + \frac{\lambda}{c_1^2\rho} + \frac{2\mu}{c_1^2\rho}\right) \end{aligned} \tag{1.126}$$

$$= \alpha\left(1 + \frac{\lambda + 2\mu}{\rho}\frac{1}{c_1^2}\right) = \alpha\left(1 + \frac{\lambda + 2\mu}{\rho}\frac{\rho}{\lambda + 2\mu}\right) = 2\alpha,$$

$$c = \pm lc_1, \quad \vec{v}_d = \alpha\hat{n}, \quad \boldsymbol{\sigma}_d = \mp\frac{\alpha}{c_1}(\lambda\mathbf{I} + 2\mu\mathbf{N}_{00}),$$

$$\vec{v}_c = \hat{n} \quad \text{and} \quad \boldsymbol{\sigma}_c = \mp\frac{1}{c_1\rho}\mathbf{N}_{00}, \tag{1.127}$$

where the length of the eigenvector from the adjoint problem is fixed arbitrarily and $\alpha = \frac{1}{2}$.

For the second group,

$$c = \pm lc_2, \quad \hat{n} \cdot \vec{v}_d = 0, \quad \boldsymbol{\sigma}_d = \mp\frac{\mu}{c_2}\left(\hat{n} \otimes \vec{v} + \vec{v} \otimes \hat{n}\right),$$

$$\hat{n} \cdot \vec{v}_c = 0 \quad \text{and} \quad \boldsymbol{\sigma}_c = \mp\frac{1}{2c_2\rho}\left(\hat{n} \otimes \vec{v} + \vec{v} \otimes \hat{n}\right), \tag{1.128}$$

we suppose that $\vec{v}_d = \alpha\hat{n}_k$ and $\vec{v}_c = \hat{n}_k$, where $k = 1$ and 2. Then, using equation (1.113), we obtain

$$\begin{aligned} \boldsymbol{\sigma}_d &= \mp\frac{2\mu}{c_2}\left((\hat{n} \cdot \vec{v})\mathbf{N}_{00} + (\hat{n}_1 \cdot \vec{v})\mathbf{N}_{01} + (\hat{n}_2 \cdot \vec{v})\mathbf{N}_{02}\right) \\ &= \mp\frac{2\mu\alpha}{c_2}\left((\hat{n} \cdot \hat{n}_k)\mathbf{N}_{00} + (\hat{n}_1 \cdot \hat{n}_k)\mathbf{N}_{01} + (\hat{n}_2 \cdot \hat{n}_k)\mathbf{N}_{02}\right) \\ &= \mp\frac{2\mu\alpha}{c_2}\mathbf{N}_{0k} \end{aligned} \tag{1.129}$$

and

$$\boldsymbol{\sigma}_c = \mp\frac{2}{2c_2\rho}\left((\hat{n}\cdot\vec{v})\mathbf{N}_{00} + (\hat{n}_1\cdot\vec{v})\mathbf{N}_{01} + (\hat{n}_2\cdot\vec{v})\mathbf{N}_{02}\right)$$
$$= \mp\frac{1}{c_2\rho}\left((\hat{n}\cdot\hat{n}_k)\mathbf{N}_{00} + (\hat{n}_1\cdot\hat{n}_k)\mathbf{N}_{01} + (\hat{n}_2\cdot\hat{n}_k)\mathbf{N}_{02}\right)$$
$$= \mp\frac{1}{c_2\rho}\mathbf{N}_{0k} \tag{1.130}$$

Using the normalization condition and equation (1.106) yields

$$1 = \vec{v}_d\cdot\vec{v}_c + \boldsymbol{\sigma}_d \div \boldsymbol{\sigma}_c$$
$$= \alpha\hat{n}_k\cdot\hat{n}_k + \left(\mp\frac{2\mu\alpha}{c_2}\mathbf{N}_{0k}\right)\div\left(\mp\frac{1}{c_2\rho}\mathbf{N}_{0k}\right)$$
$$= \alpha\left(1+\frac{2\mu}{c_2{}^2\rho}\mathbf{N}_{0k}\div\mathbf{N}_{0k}\right) = \alpha\left(1+2\cdot\frac{1}{2}\right) = 2\alpha, \tag{1.131}$$

$$\vec{v}_c = \hat{n}_k,\quad \boldsymbol{\sigma}_c = \mp\frac{1}{2c_2\rho}\mathbf{N}_{0k},\quad \vec{v}_d = \alpha\hat{n}_k,\quad \text{and}$$
$$\boldsymbol{\sigma}_d = \mp\frac{\mu\alpha}{c_2}\mathbf{N}_{0k}, \tag{1.132}$$

where $k = 1$ and 2, while $\alpha = \dfrac{1}{2}$.

1.4.5. Normalization of zero eigenvalues

In this section, we normalize the third group of eigenvectors. (There are three vectors in this group that correspond to the zero eigenvalues.)

The vectors of $\{0, \boldsymbol{\sigma}_{c,i}\}$, where $i = 1, 2$, and 3, i.e.,

$$\boldsymbol{\sigma}_{c,1} = \mathbf{N}_{12},\quad \boldsymbol{\sigma}_{c,2} = \mathbf{N}_{11} - \mathbf{N}_{22},\quad \text{and}$$
$$\boldsymbol{\sigma}_{c,3} = \mathbf{N}_{11} + \mathbf{N}_{22} - \frac{2\lambda}{\lambda+2\mu}\mathbf{N}_{00}, \tag{1.133}$$

satisfy the condition for the eigenvector of the adjoint problem from equation (1.104),

$$\lambda(\boldsymbol{\sigma}\div\mathbf{I})\hat{n} + 2\mu(\boldsymbol{\sigma}\cdot\hat{n}) = 0. \tag{1.134}$$

Indeed, using equations (1.110) and (1.108), we can obtain

$$\lambda(\boldsymbol{\sigma}_{c,1}\div\mathbf{I})\hat{n} + 2\mu(\boldsymbol{\sigma}_{c,1}\cdot\hat{n}) = \lambda(\mathbf{N}_{12}\div\mathbf{I})\hat{n} + 2\mu(\mathbf{N}_{12}\cdot\hat{n})$$
$$= \lambda\delta^{12}\hat{n} + 2\mu\cdot 0 = 0, \tag{1.135}$$

$$\lambda(\boldsymbol{\sigma}_{c,2}\div\mathbf{I})\hat{n} + 2\mu(\boldsymbol{\sigma}_{c,2}\cdot\hat{n})$$
$$= \lambda((\mathbf{N}_{11}-\mathbf{N}_{22})\div\mathbf{I})\hat{n} + 2\mu((\mathbf{N}_{11}-\mathbf{N}_{22})\cdot\hat{n})$$
$$= \lambda(\mathbf{N}_{11}\div\mathbf{I})\hat{n} - \lambda(\mathbf{N}_{22}\div\mathbf{I})\hat{n} + 2\mu(\mathbf{N}_{11}\cdot\hat{n}) - 2\mu(\mathbf{N}_{22}\cdot\hat{n})$$
$$= \lambda\delta^{11}\hat{n} - \lambda\delta^{22}\hat{n} + 2\mu\cdot 0 - 2\mu\cdot 0 = 0, \tag{1.136}$$

and

$$\lambda(\boldsymbol{\sigma}_{c,3}\div\mathbf{I})\hat{n} + 2\mu(\boldsymbol{\sigma}_{c,3}\cdot\hat{n})$$
$$= \lambda\left(\left(\mathbf{N}_{11}+\mathbf{N}_{22}-\frac{2\lambda}{\lambda+2\mu}\mathbf{N}_{00}\right)\div\mathbf{I}\right)\hat{n}$$
$$+ 2\mu\left(\left(\mathbf{N}_{11}+\mathbf{N}_{22}-\frac{2\lambda}{\lambda+2\mu}\mathbf{N}_{00}\right)\cdot\hat{n}\right) = \lambda(\mathbf{N}_{11}\div\mathbf{I})\hat{n}$$
$$+ \lambda(\mathbf{N}_{22}\div\mathbf{I})\hat{n} - \frac{2\lambda^2}{\lambda+2\mu}(\mathbf{N}_{00}\div\mathbf{I})\hat{n} + 2\mu(\mathbf{N}_{11}\cdot\hat{n})$$
$$+ 2\mu(\mathbf{N}_{22}\cdot\hat{n}) - \frac{4\lambda\mu}{\lambda+2\mu}(\mathbf{N}_{00}\cdot\hat{n}) + \frac{2\lambda^2}{\lambda+2\mu}(\mathbf{N}_{00}\div\mathbf{I})\hat{n}$$
$$= \lambda\delta^{11}\hat{n} + \lambda\delta^{22}\hat{n} - \frac{2\lambda^2}{\lambda+2\mu}\delta^{00}\hat{n} + 2\mu\cdot 0 + 2\mu\cdot 0$$
$$- \frac{4\lambda\mu}{\lambda+2\mu}\left(\frac{\hat{n}}{2}(\hat{n}\cdot\hat{n}) + \frac{\hat{n}}{2}(\hat{n}\cdot\hat{n})\right)$$
$$= \left(2\lambda - \frac{2\lambda^2}{\lambda+2\mu} - \frac{4\lambda\mu}{\lambda+2\mu}\right)\hat{n}$$
$$= \frac{2\lambda^2 + 4\lambda\mu - 2\lambda^2 - 4\lambda\mu}{\lambda+2\mu}\hat{n} = 0. \tag{1.137}$$

The system of the vectors biorthogonal to these vectors and appropriate for the forward problem of equation (1.60), i.e., $\boldsymbol{\sigma}\cdot\hat{n} = 0$, is

$$\boldsymbol{\sigma}_{d,1} = 2\mathbf{N}_{12},\quad \boldsymbol{\sigma}_{d,2} = \frac{1}{2}(\mathbf{N}_{11}-\mathbf{N}_{22}),\quad \text{and}$$
$$\boldsymbol{\sigma}_{d,3} = \frac{1}{2}(\mathbf{N}_{11}+\mathbf{N}_{22}). \tag{1.138}$$

Indeed, using equation (1.60), we can obtain

$$\boldsymbol{\sigma}_{d,1}\cdot\hat{n} = 2(\mathbf{N}_{12}\cdot\hat{n}) = 0 \tag{1.139}$$

and

$$\boldsymbol{\sigma}_{d,2,3}\cdot\hat{n} = \frac{1}{2}((\mathbf{N}_{11}\mp\mathbf{N}_{22})\cdot\hat{n}) = 0. \tag{1.140}$$

The coefficients are selected so that $\forall i, j$ $\boldsymbol{\sigma}_{d,i} \div \boldsymbol{\sigma}_{c,j} = \delta^{ij}$, where $i, j = 1, 2$, and 3. Indeed,

$$\boldsymbol{\sigma}_{d,1} \div \boldsymbol{\sigma}_{c,1} = 2(\mathbf{N}_{12} \div \mathbf{N}_{12}) = 1 = \delta^{11}, \quad (1.141)$$

$$\begin{aligned}\boldsymbol{\sigma}_{d,2} \div \boldsymbol{\sigma}_{c,2} &= \frac{1}{2}(\mathbf{N}_{11} - \mathbf{N}_{22}) \div (\mathbf{N}_{11} - \mathbf{N}_{22}) \\ &= \frac{1}{2} \cdot 2 = 1 = \delta^{22},\end{aligned} \quad (1.142)$$

$$\begin{aligned}\boldsymbol{\sigma}_{d,3} \div \boldsymbol{\sigma}_{c,3} &= \frac{1}{2}(\mathbf{N}_{11} + \mathbf{N}_{22}) \div \left(\mathbf{N}_{11} + \mathbf{N}_{22} - \frac{2\lambda}{\lambda + 2\mu}\mathbf{N}_{00} \right) \\ &= \frac{1}{2}(\mathbf{N}_{11} + \mathbf{N}_{22}) \div (\mathbf{N}_{11} + \mathbf{N}_{22}) \\ &\quad - \frac{2\lambda}{\lambda + 2\mu}\frac{1}{2}(\mathbf{N}_{11} + \mathbf{N}_{22}) \div \mathbf{N}_{00} \\ &= \frac{1}{2} \cdot 2 - \frac{2\lambda}{\lambda + 2\mu}\frac{1}{2}(0 - 0) = 1 = \delta^{33},\end{aligned} \quad (1.143)$$

$$\boldsymbol{\sigma}_{d,1} \div \boldsymbol{\sigma}_{c,2} = 2(\mathbf{N}_{12} \div (\mathbf{N}_{11} - \mathbf{N}_{22})) = 0 = \delta^{12}, \quad (1.144)$$

$$\boldsymbol{\sigma}_{d,2} \div \boldsymbol{\sigma}_{c,1} = \mathbf{N}_{12} \div \left(\frac{1}{2}(\mathbf{N}_{11} - \mathbf{N}_{22}) \right) = \frac{1}{2} \cdot 0 = 0 = \delta^{21}, \quad (1.145)$$

$$\begin{aligned}\boldsymbol{\sigma}_{d,1} \div \boldsymbol{\sigma}_{c,3} &= 2\mathbf{N}_{12} \div \left(\mathbf{N}_{11} + \mathbf{N}_{22} - \frac{2\lambda}{\lambda + 2\mu}\mathbf{N}_{00} \right) \\ &= 2(\mathbf{N}_{12} \div (\mathbf{N}_{11} + \mathbf{N}_{22})) - 2\frac{2\lambda}{\lambda + 2\mu}(\mathbf{N}_{12} \div \mathbf{N}_{00}) \\ &= 0 = \delta^{13},\end{aligned} \quad (1.146)$$

$$\begin{aligned}\boldsymbol{\sigma}_{d,3} \div \boldsymbol{\sigma}_{c,1} &= \frac{1}{2}(\mathbf{N}_{11} + \mathbf{N}_{22}) \div (\mathbf{N}_{11} - \mathbf{N}_{22}) \\ &= \frac{1}{2}\big(\mathbf{N}_{11} \div \mathbf{N}_{11} - \mathbf{N}_{22} \div \mathbf{N}_{22} + \mathbf{N}_{22} \div \mathbf{N}_{11} \\ &\quad - \mathbf{N}_{11} \div \mathbf{N}_{22}\big) \\ &= \frac{1}{2}(1 - 1 + 0 - 0) = 0 = \delta^{31},\end{aligned} \quad (1.147)$$

$$\begin{aligned}\boldsymbol{\sigma}_{d,2} \div \boldsymbol{\sigma}_{c,3} &= \frac{1}{2}(\mathbf{N}_{11} - \mathbf{N}_{22}) \div \left(\mathbf{N}_{11} + \mathbf{N}_{22} - \frac{2\lambda}{\lambda + 2\mu}\mathbf{N}_{00} \right) \\ &= \frac{1}{2}\Bigg(\mathbf{N}_{11} \div \mathbf{N}_{11} + \mathbf{N}_{11} \div \mathbf{N}_{22} \\ &\quad - \frac{2\lambda}{\lambda + 2\mu}\mathbf{N}_{11} \div \mathbf{N}_{00} - \mathbf{N}_{22} \div \mathbf{N}_{11} \\ &\quad - \mathbf{N}_{22} \div \mathbf{N}_{22} + \frac{2\lambda}{\lambda + 2\mu}\mathbf{N}_{22} \div \mathbf{N}_{00} \Bigg) \\ &= \frac{1}{2}\left(1 + 0 - \frac{2\lambda}{\lambda + 2\mu}0 - 0 - 1 + \frac{2\lambda}{\lambda + 2\mu}0 \right) \\ &= 0 = \delta^{23},\end{aligned} \quad (1.148)$$

and

$$\begin{aligned}\boldsymbol{\sigma}_{d,3} \div \boldsymbol{\sigma}_{c,2} &= \frac{1}{2}(\mathbf{N}_{11} + \mathbf{N}_{22}) \div (\mathbf{N}_{11} - \mathbf{N}_{22}) \\ &= \frac{1}{2}\big(\mathbf{N}_{11} \div \mathbf{N}_{11} + \mathbf{N}_{22} \div \mathbf{N}_{11} - \mathbf{N}_{11} \div \mathbf{N}_{22} \\ &\quad - \mathbf{N}_{22} \div \mathbf{N}_{22}\big) \\ &= \frac{1}{2}(1 + 0 - 0 - 1) = 0 = \delta^{32}.\end{aligned} \quad (1.149)$$

This concludes the normalization of all eigenvectors. The next section summarizes our findings.

1.4.6. Matrices Λ, Ω, and Ω^{-1}

Now we can summarize the previous sections and represent the matrices into which $\mathbf{A}$ is expanded in the following explicit form

$$\begin{gathered}\mathbf{A} = \boldsymbol{\Omega}^{-1}\boldsymbol{\Lambda}\boldsymbol{\Omega} \\ \boldsymbol{\Lambda} = l \cdot \mathrm{diag}\{c_1, -c_1, c_2, -c_2, c_2, -c_2, 0, 0, 0\},\end{gathered} \quad (1.150)$$

where

$$\begin{bmatrix}\omega_1\\ \omega_2\\ \omega_3\\ \omega_4\\ \omega_5\\ \omega_6\\ \omega_7\\ \omega_8\\ \omega_9\end{bmatrix} = \boldsymbol{\Omega}\begin{bmatrix}\vec{v}\\ \boldsymbol{\sigma}\end{bmatrix} = \begin{bmatrix}\hat{n}\cdot\vec{v} - \frac{1}{c_1\rho}\mathbf{N}_{00} \div \boldsymbol{\sigma}\\ \hat{n}\cdot\vec{v} + \frac{1}{c_1\rho}\mathbf{N}_{00} \div \boldsymbol{\sigma}\\ \hat{n}_1\cdot\vec{v} - \frac{1}{c_2\rho}\mathbf{N}_{01} \div \boldsymbol{\sigma}\\ \hat{n}_1\cdot\vec{v} + \frac{1}{c_2\rho}\mathbf{N}_{01} \div \boldsymbol{\sigma}\\ \hat{n}_2\cdot\vec{v} - \frac{1}{c_2\rho}\mathbf{N}_{02} \div \boldsymbol{\sigma}\\ \hat{n}_2\cdot\vec{v} + \frac{1}{c_2\rho}\mathbf{N}_{02} \div \boldsymbol{\sigma}\\ \mathbf{N}_{12} \div \boldsymbol{\sigma}\\ (\mathbf{N}_{11} - \mathbf{N}_{22}) \div \boldsymbol{\sigma}\\ \left(\mathbf{N}_{11} + \mathbf{N}_{22} - \frac{2\lambda}{\lambda + 2\mu}\mathbf{N}_{00}\right) \div \boldsymbol{\sigma}\end{bmatrix}. \tag{1.151}$$

This equation defines the action of matrix $\boldsymbol{\Omega}$ on an arbitrary vector. Unfortunately, the study of the eigenvectors of system (1.28) allows us to write only the effect of matrix $\boldsymbol{\Omega}^{-1}$ on the vector to the left, $\vec{\omega}^{\mathrm{T}}\boldsymbol{\Omega}^{-1}$, but not to the right, $\boldsymbol{\Omega}^{-1}\vec{\omega}$. For the latter, additional efforts are required.

To determine $\vec{u} = \boldsymbol{\Omega}^{-1}\vec{\omega}$, we solve the system of equations $\boldsymbol{\Omega}\vec{u} = \vec{\omega}$. After summing the first three pairs of rows in equation (1.151), we have

$$\hat{n}\cdot\vec{v} = \frac{\omega_1 + \omega_2}{2}, \quad \hat{n}_1\cdot\vec{v} = \frac{\omega_3 + \omega_4}{2}, \quad \text{and}$$
$$\hat{n}_2\cdot\vec{v} = \frac{\omega_5 + \omega_6}{2}. \tag{1.152}$$

Substituting this into equation (1.112) yields

$$\vec{v} = \frac{\omega_1 + \omega_2}{2}\hat{n} + \frac{\omega_3 + \omega_4}{2}\hat{n}_1 + \frac{\omega_5 + \omega_6}{2}\hat{n}_2. \tag{1.153}$$

Now, we subtract in pairs the same rows of the system of equations,

$$\mathbf{N}_{00} \div \boldsymbol{\sigma} = c_1\rho\frac{\omega_2 - \omega_1}{2}, \quad \mathbf{N}_{01} \div \boldsymbol{\sigma} = c_2\rho\frac{\omega_4 - \omega_3}{2}, \quad \text{and}$$
$$\mathbf{N}_{02} \div \boldsymbol{\sigma} = c_2\rho\frac{\omega_6 - \omega_5}{2}. \tag{1.154}$$

The last rows yield

$$\mathbf{N}_{12} \div \boldsymbol{\sigma} = \omega_7, \quad (\mathbf{N}_{11} - \mathbf{N}_{22}) \div \boldsymbol{\sigma} = \omega_8, \quad \text{and}$$
$$(\mathbf{N}_{11} + \mathbf{N}_{22}) \div \boldsymbol{\sigma} = \omega_9 + \frac{2\lambda}{\lambda + 2\mu}(\mathbf{N}_{00} \div \boldsymbol{\sigma}). \tag{1.155}$$

Also we substitute this into equation (1.120) and change the sum $\mathbf{N}_{11} + \mathbf{N}_{22}$ according to formula (1.111). Then,

$$\begin{aligned}\boldsymbol{\sigma} &= \left(c_1\rho\frac{\omega_2-\omega_1}{2}\right)\mathbf{N}_{00} + 2\left(c_2\rho\frac{\omega_4-\omega_3}{2}\right)\mathbf{N}_{01}\\ &\quad + 2\left(c_2\rho\frac{\omega_6-\omega_5}{2}\right)\mathbf{N}_{02} + 2(\omega_7)\mathbf{N}_{12} + \frac{1}{2}(\omega_8)(\mathbf{N}_{11}-\mathbf{N}_{22})\\ &\quad + \frac{1}{2}\left(\omega_9 + \frac{2\lambda}{\lambda+2\mu}c_1\rho\frac{\omega_2-\omega_1}{2}\right)(\mathbf{I}-\mathbf{N}_{00})\\ &= \frac{1}{2}\Big(\rho(\omega_2-\omega_1)\big((c_1-c_3)\mathbf{N}_{00}+c_3\mathbf{I}\big) + 2c_2\rho(\omega_4-\omega_3)\mathbf{N}_{01}\Big)\\ &\quad + \frac{1}{2}(2c_2\rho(\omega_6-\omega_5)\mathbf{N}_{02} + 4\omega_7\mathbf{N}_{12}\\ &\quad + \omega_8(\mathbf{N}_{11}-\mathbf{N}_{22}) + \omega_9(\mathbf{I}-\mathbf{N}_{00})),\end{aligned} \tag{1.156}$$

where $c_3 = \frac{\lambda}{\lambda + 2\mu}c_1$, and

$$\begin{bmatrix}\vec{v}\\ \boldsymbol{\sigma}\end{bmatrix} = \boldsymbol{\Omega}^{-1}\vec{\omega} = \frac{1}{2}\begin{bmatrix}(\omega_1+\omega_2)\hat{n} + (\omega_3+\omega_4)\hat{n}_1 + (\omega_5+\omega_6)\hat{n}_2\\ \\ \rho(\omega_2-\omega_1)((c_1-c_3)\mathbf{N}_{00}+c_3\mathbf{I})\\ +2c_2\rho(\omega_4-\omega_3)\mathbf{N}_{01}\\ +2c_2\rho(\omega_6-\omega_5)\mathbf{N}_{02} + 4\omega_7\mathbf{N}_{12}\\ +\omega_8(\mathbf{N}_{11}-\mathbf{N}_{22}) + \omega_9(\mathbf{I}-\mathbf{N}_{00})\end{bmatrix}. \tag{1.157}$$

Next, we need to operate with the separate columns of matrix $\boldsymbol{\Omega}^{-1}$. To obtain each of them, it suffices to multiply the matrix by the corresponding unit vector,

$$\vec{\omega}^{\,i,*} \equiv \boldsymbol{\Omega}^{-1}\hat{e}_i. \tag{1.158}$$

The superscripted asterisk represents the columns of inverse matrix $\boldsymbol{\Omega}^{-1}$. Thus, matrix $\boldsymbol{\Omega}^{-1}$ can be written coordinate-wise as

$$\boldsymbol{\Omega}^{-1} = \frac{1}{2}\left[\begin{matrix}\hat{n} & \hat{n} & \hat{n}_1 & \hat{n}_1 & \hat{n}_2 & \hat{n}_2 & 0 & 0 & 0\\ -\rho\big((c_1-c_3)\mathbf{N}_{00}+c_3\mathbf{I}\big) & \rho\big((c_1-c_3)\mathbf{N}_{00}+c_3\mathbf{I}\big) & -2c_2\rho\mathbf{N}_{01} & 2c_2\rho\mathbf{N}_{01} & -2c_2\rho\mathbf{N}_{02} & 2c_2\rho\mathbf{N}_{02} & 4\mathbf{N}_{12} & \mathbf{N}_{11}-\mathbf{N}_{22} & \mathbf{I}-\mathbf{N}_{00}\end{matrix}\right]. \tag{1.159}$$

This concludes the separation of matrix **A** into three matrices $\mathbf{A} = \mathbf{\Omega}^{-1}\mathbf{\Lambda}\mathbf{\Omega}$. These equations define how each matrix is applied to an arbitrary vector.

1.5. Coordinate-wise split

Equation (1.19) is written in a convenient form that allows us to develop a difference scheme for integration in three-dimensional (3D) space if one-dimensional (1D) schemes are available for the system of equations

$$\dot{\vec{u}} + \mathbf{A}_j \frac{\partial \vec{u}}{\partial \hat{\xi}_j} = 0. \qquad (1.160)$$

The approach that consists of combining the transition operators for 1D system (1.160) in order to derive difference schemes for a 3D system (1.19) is called the coordinate-wise split method. Here, we describe some basic methods for building such split schemes, and each of these methods will be used in subsequent chapters.

It is said that the problem $L_\tau(u^\tau) = F_\tau$ approximates the problem $L(u) = F$ if the misfit $r_\tau \equiv L_\tau(u^\tau) - F_\tau$ vanishes, i.e., $\|r_\tau\| \to 0$, when $\tau \to 0$. Moreover, if the misfit satisfies the inequality $\|r_\tau\| \le C\tau^p$ with $C \ne C(\tau)$, it is said that a p-order approximation occurs. A stable difference scheme having a p-order approximation also has a p-order convergence.

In order for the schemes to be stable in all three directions, the integration step should satisfy the Courant condition

$$\tau < \min_j \tau_j^{max} = \frac{h}{c_j^{max}} = \frac{h}{c_1 l_j}, \qquad (1.161)$$

where h is the minimum grid spacing (i.e., the minimum height of all of the triangles and tetrahedrons of an unstructured computational grid in 2D and 3D cases, respectively), c_j^{max} is the maximum absolute value of the eigenvalues of $\mathbf{A}_j$, and c_1 is the velocity of pressure waves from equation (1.49).

Let us introduce the operator $\mathbf{O}(\tau^p)\vec{u} = O(\tau^p)$ (in Big O notation).

For the original system of equations

$$\partial_t \vec{u} + \mathbf{A}_1 \frac{\partial \vec{u}}{\partial \hat{\xi}_1} + \mathbf{A}_2 \frac{\partial \vec{u}}{\partial \hat{\xi}_2} + \mathbf{A}_3 \frac{\partial \vec{u}}{\partial \hat{\xi}_3} = 0 \qquad (1.162)$$

to be solved exactly according to the formula for the Taylor series, the formula

$$\vec{u}\left(\hat{\xi}_1, \hat{\xi}_2, \hat{\xi}_3, t + \tau\right) = \vec{u}\left(\hat{\xi}_1, \hat{\xi}_2, \hat{\xi}_3, t\right) + \tau \frac{\partial \vec{u}\left(\hat{\xi}_1, \hat{\xi}_2, \hat{\xi}_3, t\right)}{\partial t} + \frac{\tau^2}{2} \frac{\partial^2 \vec{u}\left(\hat{\xi}_1, \hat{\xi}_2, \hat{\xi}_3, t\right)}{\partial t^2} + O\left(\tau^3\right) \qquad (1.163)$$

should be valid.

Let us introduce the corresponding operator,

$$\mathbf{P} = \mathbf{I} + \tau\left(\frac{\partial}{\partial t}\right)_0 + \frac{\tau^2}{2}\left(\frac{\partial}{\partial t}\right)_0^2 + \mathbf{O}\left(\tau^3\right). \qquad (1.164)$$

Moreover, because the original system of equations (1.162) is satisfied, then

$$\left(\frac{\partial}{\partial t}\right)_0 = -\mathbf{A}_1\left(\frac{\partial}{\partial \hat{\xi}_1}\right)_0 - \mathbf{A}_2\left(\frac{\partial}{\partial \hat{\xi}_2}\right)_0 - \mathbf{A}_3\left(\frac{\partial}{\partial \hat{\xi}_3}\right)_0. \qquad (1.165)$$

The operator **P** can be expressed as

$$\begin{aligned}\mathbf{P} &= \mathbf{I} + \tau\left(\frac{\partial}{\partial t}\right)_0 + \frac{\tau^2}{2}\left(\frac{\partial}{\partial t}\right)_0^2 + \mathbf{O}\left(\tau^3\right) \\ &= \mathbf{I} - \tau\mathbf{A}_1\left(\frac{\partial}{\partial \hat{\xi}_1}\right)_0 - \tau\mathbf{A}_2\left(\frac{\partial}{\partial \hat{\xi}_2}\right)_0 - \tau\mathbf{A}_3\left(\frac{\partial}{\partial \hat{\xi}_3}\right)_0 \\ &\quad + \frac{\tau^2}{2}\left(\mathbf{A}_1\left(\frac{\partial}{\partial \hat{\xi}_1}\right)_0 + \mathbf{A}_2\left(\frac{\partial}{\partial \hat{\xi}_2}\right)_0 + \mathbf{A}_3\left(\frac{\partial}{\partial \hat{\xi}_3}\right)_0\right)^2 \\ &\quad + \mathbf{O}\left(\tau^3\right) = \mathbf{I} + \tau\left(\frac{\partial}{\partial t}\right)_0 + \frac{\tau^2}{2}\left(\frac{\partial}{\partial t}\right)_0^2 + \mathbf{O}\left(\tau^3\right).\end{aligned} \qquad (1.166)$$

For each system (1.160), consider the operator

$$\mathbf{M}_j = \mathbf{I} + \tau\left(\frac{\partial}{\partial t}\right)_j + \frac{\tau^2}{2}\left(\frac{\partial}{\partial t}\right)_j^2 + \mathbf{O}\left(\tau^3\right), \qquad (1.167)$$

which provides a second-order approximation of the system in time. Here, according to equation (1.160),

$$\left(\frac{\partial}{\partial t}\right)_j = -\mathbf{A}_j\left(\frac{\partial}{\partial \hat{\xi}_j}\right)_j. \qquad (1.168)$$

Consider the following superposition of operators,

$$\mathbf{M}_i\mathbf{M}_j\mathbf{M}_k = \mathbf{I} + \tau\left(\frac{\partial}{\partial t}\right)_i + \tau\left(\frac{\partial}{\partial t}\right)_j + \tau\left(\frac{\partial}{\partial t}\right)_k + \frac{\tau^2}{2}\left(\frac{\partial}{\partial t}\right)_i^2 + \frac{\tau^2}{2}\left(\frac{\partial}{\partial t}\right)_j^2 + \frac{\tau^2}{2}\left(\frac{\partial}{\partial t}\right)_k^2 + \tau^2\left(\frac{\partial}{\partial t}\right)_j\left(\frac{\partial}{\partial t}\right)_k + \tau^2\left(\frac{\partial}{\partial t}\right)_i\left(\frac{\partial}{\partial t}\right)_k + \tau^2\left(\frac{\partial}{\partial t}\right)_i\left(\frac{\partial}{\partial t}\right)_j + \mathbf{O}(\tau^3). \tag{1.169}$$

This scheme provides a second-order approximation in time.

The implementation of this algorithm, however, requires data storage at three time levels. Practice has shown that, if the simplified superposition of operators $\mathbf{M}_3\mathbf{M}_2\mathbf{M}_1$ is applied, the corresponding scheme produces results very similar to second-order approximation in time. The only significant drawback of this approach is the asymmetry of the resulting scheme, which, is usually not critical.

1.6. Grid-characteristic schemes

Among all of the difference schemes for hyperbolic equations (1.160), i.e., for those whose matrix has a complete set of eigenvectors and only real eigenvalues,

$$\dot{\vec{u}} + \mathbf{\Omega}^{-1}\mathbf{\Lambda}\mathbf{\Omega}\frac{\partial \vec{u}}{\partial \hat{\xi}} = 0, \tag{1.170}$$

it is customary to distinguish the so-called grid-characteristic schemes (Magomedov and Kholodov, 1988), which essentially take into consideration the peculiarities of such equations. The specific feature of interest of the hyperbolic equations lies in their characteristic hypersurfaces, which reduce to lines in one dimension. On this surface, the equations contain no derivatives leading to the solution.

It is apparent that, after the change of variables, system (1.170) breaks down into a number of independent scalar transport equations,

$$\dot{\vec{v}} + \mathbf{\Lambda}\frac{\partial \vec{v}}{\partial \hat{\xi}} = 0(\vec{v} \equiv \mathbf{\Omega}\,\vec{u}). \tag{1.171}$$

These transport equations are solved by a well-known method (Figure 1.1). From node m of time level $n+1$

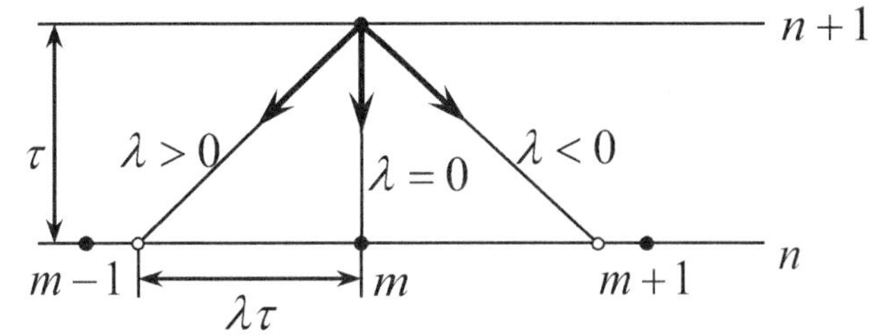

Figure 1.1. Illustration of solving the transport equations.

where it is required to obtain a solution, the characteristics are extended down. From the point of intersection of a characteristic with time level n, the corresponding component of vector $\vec{\omega}$ is transferred to the point ξ_m^{n+1}:

$$v_i^{n+1}(\xi_m) = v_i^n(\xi_m - \lambda_i\tau). \tag{1.172}$$

From this point forward, we will omit $\hat{\xi}_m$ and simply write

$$v_i^{n+1} = v_i^n(-\lambda_i\tau) \tag{1.173}$$

If no characteristic passes exactly through the current node, some methods for reconstruction of the solution are applied, i.e., interpolation. The degree of approximation of a grid-characteristic difference scheme is determined by the choice of a method for reconstruction of the solution at the previous level. Magomedov and Kholodov (1988) derive grid-characteristic schemes of the first (Courant-Isaacson-Rees) and second (Lax-Wendroff) orders.

After all of the components $\vec{v}$ have been transferred, the solution itself can be restored as

$$\vec{u}^{n+1} = \mathbf{\Omega}^{-1}\vec{v}^{n+1}. \tag{1.174}$$

In some software implementations of the grid-characteristic method, a matrix is directly inverted, which is equivalent to solving a system of equations. We, however, offer a different approach. Let us substitute one of the last expressions into another. Then,

$$\vec{u}^{n+1} = \mathbf{\Omega}^{-1}\begin{bmatrix} \cdots \\ v_i^n(-\lambda_i\tau) \\ \cdots \end{bmatrix} = \mathbf{\Omega}^{-1}\begin{bmatrix} \cdots \\ \vec{\omega}^i \cdot \vec{u}^n(-\lambda_i\tau) \\ \cdots \end{bmatrix} = \sum_i \left(\vec{\omega}^i \cdot \vec{u}^n(-\lambda_i\tau)\right)\vec{\omega}^{*,i} = \sum_i \left(\vec{\omega}^i \cdot \vec{\omega}^{*,i}\right)\vec{u}^n(-\lambda_i\tau), \tag{1.175}$$

where $\vec{\omega}^i$ is the ith row of $\mathbf{\Omega}$ and $\vec{\omega}^{*,i}$ is the ith column of $\mathbf{\Omega}^{-1}$. Denoting $\mathbf{X}^i = \vec{\omega}^{*,i}\vec{\omega}^i$, we finally obtain

$$\vec{u}^{n+1} = \sum_i \mathbf{X}^i \vec{u}^n(-\lambda_i \tau). \tag{1.176}$$

Alternatively, we can bring together the sum of all elements associated with the same eigenvalue:

$$\mathbf{X}^c = \sum_{\lambda_i = c} \vec{\omega}^{*,i}\vec{\omega}^i. \tag{1.177}$$

In as much as $\mathbf{I} = (\mathbf{\Omega}^{-1})^{\mathrm{T}}\mathbf{\Omega}^{\mathrm{T}} = \Sigma_i \mathbf{X}^i$, then $\Sigma_{\lambda_i=0}\mathbf{X}^i = \mathbf{I} - \Sigma_{\lambda_i \neq 0}\mathbf{X}^i$, and equation (1.177) can take the form

$$\begin{aligned}\vec{u}^{n+1} &= \sum_i \mathbf{X}^i \vec{u}^n(-\lambda_i\tau) = \sum_{\lambda_i\neq 0}\mathbf{X}^i \vec{u}^n(-\lambda_i\tau) + \sum_{\lambda_i=0}\mathbf{X}^i \vec{u}^n(-\lambda_i\tau)\\ &= \sum_{\lambda_i\neq 0}\mathbf{X}^i \vec{u}^n(-\lambda_i\tau) + \sum_{\lambda_i=0}\mathbf{X}^i \vec{u}^n\\ &= \mathbf{I}\vec{u}^n - \sum_{\lambda_i\neq 0}\mathbf{X}^i \vec{u}^n + \sum_{\lambda_i\neq 0}\mathbf{X}^i \vec{u}^n(-\lambda_i\tau),\end{aligned} \tag{1.178}$$

or the form which often is more convenient to use,

$$\vec{u}^{n+1} = \vec{u}^n + \sum_{\lambda_i\neq 0}\mathbf{X}^i\left(\vec{u}^n(-\lambda_i\tau) - \vec{u}^n\right). \tag{1.179}$$

We emphasize once again that this formula is valid for any grid-characteristic scheme, regardless of the order of reconstruction of a function at the previous time level.

Deriving the action of the matrices $\mathbf{X}$ on the vector is not difficult if we refer to expressions (1.159) and (1.151):

$$\mathbf{X}^{\pm c2}\begin{bmatrix}\vec{v}\\ \boldsymbol{\sigma}\end{bmatrix} = \frac{1}{2}\begin{bmatrix}(\hat{n}_1\cdot\vec{v})\hat{n}_1 \mp \frac{1}{c_2\rho}(\mathbf{N}_{01}\div\boldsymbol{\sigma})\hat{n}_1 + (\hat{n}_2\cdot\vec{v})\hat{n}_2\\ \mp\frac{1}{c_2\rho}(\mathbf{N}_{02}\div\boldsymbol{\sigma})\hat{n}_2\\ \mp 2c_2\rho(\hat{n}_1\cdot\vec{v})\mathbf{N}_{01} + 2(\mathbf{N}_{01}\div\boldsymbol{\sigma})\mathbf{N}_{01}\\ \mp 2c_2\rho(\hat{n}_2\cdot\vec{v})\mathbf{N}_{02} + 2(\mathbf{N}_{02}\div\boldsymbol{\sigma})\mathbf{N}_{02}\end{bmatrix}$$

$$= \frac{1}{2}\begin{bmatrix}\vec{v} - (\hat{n}\cdot\vec{v})\hat{n} \mp \frac{1}{c_2\rho}\left(\boldsymbol{\sigma}\cdot\hat{n} - (\mathbf{N}_{00}\div\boldsymbol{\sigma})\hat{n}\right)\hat{n}_1\\ \mp c_2\rho\left(\hat{n}\otimes\vec{v} + \vec{v}\otimes\hat{n} - 2(\hat{n}\cdot\vec{v})\mathbf{N}_{00}\right)\\ + 2\left(\hat{n}\otimes(\boldsymbol{\sigma}\cdot\hat{n}) + (\boldsymbol{\sigma}\cdot\hat{n})\otimes\hat{n}\right.\\ \left. - 2(\boldsymbol{\sigma}\div\mathbf{N}_{00})\mathbf{N}_{00}\right)\end{bmatrix}. \tag{1.180}$$

The sum of the characteristic matrices for positive (negative) eigenvalues is determined by adding the last two expressions:

$$\mathbf{X}^{\pm}\begin{bmatrix}\vec{v}\\ \boldsymbol{\sigma}\end{bmatrix} = \mathbf{X}^{\pm c2}\begin{bmatrix}\vec{v}\\ \boldsymbol{\sigma}\end{bmatrix} + \mathbf{X}^{\pm c1}\begin{bmatrix}\vec{v}\\ \boldsymbol{\sigma}\end{bmatrix}$$

$$= \frac{1}{2}\begin{bmatrix}\vec{v} \mp \frac{1}{c_2\rho}\left(\boldsymbol{\sigma}\cdot\hat{n} - (\mathbf{N}_{00}\div\boldsymbol{\sigma})\hat{n}\right)\hat{n} \mp \frac{1}{c_1\rho}(\mathbf{N}_{00}\div\boldsymbol{\sigma})\hat{n}\\ \mp c_2\rho\left(\hat{n}\otimes\vec{v} + \vec{v}\otimes\hat{n} - 2(\hat{n}\cdot\vec{v})\mathbf{N}_{00}\right)\\ \mp\rho(\hat{n}\cdot\vec{v})\left((c_1 - c_3)\mathbf{N}_{00} + c_3\mathbf{I}\right)\\ + 2\left(\hat{n}\otimes(\boldsymbol{\sigma}\cdot\hat{n}) + (\boldsymbol{\sigma}\cdot\hat{n})\otimes\hat{n} - 2(\boldsymbol{\sigma}\div\mathbf{N}_{00})\mathbf{N}_{00}\right)\\ + \frac{1}{\lambda + 2\mu}(\mathbf{N}_{00}\div\boldsymbol{\sigma})(2\mu\mathbf{N}_{00} + \lambda\mathbf{I})\end{bmatrix}. \tag{1.181}$$

1.7. Computation at the boundary of integration domain

Formulae (1.176) and (1.179) cannot be applied directly in the vicinity of a boundary because they contain values of $\vec{u}^n(-\lambda_i\tau)$ that are outside of the domain of integration. A correct solution of the problem at the boundary requires a statement of boundary conditions, their number equal to the number of outgoing characteristics (Magomedov and Kholodov, 1988). For the left-side boundary, the outgoing characteristics are those associated with positive eigenvalues, whereas for the right-side boundary, the outgoing characteristics are those associated with negative eigenvalues (Figure 1.1). The linear boundary conditions can be written in the general form

$$\mathbf{B}\vec{u}^{n+1} = \vec{b}, \tag{1.182}$$

where $\mathbf{B}$ is a rectangular matrix and $\vec{b}$ is a column vector.

We divide the entire set of eigenvalues of $\mathbf{A}$ into two groups. These groups depend on the relation of the characteristics to the integration domain at the previous time level, λ^{in} and λ^{out}. We assume that equation (1.76) also holds true for the boundary nodes if the values of $\vec{u}^n(-\lambda_i\tau)$, $\lambda_i \in \lambda^{out}$, are specifically chosen so that equation (1.182) is satisfied,

$$\mathbf{B}\sum_i \mathbf{X}^i\vec{u}^n(-\lambda_i\tau) = \vec{b}, \tag{1.183}$$

which is equivalent to

$$\mathbf{B}\left(\sum_{\lambda_i \in \lambda^{in}} \mathbf{X}^i \vec{u}^n(-\lambda_i \tau) + \sum_{\lambda_i \in \lambda^{out}} \alpha_i \vec{\omega}^{*,i}\right) = \vec{b}, \tag{1.184}$$

where α_i are some coefficients that provide this equality. After denoting the first sum in this equation as $\vec{u}^{in}$ and collecting all of the columns $\vec{\omega}^{*,i}$ and all of the coefficients α_i from the second sum in the rectangular matrix $\mathbf{\Omega}^{*,out}$ and the vector $\vec{\alpha}$, respectively, we have

$$\mathbf{B}\left[\vec{u}^{in} + \mathbf{\Omega}^{*,out}\vec{\alpha}\right] = \vec{b}. \tag{1.185}$$

There is no need to calculate the value of vector $\vec{\alpha}$, so we exclude it from this expression:

$$\vec{\alpha} = \left(\mathbf{B\Omega}^{*,out}\right)^{-1}\left(\vec{b} - \mathbf{B}\vec{u}^{in}\right). \tag{1.186}$$

Substituting this into equation (1.176) yields

$$\vec{u}^{n+1} = \vec{u}^{in} + \mathbf{\Omega}^{*,out}\left(\mathbf{B\Omega}^{*,out}\right)^{-1}\left(\vec{b} - \mathbf{B}\vec{u}^{in}\right). \tag{1.187}$$

Thus, we have a generalized representation of the method for calculating a boundary node, at least one outgoing characteristic of which is out of the domain of integration at the previous time level. This formula is valid for any order of accuracy of reconstructing a function at time level n and for any set of boundary conditions $\{\mathbf{B},\vec{b}\}$.

Expression (1.186) sets the criterion for a correct setting of a boundary condition, namely non-singularity of matrix $\mathbf{B\Omega}^{*,out}$.

In the subsequent sections, we will demonstrate how this method is simplified for linear elasticity equations, with the definition of the matrix from equation (1.27) and for specific boundary conditions.

1.7.1. Given external force

In this condition, the density of an external force that acts on the surface of a body and is exerted by other objects is given. When the force is zero, the surface is said to be free. This condition can be written formally as

$$\boldsymbol{\sigma} \cdot \vec{p} = \vec{f}, \tag{1.188}$$

where $\vec{p}$ denotes the outward unit normal to the body surface and $\vec{f}$ is the surface density of external forces. Expression (1.188) contains three scalar conditions that compensate for the lack of data for three outgoing characteristics. Reducing the boundary conditions to the form of equation (1.182), we obtain

$$\mathbf{B}\begin{bmatrix}\vec{v}\\ \boldsymbol{\sigma}\end{bmatrix} = \boldsymbol{\sigma} \cdot \vec{p} \quad \text{with } \vec{b} = \vec{f}. \tag{1.189}$$

Substituting this into equation (1.187) yields

$$\vec{u}^{n+1} = \vec{u}^{in} - \mathbf{\Omega}^{*,out}\left(\mathbf{B\Omega}^{*,out}\right)^{-1}\left(\boldsymbol{\sigma}^{in} \cdot \vec{p} - \vec{f}\right), \tag{1.190}$$

where $\vec{u}^{in} = \{\vec{v}^{in}, \boldsymbol{\sigma}^{in}\}$.

The effect of $\mathbf{\Omega}^{*,out}$ is derived from equation (1.154):

$$\mathbf{\Omega}^{*,out}\begin{bmatrix}\omega_1\\ \omega_2\\ \omega_3\end{bmatrix} = \frac{1}{2}\begin{bmatrix}\omega_1\hat{n} + \omega_2\hat{n}_1 + \omega_3\hat{n}_2\\ \mp\big(\rho\omega_1((c_1 - c_3)\mathbf{N}_{00} + c_3\mathbf{I})\\ + 2c_2\rho\omega_2\mathbf{N}_{01} + 2c_2\rho\omega_3\mathbf{N}_{02}\big)\end{bmatrix} \tag{1.191}$$

and

$$\mathbf{B\Omega}^{*,out}\begin{bmatrix}\omega_1\\ \omega_2\\ \omega_3\end{bmatrix} = \mp\frac{\rho}{2}\Big(\omega_1\left((c_1 - c_3)\mathbf{N}_{00} + c_3\mathbf{I}\right) + 2c_2\omega_2\mathbf{N}_{01} + 2c_2\omega_3\mathbf{N}_{02}\Big) \cdot \vec{p} \tag{1.192}$$

where the negative sign refers to the left-side boundary of the domain of integration, and the positive sign to the right-side boundary.

1.7.2. Given surface velocity

In this condition, it is supposed that, at the surface of a body, a velocity is given. For instance, it can be the velocity of an external object that is in contact with the body being considered. When the velocity is zero, the surface is fixed. This condition can be written formally as $\vec{v} = \vec{V}$ where $\vec{V}$ is the given velocity. This expression contains three scalar conditions that compensate for the lack of data for three outgoing characteristics. Reducing the boundary conditions to the form of equation (1.182) yields

$$\mathbf{B}\begin{bmatrix}\vec{v}\\ \boldsymbol{\sigma}\end{bmatrix} = \vec{v} \quad \text{with } \vec{b} = \vec{V}. \tag{1.193}$$

By analogy with the case of given surface forces, we obtain

$$\Omega^{*,out}\begin{bmatrix}\omega_1\\ \omega_2\\ \omega_3\end{bmatrix}=\frac{1}{2}\begin{bmatrix}\omega_1\hat{n}+\omega_2\hat{n}_1+\omega_3\hat{n}_2\\ \mp\left(\rho\omega_1((c_1-c_3)\mathbf{N}_{00}+c_3\mathbf{I})\right.\\ \left.+2c_2\rho\omega_2\mathbf{N}_{01}+2c_2\rho\omega_3\mathbf{N}_{02}\right)\end{bmatrix} \tag{1.194}$$

and

$$\mathbf{B}\Omega^{*,out}\begin{bmatrix}\omega_1\\ \omega_2\\ \omega_3\end{bmatrix}=\frac{1}{2}\begin{bmatrix}\omega_1\hat{n}+\omega_2\hat{n}_1+\omega_3\hat{n}_2\\ 0\end{bmatrix}. \tag{1.195}$$

This gives us the equation for a boundary condition for a given surface velocity.

1.7.3. Mixed conditions

In the first mixed condition, we assume that the normal component of a surface velocity and the tangential component of an external force are given. This condition occurs when the body considered is in contact with an external object. The tangential force component is zero in the absence of friction and otherwise is proportional to the sliding velocity.

Let us denote the given tangential force by $\vec{f}_\tau$, with the normal component of this vector being arbitrary, and denote normal velocity by V_p. Then, the mixed condition can be written formally as

$$\begin{aligned}&\boldsymbol{\sigma}\cdot\vec{p}=\vec{f},\quad \vec{v}\cdot\vec{p}=V_p,\quad \text{and}\\ &\vec{f}=\vec{f}_\tau-\left(\vec{f}_\tau\cdot\vec{p}\right)\vec{p}+\left(\vec{f}\cdot\vec{p}\right)\vec{p}.\end{aligned} \tag{1.196}$$

Solving (1.196) for $\vec{f}$, one could use equation (1.190) in conjunction with (1.191) and (1.192) to obtain $\vec{u}^{n+1}$.

For completeness, we describe another mixed condition when the tangential component of a surface velocity and the normal component of an external force are given. If both values are zero, we have a "rough" free boundary, any point of which is allowed to move only perpendicularly, but the normal outward force is equal to zero.

Let us denote the given tangential velocity by $\vec{V}_\tau$, with the normal component of this vector being arbitrary, and denote surface density of normal force by f_p. Then, the second mixed condition can be written formally as

$$\vec{p}\cdot\boldsymbol{\sigma}\cdot\vec{p}=f_p\quad\text{and}\quad \vec{v}=\vec{V}_\tau-\left(\vec{V}_\tau\cdot\vec{p}\right)\vec{p}+(\vec{v}\cdot\vec{p})\vec{p}. \tag{1.197}$$

Solving (1.197) for v, one could use equation (1.190) in conjunction with (1.194) and (1.195) to obtain $\vec{u}^{n+1}$.

1.8. Contact between two bodies

Suppose that two bodies, e.g., a and b, possibly of different materials, touch one another at a given point on a surface with the normal $\vec{p}$ external to body a. Furthermore, two surface nodes of meshes constructed in bodies a and b are identical (Figure 1.2). For these nodes, it is necessary to construct correctors similar to those we implemented in Section 1.7. They will replace any missing characteristics (those that are out of the integration domain at the previous time level) and account for the features of a particular type of contact between the bodies. Different types of contacts will be discussed next.

1.8.1. Bonded contact

When the two bodies are in bonded contact, the velocities of the adjoining surface points must be the same, $\vec{v}_a=\vec{v}_b=\vec{V}$, and the external forces acting on each body must be equal and of opposite sign, $\vec{f}_a=-\vec{f}_b$. The external forces acting on both bodies in the next step are, respectively,

$$\vec{f}_a=\boldsymbol{\sigma}_a^{n+1}\cdot\vec{p}\quad\text{and}\quad\vec{f}_b=\boldsymbol{\sigma}_b^{n+1}\cdot(-\vec{p}). \tag{1.198}$$

1.8.2. Free sliding

When the two bodies slide freely along the common interface, the normal components of velocity must coin-

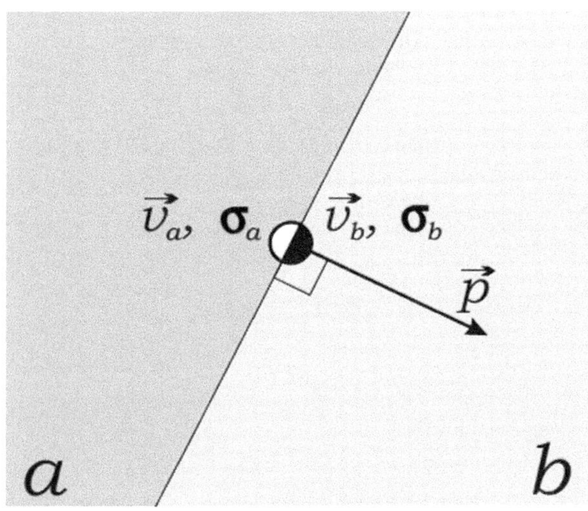

Figure 1.2. Contact between two media.

cide, $\vec{v}_a \cdot \vec{p} = \vec{v}_b \cdot \vec{p} = \vec{V}_p$, and the surface density of normal components of force must compensate for each other because $f_p^a = -f_p^b$, while both of the tangential components of force vanish, $f_\tau^a = f_\tau^b = 0$.

1.9. A case of translational symmetry along one axis

In many practical problems, a translational symmetry along one axis exists. Sometimes symmetry may not be absolute but could be considered a good approximation. For example, in exploration seismology, fractures are not infinite along their long dimension but may be approximated as having infinite length. The primary advantage of such an approximation is that it effectively converts the problem from 3D to 2D and thus may be solved more inexpensively (using less computing resources). This section will repeat the same steps that we used for the full 3D system.

The 3D system of equations (1.9),

$$\rho \dot{\vec{v}} = \boldsymbol{\nabla} \cdot \boldsymbol{\sigma} \quad \text{and} \quad \dot{\boldsymbol{\sigma}} = \lambda(\boldsymbol{\nabla} \cdot \vec{v})\mathbf{I} + \mu\left(\boldsymbol{\nabla} \otimes \vec{v} + (\boldsymbol{\nabla} \otimes \vec{v})^{\mathrm{T}}\right), \tag{1.199}$$

can be represented coordinate-wise as

$$\begin{aligned}
\rho \dot{v}^1 &= \nabla_1 \sigma^{11} + \nabla_2 \sigma^{12} + \nabla_3 \sigma^{13},\\
\rho \dot{v}^2 &= \nabla_1 \sigma^{12} + \nabla_2 \sigma^{22} + \nabla_3 \sigma^{23},\\
\rho \dot{v}^3 &= \nabla_1 \sigma^{13} + \nabla_2 \sigma^{23} + \nabla_3 \sigma^{33},\\
\dot{\sigma}^{11} &= \lambda\left(\nabla_1 v^1 + \nabla_2 v^2 + \nabla_3 v^3\right) + 2\mu \nabla_1 v^1,\\
\dot{\sigma}^{22} &= \lambda\left(\nabla_1 v^1 + \nabla_2 v^2 + \nabla_3 v^3\right) + 2\mu \nabla_2 v^2,\\
\dot{\sigma}^{33} &= \lambda\left(\nabla_1 v^1 + \nabla_2 v^2 + \nabla_3 v^3\right) + 2\mu \nabla_3 v^3,\\
\dot{\sigma}^{23} &= \mu\left(\nabla_2 v^3 + \nabla_3 v^2\right),\\
\dot{\sigma}^{13} &= \mu\left(\nabla_1 v^3 + \nabla_3 v^1\right),\\
\dot{\sigma}^{12} &= \mu\left(\nabla_1 v^2 + \nabla_2 v^1\right).
\end{aligned} \tag{1.200}$$

Suppose that there is translational symmetry along an axis, e.g., x_3. Then, $\nabla_3 \equiv 0$, and also

$$\begin{aligned}
\rho \dot{v}^1 &= \nabla_1 \sigma^{11} + \nabla_2 \sigma^{12},\\
\rho \dot{v}^2 &= \nabla_1 \sigma^{12} + \nabla_2 \sigma^{22},\\
\rho \dot{v}^3 &= \nabla_1 \sigma^{13} + \nabla_2 \sigma^{23}\\
\dot{\sigma}^{11} &= \lambda\left(\nabla_1 v^1 + \nabla_2 v^2\right) + 2\mu \nabla_1 v^1,\\
\dot{\sigma}^{22} &= \lambda\left(\nabla_1 v^1 + \nabla_2 v^2\right) + 2\mu \nabla_2 v^2,\\
\dot{\sigma}^{33} &= \lambda\left(\nabla_1 v^1 + \nabla_2 v^2\right),\\
\dot{\sigma}^{23} &= \mu \nabla_2 v^3,\\
\dot{\sigma}^{13} &= \mu \nabla_1 v^3,\\
\dot{\sigma}^{12} &= \mu\left(\nabla_1 v^2 + \nabla_2 v^1\right).
\end{aligned} \tag{1.201}$$

We seek a solution in the form $v^3 = const$, $\sigma^{23} = const$, and $\sigma^{13} = const$. Then,

$$\begin{aligned}
\rho \dot{v}^1 &= \nabla_1 \sigma^{11} + \nabla_2 \sigma^{12},\\
\rho \dot{v}^2 &= \nabla_1 \sigma^{12} + \nabla_2 \sigma^{22},\\
\dot{\sigma}^{11} &= \lambda\left(\nabla_1 v^1 + \nabla_2 v^2\right) + 2\mu \nabla_1 v^1,\\
\dot{\sigma}^{22} &= \lambda\left(\nabla_1 v^1 + \nabla_2 v^2\right) + 2\mu \nabla_2 v^2,\\
\dot{\sigma}^{33} &= \lambda\left(\nabla_1 v^1 + \nabla_2 v^2\right),\\
\dot{\sigma}^{12} &= \mu\left(\nabla_1 v^2 + \nabla_2 v^1\right).
\end{aligned} \tag{1.202}$$

The equation $\dot{\sigma}^{33} = \lambda\left(\nabla_1 v^1 + \nabla_2 v^2\right)$ can be dropped, because σ^{33} is absent on the right sides of the other equations. As a result, we obtain the system

$$\begin{aligned}
\rho \dot{v}^1 &= \nabla_1 \sigma^{11} + \nabla_2 \sigma^{12},\\
\rho \dot{v}^2 &= \nabla_1 \sigma^{12} + \nabla_2 \sigma^{22},\\
\dot{\sigma}^{11} &= \lambda\left(\nabla_1 v^1 + \nabla_2 v^2\right) + 2\mu \nabla_1 v^1,\\
\dot{\sigma}^{22} &= \lambda\left(\nabla_1 v^1 + \nabla_2 v^2\right) + 2\mu \nabla_2 v^2,\\
\dot{\sigma}^{12} &= \mu\left(\nabla_1 v^2 + \nabla_2 v^1\right).
\end{aligned} \tag{1.203}$$

Now introduce the concepts of 2D vectors and a 2D symmetric tensor of the second rank, which are used next. As a result, equation (1.203) can be written in vector form as

$$\begin{cases} \rho\dot{\vec{v}} = \nabla \cdot \boldsymbol{\sigma} \\ \dot{\boldsymbol{\sigma}} = \lambda(\nabla \cdot \vec{v})\mathbf{I} + \mu\left(\nabla \otimes \vec{v} + (\nabla \otimes \vec{v})^{\mathrm{T}}\right), \end{cases} \quad (1.204)$$

which coincides with the 3D system (1.9).

1.9.1. Selection of coordinate system

In a software implementation of this process, as well as in the previous 3D case, it is convenient to store the state of a medium (in the case being considered, the components of velocity and stress) in a fixed Cartesian coordinate system $(\hat{x}_1, \hat{x}_2)$. At the same time, derivatives are approximately computed in directions $\hat{\xi}_1$ and $\hat{\xi}_2$, selected according to the local configuration of grid nodes.

Let us derive an expression for the components of gradient ∇ in the new coordinate system. For this task, consider the following expressions.

By definition, the derivative in the direction $\hat{\xi}_j$ is

$$\frac{\partial \hat{x}_i}{\partial \hat{\xi}_j} = \left(\hat{\xi}_j \cdot \nabla\right)\hat{x}_i. \quad (1.205)$$

Represent equation (1.205) as the sum

$$\frac{\partial \hat{x}_i}{\partial \hat{\xi}_j} = \left(\hat{\xi}_j \cdot \nabla\right)\hat{x}_i = \sum_k \left(\hat{\xi}_j\right)^k \frac{\partial \hat{x}_i}{\partial \hat{x}_k} = \sum_k \left(\hat{\xi}_j\right)^k \delta^{ik} = \left(\hat{\xi}_j\right)^k, \quad (1.206)$$

which can be written in matrix notation as

$$\frac{\partial(\hat{x}_1, \hat{x}_2)}{\partial(\hat{\xi}_1, \hat{\xi}_2)} = \begin{bmatrix} (\hat{\xi}_1)^1 & (\hat{\xi}_2)^1 \\ (\hat{\xi}_1)^2 & (\hat{\xi}_2)^2 \end{bmatrix} = \begin{bmatrix} \hat{\xi}_1 & \hat{\xi}_2 \end{bmatrix}, \quad (1.207)$$

where

$$\frac{\partial(a^1, a^2)}{\partial(b^1, b^2)} = \begin{bmatrix} \dfrac{\partial a^1}{\partial b^1} & \dfrac{\partial a^1}{\partial b^2} \\ \dfrac{\partial a^2}{\partial b^1} & \dfrac{\partial a^2}{\partial b^2} \end{bmatrix}. \quad (1.208)$$

Introduce the notation

$$\begin{bmatrix} \vec{q}_1^{\,\mathrm{T}} \\ \vec{q}_2^{\,\mathrm{T}} \end{bmatrix} = \begin{bmatrix} (q_1)^1 & (q_1)^2 \\ (q_2)^1 & (q_2)^2 \end{bmatrix} = \frac{\partial(\hat{\xi}_1, \hat{\xi}_2)}{\partial(\hat{x}_1, \hat{x}_2)} = \left(\frac{\partial(\hat{x}_1, \hat{x}_2)}{\partial(\hat{\xi}_1, \hat{\xi}_2)}\right)^{-1}, \quad (1.209)$$

where $\vec{q}_i \cdot \hat{\xi}_j = \delta^{ij}$.

Thus, $(\vec{q}_1, \vec{q}_2)$ is the basis biorthogonal relative to $\hat{\xi}_1, \hat{\xi}_2$.

In the 3D case, the vectors of the biorthogonal basis can be expressed as

$$\vec{p}_1 = \frac{\vec{\eta}_2 \times \vec{\eta}_3}{(\vec{\eta}_1, \vec{\eta}_2, \vec{\eta}_3)}, \quad \vec{p}_2 = \frac{\vec{\eta}_3 \times \vec{\eta}_1}{(\vec{\eta}_1, \vec{\eta}_2, \vec{\eta}_3)}, \quad \text{and} \quad \vec{p}_3 = \frac{\vec{\eta}_1 \times \vec{\eta}_2}{(\vec{\eta}_1, \vec{\eta}_2, \vec{\eta}_3)}, \quad (1.210)$$

where the numerators are cross (vector) products, and the denominators are mixed (scalar triple) products.

Introducing the notations

$$\vec{\eta}_1 = \begin{bmatrix} (\hat{\xi}_1)^1 \\ (\hat{\xi}_1)^2 \\ 0 \end{bmatrix}, \quad \vec{\eta}_2 = \begin{bmatrix} (\hat{\xi}_2)^1 \\ (\hat{\xi}_2)^2 \\ 0 \end{bmatrix}, \quad \text{and} \quad \vec{\eta}_3 = \begin{bmatrix} 0 \\ 0 \\ 1 \end{bmatrix}, \quad (1.211)$$

we obtain

$$(\vec{\eta}_1, \vec{\eta}_2, \vec{\eta}_3) = \vec{\eta}_1 \cdot (\vec{\eta}_2 \times \vec{\eta}_3) = \begin{bmatrix} (\hat{\xi}_1)^1 \\ (\hat{\xi}_1)^2 \\ 0 \end{bmatrix} \cdot \begin{bmatrix} (\hat{\xi}_2)^2 \\ -(\hat{\xi}_2)^1 \\ 0 \end{bmatrix} \times \begin{bmatrix} 0 \\ 0 \\ 1 \end{bmatrix}$$
$$= \begin{bmatrix} (\hat{\xi}_1)^1 \\ (\hat{\xi}_1)^2 \\ 0 \end{bmatrix} \cdot \begin{bmatrix} (\hat{\xi}_2)^2 \\ -(\hat{\xi}_2)^1 \\ 0 \end{bmatrix} = (\hat{\xi}_1)^1 (\hat{\xi}_2)^2 - (\hat{\xi}_1)^2 (\hat{\xi}_2)^1, \quad (1.212)$$

$$\vec{\eta}_2 \times \vec{\eta}_3 = \begin{bmatrix} (\hat{\xi}_2)^1 \\ (\hat{\xi}_2)^2 \\ 0 \end{bmatrix} \times \begin{bmatrix} 0 \\ 0 \\ 1 \end{bmatrix} = \begin{bmatrix} (\hat{\xi}_2)^2 \\ (\hat{\xi}_2)^1 \\ 0 \end{bmatrix}, \quad (1.213)$$

$$\vec{\eta}_3 \times \vec{\eta}_1 = \begin{bmatrix} 0 \\ 0 \\ 1 \end{bmatrix} \times \begin{bmatrix} (\hat{\xi}_1)^1 \\ (\hat{\xi}_1)^2 \\ 0 \end{bmatrix} = \begin{bmatrix} -(\hat{\xi}_1)^2 \\ (\hat{\xi}_1)^1 \\ 0 \end{bmatrix}, \quad (1.214)$$

and

$$\vec{\eta}_1 \times \vec{\eta}_2 = \begin{bmatrix} (\hat{\xi}_1)^1 \\ (\hat{\xi}_1)^2 \\ 0 \end{bmatrix} \times \begin{bmatrix} (\hat{\xi}_2)^1 \\ (\hat{\xi}_2)^2 \\ 0 \end{bmatrix} = \begin{bmatrix} 0 \\ 0 \\ (\hat{\xi}_1)^1 (\hat{\xi}_2)^2 - (\hat{\xi}_1)^2 (\hat{\xi}_2)^1 \end{bmatrix} \tag{1.215}$$

Inasmuch as the vectors $\vec{p}_j$ are

$$\vec{p}_1 = \frac{\vec{\eta}_2 \times \vec{\eta}_3}{(\vec{\eta}_1, \vec{\eta}_2, \vec{\eta}_3)} = \begin{bmatrix} \dfrac{(\hat{\xi}_2)^2}{(\hat{\xi}_1)^1 (\hat{\xi}_2)^2 - (\hat{\xi}_1)^2 (\hat{\xi}_2)^1} \\ -\dfrac{(\hat{\xi}_2)^1}{(\hat{\xi}_1)^1 (\hat{\xi}_2)^2 - (\hat{\xi}_1)^2 (\hat{\xi}_2)^1} \\ 0 \end{bmatrix}, \tag{1.216}$$

$$\vec{\eta}_2 = \frac{\vec{\eta}_3 \times \vec{\eta}_1}{(\vec{\eta}_1, \vec{\eta}_2, \vec{\eta}_3)} = \begin{bmatrix} \dfrac{(\hat{\xi}_1)^2}{(\hat{\xi}_1)^1 (\hat{\xi}_2)^2 - (\hat{\xi}_1)^2 (\hat{\xi}_2)^1} \\ -\dfrac{(\hat{\xi}_1)^1}{(\hat{\xi}_1)^1 (\hat{\xi}_2)^2 - (\hat{\xi}_1)^2 (\hat{\xi}_2)^1} \\ 0 \end{bmatrix}, \tag{1.217}$$

and

$$\vec{p}_3 = \frac{\vec{\eta}_1 \times \vec{\eta}_2}{(\vec{\eta}_1, \vec{\eta}_2, \vec{\eta}_3)} = \begin{bmatrix} 0 \\ 0 \\ \dfrac{(\hat{\xi}_1)^1 (\hat{\xi}_2)^2 - (\hat{\xi}_1)^2 (\hat{\xi}_2)^1}{(\hat{\xi}_1)^1 (\hat{\xi}_2)^2 - (\hat{\xi}_1)^2 (\hat{\xi}_2)^1} \end{bmatrix} = \begin{bmatrix} 0 \\ 0 \\ 1 \end{bmatrix}, \tag{1.218}$$

we have

$$\vec{q}_1 = \begin{bmatrix} \dfrac{(\hat{\xi}_2)^2}{(\hat{\xi}_1)^1 (\hat{\xi}_2)^2 - (\hat{\xi}_1)^2 (\hat{\xi}_2)^1} \\ -\dfrac{(\hat{\xi}_2)^1}{(\hat{\xi}_1)^1 (\hat{\xi}_2)^2 - (\hat{\xi}_1)^2 (\hat{\xi}_2)^1} \end{bmatrix} \quad \text{and} \quad \vec{q}_2 = \begin{bmatrix} -\dfrac{(\hat{\xi}_1)^2}{(\hat{\xi}_1)^1 (\hat{\xi}_2)^2 - (\hat{\xi}_1)^2 (\hat{\xi}_2)^1} \\ \dfrac{(\hat{\xi}_1)^1}{(\hat{\xi}_1)^1 (\hat{\xi}_2)^2 - (\hat{\xi}_1)^2 (\hat{\xi}_2)^1} \end{bmatrix}. \tag{1.219}$$

To simplify the subsequent formulae significantly, let us associate each of the vectors $\vec{q}_j$ with an orthonormal basis $(\hat{n}_{j,0}, \hat{n}_{j,1})$, in which $\hat{n}_{j,0}$ is collinear to $\vec{q}_j$, i.e.,

$$\hat{n}_{j,0} = \frac{\vec{q}_j}{\|\vec{q}_j\|} \quad \text{with } l_j = \|\vec{q}_j\|, \tag{1.220}$$

whereas unit vector $\hat{n}_{j,1}$ is chosen arbitrarily. In the following, we omit the first index j and, in $\hat{n}_{j,0}$, the second index 0, if their absence is evident from the context.

Equations (1.204) are written in a form independent of the choice of a particular coordinate system. It is only necessary to specify the operators which should be substituted for the gradient components. In the Cartesian coordinate system,

$$\boldsymbol{\nabla} = \{\nabla_1, \nabla_2\} = \frac{\partial}{\partial \hat{x}}, \quad \text{where } \nabla_i = \frac{\partial}{\partial \hat{x}_i}. \tag{1.221}$$

Otherwise, we apply the change of variables formula and obtain

$$\nabla_i = \sum_j \frac{\partial \hat{\xi}_j}{\partial \hat{x}_i} \frac{\partial}{\partial \hat{\xi}_j} \quad \text{and} \quad \boldsymbol{\nabla} = \{\nabla_1, \nabla_2\}. \tag{1.222}$$

1.9.2. Notation generalization for differential equations

Let us represent the system of equations in the canonical form

$$\dot{\vec{u}} + \sum_j \left(\mathbf{A}_j \cdot \frac{\partial \vec{u}}{\partial \hat{\xi}_j} \right) = 0, \tag{1.223}$$

where vector $\vec{u}$ comprises all of the variables of the system. In our case of equations (1.204), the vector is

$$\vec{u} = \{\vec{v}, \boldsymbol{\sigma}\}^{\mathrm{T}} = \{v^1, v^2, v^3, \sigma^{11}, \sigma^{12}, \sigma^{22}\}^{\mathrm{T}}. \tag{1.224}$$

To obtain a desired expression for matrices $\mathbf{A}_j$, we repeat the derivation of the 3D case by replacing the 3D vectors and tensors with 2D vectors. As a result, we have

$$\mathbf{A}_j \begin{bmatrix} \vec{v} \\ \boldsymbol{\sigma} \end{bmatrix} = -l_j \begin{bmatrix} \dfrac{1}{\rho} (\boldsymbol{\sigma} \cdot \hat{n}_j) \\ \lambda (\hat{n}_j \cdot \vec{v}) \mathbf{I} + \mu (\hat{n}_j \otimes \vec{v} + \vec{v} \otimes \hat{n}_j) \end{bmatrix}. \tag{1.225}$$

1.9.3. Spectral study of the system

The purpose of this section is to determine all eigenvalues c and eigenvectors $\vec{u} = \{\vec{v}, \boldsymbol{\sigma}\}^{\mathrm{T}}$ of the matrices $\mathbf{A}_j$:

$$\mathbf{A}\vec{u} = c\vec{u} \tag{1.226}$$

(note that the subscript j will be omitted until the end of this section).

As in the 3D case, we introduce the diagonal matrix

$$\boldsymbol{\Lambda} = \mathrm{diag}\{c_j\} = \begin{bmatrix} c_1 & 0 & \cdots & 0 \\ 0 & c_2 & \cdots & 0 \\ \vdots & \vdots & \ddots & \vdots \\ 0 & 0 & \cdots & c_5 \end{bmatrix} \tag{1.227}$$

that comprises all eigenvalues of matrix $\mathbf{A}$, and the matrix

$$\boldsymbol{\Omega}^{-1} = \{\vec{u}_i\} = \begin{bmatrix} (\vec{u}_1)_1 & (\vec{u}_2)_1 & \cdots & (\vec{u}_5)_1 \\ (\vec{u}_1)_2 & (\vec{u}_2)_2 & \cdots & (\vec{u}_5)_2 \\ \vdots & \vdots & \ddots & \vdots \\ (\vec{u}_1)_5 & (\vec{u}_2)_5 & \cdots & (\vec{u}_5)_5 \end{bmatrix}, \tag{1.228}$$

whose columns are the eigenvectors of $\mathbf{A}$ following in the same order as the corresponding eigenvalues in $\boldsymbol{\Lambda}$.

As in the 3D case, each row in $\boldsymbol{\Omega}$ is an eigenrow of $\mathbf{A}$, with $\vec{q}^{\,\mathrm{T}}$ being the eigenrow of $\mathbf{A}$ associated with the eigenvalue Q when the equality $\vec{q}^{\,\mathrm{T}}\mathbf{A} = \vec{q}^{\,\mathrm{T}}Q$ holds true. Being transposed, the eigenrows turn into the eigenvectors of $\mathbf{A}^{\mathrm{T}}$. These eigenvectors can be determined as the solution of the adjoint problem with respect to equation (1.226),

$$\mathbf{A}^{\mathrm{T}}\vec{u} = c\vec{u}. \tag{1.229}$$

The adjoint problem has the same set of eigenvalues. And, we can determine that the eigenvectors are coordinated because

$$\boldsymbol{\Omega}^{-1}\boldsymbol{\Omega} = \mathbf{I}. \tag{1.230}$$

It also follows that different elements of $\boldsymbol{\Omega}$ have dimensions of the reciprocal of velocity and the reciprocal of stress, which, after left-multiplication by $\boldsymbol{\Omega}^{-1}$, yield dimensionless matrix $\mathbf{I}$.

1.9.4. Forward problem

By the same mathematical manipulation as in the 3D case, we arrive at the same equation for the eigenvalues,

$$(\lambda + \mu)(\hat{n} \cdot \vec{v})\hat{n} = \left(l^{-2}c^2\rho - \mu\right)\vec{v}. \tag{1.231}$$

It has three qualitatively different groups of solutions.

1) If vectors $\vec{v}$ and $\hat{n}$ are collinear and $\vec{v} \neq 0$, then $(\hat{n} \cdot \vec{v}) = \|\hat{n}\|\,\|\vec{v}\| = \|\vec{v}\| = v$ and $\vec{v} = v\hat{n}$.
   Substituting into equation (1.231) $c^2 = l^2 \dfrac{\lambda + 2\mu}{\rho}$, which follows from $(\lambda + \mu)v\hat{n} = \left(l^{-2}c^2\rho - \mu\right)v\hat{n}$ and $\lambda + \mu = l^{-2}c^2\rho - \mu$, and introducing, $c_1 = \sqrt{\dfrac{\lambda + 2\mu}{\rho}}$, which represents the velocity of pressure waves in an elastic material, we obtain the eigenpair $c = \pm lc_1$ and $\vec{v} = \alpha\hat{n}$. Also, after some easy algebra, we have

$$\begin{aligned} \sigma^{ik} &= \left(-lc^{-1}\left(\lambda(\hat{n} \cdot \vec{v})\mathbf{I} + \mu(\hat{n} \otimes \vec{v} + \vec{v} \otimes \hat{n})\right)\right)^{ik} \\ &= -lc^{-1}\left(\lambda\alpha\delta^{ik} + \mu\alpha\left(n^i n^k + n^k n^i\right)\right) \\ &= -lc^{-1}\left(\lambda\alpha\delta^{ik} + 2\mu\alpha n^i n^k\right) \\ &= -lc^{-1}\alpha\left(\lambda\delta^{ik} + 2\mu(\hat{n} \otimes \hat{n})^{ik}\right) \end{aligned} \tag{1.232}$$

and

$$\boldsymbol{\sigma} = \mp\frac{\alpha}{c_1}(\lambda\mathbf{I} + 2\mu\hat{n} \otimes \hat{n}), \tag{1.233}$$

where α is an arbitrary nonzero multiplier with a dimension of velocity.

2) If vectors $\vec{v}$ and $\hat{n}$ are orthogonal and $\vec{v} \neq 0$, the left side of equation (1.231) vanishes,

$$0 = \left(l^{-2}c^2\rho - \mu\right)\vec{v}. \tag{1.234}$$

This is valid only when the right side vanishes too,

$$l^{-2}c^2\rho - \mu = 0, \tag{1.235}$$

where $c^2 = \dfrac{\mu l^2}{\rho}$.

After introducing the notation $c_2 = \sqrt{\dfrac{\mu}{\rho}}$, which represents the velocity of shear waves in an elastic material, we can describe the set of eigenvalues and eigenvectors as

$$c = \pm lc_2, \quad \hat{n} \cdot \vec{v} = 0, \tag{1.236}$$

and

$$\begin{aligned} \boldsymbol{\sigma} &= -lc^{-1}\left(\lambda(\hat{n} \cdot \vec{v})\mathbf{I} + \mu(\hat{n} \otimes \vec{v} + \vec{v} \otimes \hat{n})\right) \\ &= -lc^{-1}\mu\left(\hat{n} \otimes \vec{v} + \vec{v} \otimes \hat{n}\right) = \mp\frac{\mu}{c_2}\left(\hat{n} \otimes \vec{v} + \vec{v} \otimes \hat{n}\right). \end{aligned} \tag{1.237}$$

The equation $\hat{n} \cdot \vec{v} = \Sigma_i n^i v^i = 0$ has one solution, as we demonstrate next.

Inasmuch as $\|\hat{n}\| = 1$, there exists a nonzero n^i, e.g., n^1. Then, $v^1 + \frac{n^2}{n^1}v^2 = 0$ and $\vec{v} = \left\{-\frac{n^2}{n^1}\alpha, \alpha\right\}$, where $\alpha \in \mathbb{R}$.

Therefore, a total of two linearly independent eigenvectors are in this set.

3) If the eigenvalue is zero and the components of the eigenvector related to velocity are zero, we obtain the following equation for the remaining components of the eigenvector,

$$c = 0, \quad \vec{v} = 0, \quad \text{and} \quad \boldsymbol{\sigma} \cdot \hat{n} = -l^{-1}c\rho\vec{v} = 0. \quad (1.238)$$

There is one solution to the system

$$(\boldsymbol{\sigma} \cdot \hat{n})^k = \sum_i n^i \sigma^{ik} = 0, \quad k = 1 \text{ and } 2. \quad (1.239)$$

Indeed, because $\|\hat{n}\| = 1$, there exists a nonzero n^i, e.g., n^1. Then, $\sigma^{1k} + \frac{n^2}{n^1}\sigma^{2k} = 0$ is valid for $k = 1$ and 2, and

$$\{\sigma^{1k}, \sigma^{2k}\} = \left\{-\frac{n^2}{n^1}\alpha_k, \alpha_k\right\}, \quad (1.240)$$

where $\alpha_k \in \mathbb{R}$.

Because tensor $\boldsymbol{\sigma}$ is symmetric, i.e., $\sigma^{ij} = \sigma^{ji}$, which yields one more equation, one solution remains and, therefore, there is one eigenvector associated with zero eigenvalue.

In items 1, 2, and 3, we determine two, two, and one eigenvectors, respectively. In total, all of the five linearly independent eigenvectors and their eigenvalues are thus determined.

1.9.5. Adjoint problem

The transpose of matrix $\mathbf{A}$ from equation (1.225) is derived from the equality $\vec{u}_1{}^{\mathrm{T}}\mathbf{A}^{\mathrm{T}}\vec{u} = \vec{u}^{\mathrm{T}}\mathbf{A}\vec{u}_1$, where $\vec{u}$ and $\vec{u}_1$ are arbitrary vectors.

Repeating the derivation of the 3D case by replacing the 3D vectors and tensors with 2D vectors allows us to obtain

$$\mathbf{A}^{\mathrm{T}}\begin{bmatrix}\vec{v}\\ \boldsymbol{\sigma}\end{bmatrix} = -l\begin{bmatrix}\lambda(\boldsymbol{\sigma} \div \mathbf{I})\hat{n} + 2\mu(\boldsymbol{\sigma}\cdot\hat{n})\\ \frac{1}{2\rho}(\vec{v}\otimes\hat{n} + \hat{n}\otimes\vec{v})\end{bmatrix}. \quad (1.241)$$

As in the 3D case, $\vec{v}$ and $\boldsymbol{\sigma}$ have dimensions of the reciprocal of velocity and the reciprocal of stress, respectively.

Following the reasoning of the 3D case, we obtain three groups of eigenvalues and eigenvectors.

1) If vectors $\vec{v}$ and $\hat{n}$ are collinear and $\vec{v} \neq 0$, then $(\hat{n}\cdot\vec{v}) = \|\hat{n}\|\|\vec{v}\| = \|\vec{v}\| = v$ and $\vec{v} = v \cdot \hat{n}$.
Here, $(\lambda+\mu)v\hat{n} = (l^{-2}c^2\rho - \mu)v\hat{n}, \lambda + \mu = l^{-2}c^2\rho - \mu$ and $c^2 = l^2\frac{\lambda + 2\mu}{\rho}$.
In this case, we obtain the eigenpair $c = \pm lc_1$ and $\vec{v} = \alpha\hat{n}$. Similarly, we have,

$$\sigma^{ik} = -\frac{l}{2\rho c}(v^i n^k + n^i v^k) = -\frac{l\alpha}{2\rho c}(n^i n^k + n^i n^k)$$
$$= \mp\frac{\alpha}{\rho c_1}(\hat{n}\otimes\hat{n})^{ik} \quad (1.242)$$

and

$$\boldsymbol{\sigma} = \mp\frac{\alpha}{\rho c_1}(\hat{n}\otimes\hat{n}), \quad (1.243)$$

where α is an arbitrary nonzero multiplier with a dimension of velocity.

2) If vectors $\vec{v}$ and $\hat{n}$ are orthogonal and $\vec{v} \neq 0$, we have

$$0 = (l^{-2}c^2\rho - \mu)\vec{v}. \quad (1.244)$$

This is valid only when the right side vanishes too,

$$l^{-2}c^2\rho - \mu = 0, \quad (1.245)$$

where $c^2 = \frac{\mu l^2}{\rho}$.

Now, we can describe the set of eigenvalues and eigenvectors

$$c = \pm lc_2, \quad \hat{n}\cdot\vec{v} = 0, \quad (1.246)$$

and

$$\boldsymbol{\sigma} = -\frac{l}{2\rho c}(\vec{v}\otimes\hat{n} + \hat{n}\otimes\vec{v})$$
$$= \mp\frac{1}{2\rho c_2}(\vec{v}\otimes\hat{n} + \hat{n}\otimes\vec{v}). \quad (1.247)$$

The equation $\hat{n}\cdot\vec{v} = \Sigma_i n^i v^i = 0$ has one solution, as we demonstrate in the following analysis.

Inasmuch as $\|\hat{n}\| = 1$, there exists a nonzero n^i, e.g., n^1. Then, $v^1 + \frac{n^2}{n^1}v^2 = 0$ and $v = \left\{-\frac{n^2}{n^1}\alpha, \alpha\right\}$, where $\alpha \in \mathbb{R}$.

Therefore, a total of two linearly independent eigenvectors $\{\vec{v}, \boldsymbol{\sigma}\}$ are in this set.

3) If the eigenvalue is zero and the components of the eigenvector related to velocity are zero, we obtain the following equation for the remaining components of the eigenvector

$$c = 0, \quad \vec{v} = 0, \quad \text{and}$$
$$0 = c\vec{v} = -l\left(\lambda(\boldsymbol{\sigma} \div \mathbf{I})\hat{n} + 2\mu(\boldsymbol{\sigma}\cdot\hat{n})\right), \quad (1.248)$$

i.e., for $\boldsymbol{\sigma}$,

$$\begin{aligned} 0 &= \lambda(\boldsymbol{\sigma} \div \mathbf{I})\hat{n} + 2\mu(\boldsymbol{\sigma}\cdot\hat{n}) \\ &= \lambda\left(\sum_i \sigma^{ii}\right)n^k + 2\mu(\boldsymbol{\sigma}\cdot\hat{n})^k \\ &= \lambda n^k \sum_i \sigma^{ii} + 2\mu\sum_i \sigma^{ik}n^i = 0. \end{aligned} \quad (1.249)$$

There are two linearly independent solutions to the system

$$\lambda n^k \sum_i \sigma^{ii} + 2\mu \sum_i \sigma^{ik}n^i = 0, \quad k = 1 \text{ and } 2, \quad (1.250)$$

To prove this, let us represent the system as

$$\begin{cases} \lambda n^1\left(\sigma^{11} + \sigma^{22}\right) + 2\mu\left(n^1\sigma^{11} + n^2\sigma^{12}\right) = 0 \\ \lambda n^2\left(\sigma^{11} + \sigma^{22}\right) + 2\mu\left(n^1\sigma^{12} + n^2\sigma^{22}\right) = 0 \end{cases}. \quad (1.251)$$

A) If $n^2 = 0$ and $n^1 = 1$, then

$$\begin{cases} \lambda\left(\sigma^{11} + \sigma^{22}\right) + 2\mu\sigma^{11} = 0 \\ 2\mu\sigma^{12} = 0 \end{cases} \quad (1.252)$$

or

$$\begin{bmatrix} \lambda + 2\mu & \lambda & 0 \\ 0 & 0 & 2\mu \end{bmatrix}\begin{bmatrix}\sigma^{11} \\ \sigma^{22} \\ \sigma^{12}\end{bmatrix} = \mathbf{B}\begin{bmatrix}\sigma^{11} \\ \sigma^{22} \\ \sigma^{12}\end{bmatrix} = 0. \quad (1.253)$$

The rank of matrix **B** is two, which is why there are two linearly independent solutions.

B) If $n^1 \neq 0$ and $n^2 \neq 0$, then

$$\begin{cases} \lambda n^1\sigma^{11} + \lambda n^1\sigma^{22} + 2\mu n^1\sigma^{11} + 2\mu n^2\sigma^{12} = 0 \\ \lambda n^2\sigma^{11} + \lambda n^2\sigma^{22} + 2\mu n^1\sigma^{12} + 2\mu n^2\sigma^{22} = 0 \end{cases} \quad (1.254)$$

or

$$\begin{bmatrix} (\lambda + 2\mu)n^1 & \lambda n^1 & 2\mu n^2 \\ \lambda n^2 & (\lambda + 2\mu)n^2 & 2\mu n^1 \end{bmatrix}\begin{bmatrix}\sigma^{11} \\ \sigma^{22} \\ \sigma^{12}\end{bmatrix} = \mathbf{B}\begin{bmatrix}\sigma^{11} \\ \sigma^{22} \\ \sigma^{12}\end{bmatrix} = 0. \quad (1.255)$$

Suppose that the first row is linearly dependent on the second. Then there exists some α, such that

$$\begin{cases} \lambda n^1 + 2\mu n^1 = \alpha\lambda n^2 \\ \lambda n^1 = \alpha\left(\lambda n^2 + 2\mu n^2\right). \\ 2\mu n^2 = \alpha 2\mu n^1 \end{cases} \quad (1.256)$$

From the last equation, it follows that $n^2 = \alpha n^1$. Substituting this into the first two equations yields

$$\begin{aligned} \lambda + 2\mu &= \alpha^2\lambda \\ \lambda &= \alpha^2(\lambda + 2\mu), \end{aligned} \quad (1.257)$$

where $\alpha^2\lambda = \alpha^{-2}\lambda$, $\alpha^4 = 1$, and $\lambda + 2\mu = \lambda$, which is impossible at a nonzero value of parameter μ. Therefore, in this case, the rank of matrix **B** is also two, and there are two linearly independent solutions. As a result, we have

$$c = \pm lc_1, \quad \vec{v} = \alpha\hat{n}, \quad \text{and} \quad \boldsymbol{\sigma} = \mp\frac{\alpha}{c_1\rho}\hat{n}\otimes\hat{n}, \quad (1.258)$$

$$\begin{aligned} &c = \pm lc_2, \quad \hat{n}\cdot\vec{v} = 0, \quad \text{and} \\ &\boldsymbol{\sigma} = \mp\frac{1}{2c_2\rho}\left(\hat{n}\otimes\vec{v} + \vec{v}\otimes\hat{n}\right), \end{aligned} \quad (1.259)$$

and

$$\begin{aligned} &c = 0, \quad \vec{v} = 0, \quad \text{and} \\ &\lambda(\boldsymbol{\sigma} \div \mathbf{I})\hat{n} + 2\mu(\boldsymbol{\sigma}\cdot\hat{n}) = 0. \end{aligned} \quad (1.260)$$

1.9.6. Important notations and relations

Next, consider what changes must be made in the formulae for $\mathbf{N}_{ij} = \mathbf{N}_{ji}$ in the 2D case.

We define the symmetric matrices $\mathbf{N}_{ij} = \mathbf{N}_{ji}$ as

$$\mathbf{N}_{ij} = \frac{1}{2}\left(\hat{n}_i\otimes\hat{n}_j + \hat{n}_j\otimes\hat{n}_i\right), \quad i, j = 0 \text{ and } 1. \quad (1.261)$$

The matrices $\mathbf{N}_{ij}$ are orthogonal to each other with respect to the operation of the double dot product:

$$\mathbf{N}_{ij} \div \mathbf{N}_{kl} = \frac{1}{2}\left(1+\delta^{ij}\right)\delta^{ik}\delta^{jl} \quad \text{for } \forall i \le j, k \le l. \quad (1.262)$$

The effect of $\mathbf{N}_{ij}$ on vector $\hat{n}_k$ for $\forall i, j, k$ appears as

$$\mathbf{N}_{ij} \cdot \hat{n}_k = \frac{\hat{n}_i}{2}\left(\hat{n}_j \cdot \hat{n}_k\right) + \frac{\hat{n}_j}{2}\left(\hat{n}_i \cdot \hat{n}_k\right). \quad (1.263)$$

In particular,

$$\begin{aligned}\mathbf{N}_{ij} \cdot \hat{n} = \mathbf{N}_{ij} \cdot \hat{n}_0 &= \frac{\hat{n}_i}{2}\left(\hat{n}_j \cdot \hat{n}_0\right) + \frac{\hat{n}_j}{2}\left(\hat{n}_i \cdot \hat{n}_0\right)\\ &= 0 \quad \text{for } i, j = 1,2.\end{aligned} \quad (1.264)$$

Another important property of the matrices $\mathbf{N}_{ij}$ is the result of their double convolution with an arbitrary symmetric matrix $\boldsymbol{\sigma}$,

$$\forall i, j: \mathbf{N}_{ij} \div \boldsymbol{\sigma} = \hat{n}_i^{\mathrm{T}} \cdot \boldsymbol{\sigma} \cdot \hat{n}_j, \quad (1.265)$$

which, for the unit matrix, is

$$\forall i, j: \mathbf{N}_{ij} \div \mathbf{I} = \hat{n}_i^{\mathrm{T}} \cdot \mathbf{I} \cdot \hat{n}_j = \hat{n}_i^{\mathrm{T}} \cdot \hat{n}_j = \left(\hat{n}_i \cdot \hat{n}_j\right) = \delta^{ij}. \quad (1.266)$$

Also,

$$\mathbf{N}_{00} + \mathbf{N}_{11} = \mathbf{I}. \quad (1.267)$$

The decomposition of an arbitrary vector $\vec{v}$ with respect to an orthonormal basis is

$$\vec{v} = \left(\hat{n} \cdot \vec{v}\right)\hat{n} + \left(\hat{n}_1 \cdot \vec{v}\right)\hat{n}_1. \quad (1.268)$$

This leads to

$$\frac{1}{2}\left(\hat{n} \otimes \vec{v} + \vec{v} \otimes \hat{n}\right) = \left(\hat{n} \cdot \vec{v}\right)\mathbf{N}_{00} + \left(\hat{n}_1 \cdot \vec{v}\right)\mathbf{N}_{01}. \quad (1.269)$$

An arbitrary symmetric matrix $\boldsymbol{\sigma}$ can be expanded as

$$\boldsymbol{\sigma} = (\mathbf{N}_{00} \div \boldsymbol{\sigma})\mathbf{N}_{00} + 2(\mathbf{N}_{01} \div \boldsymbol{\sigma})\mathbf{N}_{01} + (\mathbf{N}_{11} \div \boldsymbol{\sigma})\mathbf{N}_{11}. \quad (1.270)$$

1.9.7. Normalization of nonzero eigenvectors

The dot product of eigenvectors $\{\vec{v}_d, \boldsymbol{\sigma}_d\}$ from the forward problem, which constitute the columns of $\boldsymbol{\Omega}^{-1}$, and eigenvectors $\{\vec{v}_c, \boldsymbol{\sigma}_c\}$ from the adjoint problem, which constitute the rows of $\boldsymbol{\Omega}$, must be unity for the corresponding pairs of vectors, i.e.,

$$\vec{v}_d \cdot \vec{v}_c + \boldsymbol{\sigma}_d \div \boldsymbol{\sigma}_c = 1, \quad (1.271)$$

and zero otherwise.

Let us substitute the preceding values of the eigenvectors into this equation in order to normalize their lengths to unity. For the first group of eigenvectors

$$\begin{aligned}&c = \pm lc_1, \quad \vec{v}_d = \alpha\hat{n}, \quad \boldsymbol{\sigma}_d = \mp\frac{\alpha}{c_1}\left(\lambda\mathbf{I} + 2\mu\left(\hat{n}\otimes\hat{n}\right)\right), \quad \text{and}\\ &\vec{v}_c = \hat{n}, \quad \boldsymbol{\sigma}_c = \mp\frac{1}{c_1\rho}\left(\hat{n}\otimes\hat{n}\right),\end{aligned} \quad (1.272)$$

we have

$$\begin{aligned}1 &= \vec{v}_d\cdot\vec{v}_c + \boldsymbol{\sigma}_d \div \boldsymbol{\sigma}_c = \alpha\left(\hat{n}\cdot\hat{n}\right) + \left(\mp\frac{\alpha}{c_1}\left(\lambda\mathbf{I} + 2\mu\left(\hat{n}\otimes\hat{n}\right)\right)\right)\\ &\div\left(\mp\frac{1}{c_1\rho}\left(\hat{n}\otimes\hat{n}\right)\right) = \alpha + \frac{\alpha}{c_1^2\rho}\left(\lambda\mathbf{I} + 2\mu\left(\hat{n}\otimes\hat{n}\right)\right)\div\left(\hat{n}\otimes\hat{n}\right)\\ &= \alpha\left(1 + \frac{\lambda}{c_1^2\rho}(\mathbf{N}_{00}\div\mathbf{I}) + \frac{2\mu}{c_1^2\rho}(\mathbf{N}_{00}\div\mathbf{N}_{00})\right).\end{aligned} \quad (1.273)$$

Using equation (1.266), we obtain

$$\begin{aligned}1 &= \alpha\left(1 + \frac{\lambda}{c_1^2\rho}(\mathbf{N}_{00}\div\mathbf{I}) + \frac{2\mu}{c_1^2\rho}(\mathbf{N}_{00}\div\mathbf{N}_{00})\right)\\ &= \alpha\left(1 + \frac{\lambda}{c_1^2\rho} + \frac{2\mu}{c_1^2\rho}\right) = \alpha\left(1 + \frac{\lambda+2\mu}{\rho}\frac{1}{c_1^2}\right)\\ &= \alpha\left(1 + \frac{\lambda+2\mu}{\rho}\frac{\rho}{\lambda+2\mu}\right) = 2\alpha,\\ &c = \pm lc_1, \quad \vec{v}_d = \alpha\hat{n}, \quad \boldsymbol{\sigma}_d = \mp\frac{\alpha}{c_1}(\lambda\mathbf{I} + 2\mu\mathbf{N}_{00}),\\ &\vec{v}_c = \hat{n}, \quad \text{and} \quad \boldsymbol{\sigma}_c = \mp\frac{1}{c_1\rho}\mathbf{N}_{00},\end{aligned} \quad (1.274)$$

where the length of the eigenvector from the adjoint problem is fixed arbitrarily and $\alpha = \frac{1}{2}$.

For the second group,

$$\begin{aligned}&c = \pm lc_2, \quad \hat{n}\cdot\vec{v}_d = 0,\\ &\boldsymbol{\sigma}_d = \mp\frac{\mu}{c_2}\left(\hat{n}\otimes\vec{v} + \vec{v}\otimes\hat{n}\right), \quad \hat{n}\cdot\vec{v}_c = 0, \quad \text{and}\\ &\boldsymbol{\sigma}_c = \mp\frac{1}{2c_2\rho}\left(\hat{n}\otimes\vec{v} + \vec{v}\otimes\hat{n}\right),\end{aligned} \quad (1.275)$$

we suppose that $\vec{v}_d = \alpha\hat{n}_1$ and $\vec{v}_c = \hat{n}_1$. Then, using equation (1.269), we obtain

$$\boldsymbol{\sigma}_d = \mp\frac{2\mu}{c_2}\left((\hat{n}\cdot\vec{v})\mathbf{N}_{00} + (\hat{n}_1\cdot\vec{v})\mathbf{N}_{01}\right) = \mp\frac{2\mu\alpha}{c_2}\left((\hat{n}\cdot\hat{n}_1)\mathbf{N}_{00} + (\hat{n}_1\cdot\hat{n}_1)\mathbf{N}_{01}\right) = \mp\frac{2\mu\alpha}{c_2}\mathbf{N}_{01}$$

and

$$\boldsymbol{\sigma}_c = \mp\frac{2}{2c_2\rho}\left((\hat{n}\cdot\vec{v})\mathbf{N}_{00} + (\hat{n}_1\cdot\vec{v})\mathbf{N}_{01}\right) = \mp\frac{1}{c_2\rho}\left((\hat{n}\cdot\hat{n}_k)\mathbf{N}_{00} + (\hat{n}_1\cdot\hat{n}_1)\mathbf{N}_{01}\right) = \mp\frac{1}{c_2\rho}\mathbf{N}_{01}. \tag{1.276}$$

Using the normalization condition and equation (1.262) yields

$$\begin{aligned}1 &= \vec{v}_d\cdot\vec{v}_c + \boldsymbol{\sigma}_d \div \boldsymbol{\sigma}_c = \alpha\hat{n}_1\cdot\hat{n}_1 + \left(\mp\frac{2\mu\alpha}{c_2}\mathbf{N}_{01}\right)\div\left(\mp\frac{1}{c_2\rho}\mathbf{N}_{01}\right)\\ &= \alpha\left(1+\frac{2\mu}{c_2^{\,2}\rho}\mathbf{N}_{01}\div\mathbf{N}_{01}\right) = \alpha\left(1+2\cdot\frac{1}{2}\right) = 2\alpha,\\ \vec{v}_c &= \hat{n}_1,\quad \boldsymbol{\sigma}_c = \mp\frac{1}{2c_2\rho}\mathbf{N}_{01},\quad \vec{v}_d = \alpha\hat{n}_1,\quad \text{and}\\ \boldsymbol{\sigma}_d &= \mp\frac{\mu\alpha}{c_2}\mathbf{N}_{01},\end{aligned} \tag{1.277}$$

where $\alpha = \frac{1}{2}$.

1.9.8. Normalization of zero eigenvalues

The vector $\boldsymbol{\sigma}_c = \mathbf{N}_{11} - \frac{\lambda}{\lambda+2\mu}\mathbf{N}_{00}$ of the eigenpair $\{0,\boldsymbol{\sigma}_c\}$ satisfies the condition of the eigenvector of the adjoint problem from equation (1.260), i.e., $\lambda(\boldsymbol{\sigma}\div\mathbf{I})\hat{n} + 2\mu(\boldsymbol{\sigma}\cdot\hat{n}) = 0$.

Indeed, using equations (1.264) and (1.266), we can obtain

$$\begin{aligned}&\lambda(\boldsymbol{\sigma}_c\div\mathbf{I})\hat{n} + 2\mu(\boldsymbol{\sigma}_c\cdot\hat{n})\\ &= \lambda\left(\left(\mathbf{N}_{11}-\frac{\lambda}{\lambda+2\mu}\mathbf{N}_{00}\right)\div\mathbf{I}\right)\hat{n}\\ &\quad + 2\mu\left(\left(\mathbf{N}_{11}-\frac{\lambda}{\lambda+2\mu}\mathbf{N}_{00}\right)\cdot\hat{n}\right)\\ &= \lambda(\mathbf{N}_{11}\div\mathbf{I})\hat{n} - \frac{\lambda^2}{\lambda+2\mu}(\mathbf{N}_{00}\div\mathbf{I})\hat{n} + 2\mu(\mathbf{N}_{11}\cdot\hat{n})\\ &\quad - \frac{2\lambda\mu}{\lambda+2\mu}(\mathbf{N}_{00}\cdot\hat{n}) = \lambda\delta^{11}\hat{n} - \frac{\lambda^2}{\lambda+2\mu}\delta^{00}\hat{n} + 2\mu\cdot 0\\ &\quad - \frac{2\lambda\mu}{\lambda+2\mu}\left(\frac{\hat{n}}{2}(\hat{n}\cdot\hat{n}) + \frac{\hat{n}}{2}(\hat{n}\cdot\hat{n})\right)\\ &= \left(\lambda - \frac{\lambda^2}{\lambda+2\mu} - \frac{2\lambda\mu}{\lambda+2\mu}\right)\hat{n} = \frac{\lambda^2+2\lambda\mu-\lambda^2-2\lambda\mu}{\lambda+2\mu}\hat{n} = 0.\end{aligned} \tag{1.278}$$

The vector suitable for the forward problem $(\boldsymbol{\sigma}\cdot\hat{n} = 0)$ is $\boldsymbol{\sigma}_d = \mathbf{N}_{11}$.

Indeed, using equation (1.264), we obtain

$$\boldsymbol{\sigma}_d\cdot\hat{n} = \mathbf{N}_{11}\cdot\hat{n} = 0. \tag{1.279}$$

The coefficients are chosen so that $\boldsymbol{\sigma}_d \div \boldsymbol{\sigma}_c = 1$.

Indeed,

$$\begin{aligned}\boldsymbol{\sigma}_d\div\boldsymbol{\sigma}_c &= \mathbf{N}_{11}\div\left(\mathbf{N}_{11} - \frac{\lambda}{\lambda+2\mu}\mathbf{N}_{00}\right)\\ &= (\mathbf{N}_{11}\div\mathbf{N}_{11}) - \frac{\lambda}{\lambda+2\mu}(\mathbf{N}_{11}\div\mathbf{N}_{00}) = 1.\end{aligned} \tag{1.280}$$

1.9.9. Matrices Λ, Ω, and Ω^{-1}

Now we can summarize the previous sections and represent the matrices into which **A** is expanded in the following explicit form:

$$\boldsymbol{\Lambda} = l\cdot\mathrm{diag}\{c_1, -c_1, c_2, -c_2, 0\}, \tag{1.281}$$

$$\begin{bmatrix}\omega_1\\ \omega_2\\ \omega_3\\ \omega_4\\ \omega_5\end{bmatrix} = \mathbf{\Omega}\begin{bmatrix}\vec{v}\\ \boldsymbol{\sigma}\end{bmatrix} = \begin{bmatrix}\hat{n}\cdot\vec{v} - \dfrac{1}{c_1\rho}\mathbf{N}_{00} \div \boldsymbol{\sigma}\\ \hat{n}\cdot\vec{v} + \dfrac{1}{c_1\rho}\mathbf{N}_{00} \div \boldsymbol{\sigma}\\ \hat{n}_1\cdot\vec{v} - \dfrac{1}{c_2\rho}\mathbf{N}_{01} \div \boldsymbol{\sigma}\\ \hat{n}_1\cdot\vec{v} + \dfrac{1}{c_2\rho}\mathbf{N}_{01} \div \boldsymbol{\sigma}\\ \left(\mathbf{N}_{11} - \dfrac{\lambda}{\lambda+2\mu}\mathbf{N}_{00}\right) \div \boldsymbol{\sigma}\end{bmatrix}, \quad (1.282)$$

and

$$\begin{bmatrix}\vec{v}\\ \boldsymbol{\sigma}\end{bmatrix} = \mathbf{\Omega}^{-1}\vec{\omega} = \frac{1}{2}\begin{bmatrix}(\omega_1+\omega_2)\hat{n} + (\omega_3+\omega_4)\hat{n}_1\\ (\omega_2-\omega_1)\left(\rho(c_1-c_3)\mathbf{N}_{00} + \rho c_3\mathbf{I}\right)\\ +2\rho c_2(\omega_4-\omega_3)\mathbf{N}_{01} + 2\omega_5(\mathbf{I}-\mathbf{N}_{00})\end{bmatrix}, \quad (1.283)$$

where $c_3 = \dfrac{\lambda}{\lambda+2\mu}c_1$.

Next, we need to address separate columns of $\mathbf{\Omega}^{-1}$, which are

$$\mathbf{\Omega}^{-1} = \frac{1}{2}\begin{bmatrix}\hat{n} & \hat{n} & \hat{n}_1 & \hat{n}_1 & 0\\ -\rho(c_1-c_3)\mathbf{N}_{00}+\rho c_3\mathbf{I} & \rho(c_1-c_3)\mathbf{N}_{00}+\rho c_3\mathbf{I} & -2\rho c_2\mathbf{N}_{01} & 2\rho c_2\mathbf{N}_{01} & 2(\mathbf{I}-\mathbf{N}_{00})\end{bmatrix}. \quad (1.284)$$

1.9.10. Coordinate-wise split

For the final system of equations,

$$\partial_t\vec{u} + \mathbf{A}_1\frac{\partial\vec{u}}{\partial\hat{\xi}_1} + \mathbf{A}_2\frac{\partial\vec{u}}{\partial\hat{\xi}_2} = 0, \quad (1.285)$$

to be solved exactly according to the formula for the Fourier series, the formula

$$\vec{u}\left(\hat{\xi}_1,\hat{\xi}_2,t+\tau\right) = \vec{u}\left(\hat{\xi}_1,\hat{\xi}_2,t\right) + \tau\frac{\partial\vec{u}\left(\hat{\xi}_1,\hat{\xi}_2,t\right)}{\partial t} + \frac{\tau^2}{2}\frac{\partial^2\vec{u}\left(\hat{\xi}_1,\hat{\xi}_2,t\right)}{\partial t^2} + O\left(\tau^3\right) \quad (1.286)$$

should be valid.

Let us introduce the corresponding operator

$$\mathbf{P} = \mathbf{I} + \tau\left(\frac{\partial}{\partial t}\right)_0 + \frac{\tau^2}{2}\left(\frac{\partial}{\partial t}\right)_0^2 + \mathbf{O}\left(\tau^3\right). \quad (1.287)$$

For each system, consider the operator

$$\mathbf{M}_j = \mathbf{I} + \tau\left(\frac{\partial}{\partial t}\right)_j + \frac{\tau^2}{2}\left(\frac{\partial}{\partial t}\right)_j^2 + \mathbf{O}\left(\tau^3\right). \quad (1.288)$$

This scheme $\mathbf{M}_i\mathbf{M}_j$ provides a second-order approximation in time. The implementation of this algorithm, however, requires data storage at three time levels. Practice has shown that, if the simplified superposition of operators $\mathbf{M}_2\mathbf{M}_1$ is applied, the corresponding scheme produces results very close to a second-order approximation in time. The only significant drawback of this approach is the asymmetry of the resulting scheme, which, is usually not critical.

1.9.11. Given external force

In this condition, the density of an external force that acts on the surface of a body and is exerted by other objects is given. When the force is zero, the surface is said to be free. This condition can be written formally as

$$\boldsymbol{\sigma}\cdot\vec{p} = \vec{f}, \quad (1.289)$$

where $\vec{p}$ denotes the outward unit normal to the body surface and $\vec{f}$ is the surface density of external forces. This expression contains two scalar conditions that compensate for the lack of data for two outgoing characteristics. Reducing the boundary conditions to the canonical form, we obtain

$$\mathbf{B}\begin{bmatrix}\vec{v}\\ \boldsymbol{\sigma}\end{bmatrix} = \boldsymbol{\sigma}\cdot\vec{p} \quad \text{with } \vec{b} = \vec{f}. \quad (1.290)$$

Then,

$$\begin{aligned}\vec{u}^{n+1} &= \vec{u}^{in} + \mathbf{\Omega}^{*,out}\left(\mathbf{B}\mathbf{\Omega}^{*,out}\right)^{-1}\left(\vec{b} - \mathbf{B}\vec{u}^{in}\right)\\ &= \vec{u}^{in} - \mathbf{\Omega}^{*,out}\left(\mathbf{B}\mathbf{\Omega}^{*,out}\right)^{-1}\left(\boldsymbol{\sigma}^{in}\cdot\vec{p} - \vec{f}\right),\end{aligned} \quad (1.291)$$

where $\vec{u}^{in} = \left\{\vec{v}^{in}, \boldsymbol{\sigma}^{in}\right\}$.

1.9.12. Given surface velocity

In this condition, it is assumed that, at the surface of a body, a velocity is given. For example, it could be the velocity of an external body that is in contact with the body being considered. When the velocity is zero, the body surface is fixed. This condition can be written formally as $\vec{v} = \vec{V}$, where $\vec{V}$ is the given velocity. This expression contains three scalar conditions that compensate for the lack of data for three outgoing characteristics. Reducing the boundary conditions to the canonical form yields

$$\mathbf{B}\begin{bmatrix} \vec{v} \\ \boldsymbol{\sigma} \end{bmatrix} = \vec{v} \quad \text{with } \vec{b} = \vec{V}. \tag{1.292}$$

Chapter 2

Parallel Implementation of the Numerical Method on Unstructured Grids

2.1. The grid-characteristic method on unstructured grids

Next we consider a numerical method for the solution of a scalar transport equation in the form

$$\frac{\partial u}{\partial t} + \lambda \frac{\partial u}{\partial \xi_i} = 0. \tag{2.1}$$

Let a triangular or tetrahedral grid be introduced into a domain of integration. As the grid nodes at which the approximately calculated values of a function are kept, we can choose the structural elements of the grid, such as triangles, vertices, and edges. The choice of these determines the degree of approximation. For instance, in a first-order method, we use only the vertices of the grid as reference points for linear interpolation. For interpolation in a second-order method, we use the vertices and the midpoints of the edges. In turn, in a third-order method, the vertices, the pairs of points that divide each edge into equal parts, and the centers of mass of the triangles are used, and so on. We want to obtain the values at the nodes for the next time level $t^{n+1} = t^n + \tau^n$ using the known values at the nodes for the level t^n.

Let us draw the characteristic

$$\left(\vec{r}, t\right) + \tau\left(\lambda \hat{\xi}_i, 1\right) = \left(\vec{r}_k, t^{n+1}\right) \tag{2.2}$$

through the kth node for the upper time level, where τ is the parameter of the straight line along which the solution u is constant and $\hat{\xi}_i$ is a unit vector. This characteristic crosses the lower time level at the point $\left(\vec{r}_k - \lambda\tau^n \hat{\xi}_i, t^n\right)$ (Figure 2.1).

If this point is within the domain of integration, it belongs to a triangle of the grid, and the value at this point can be determined by interpolation inside the triangle. The point $\vec{r}_k - \lambda\tau^n \hat{\xi}_i$ is located at the distance $\lambda\tau^n$

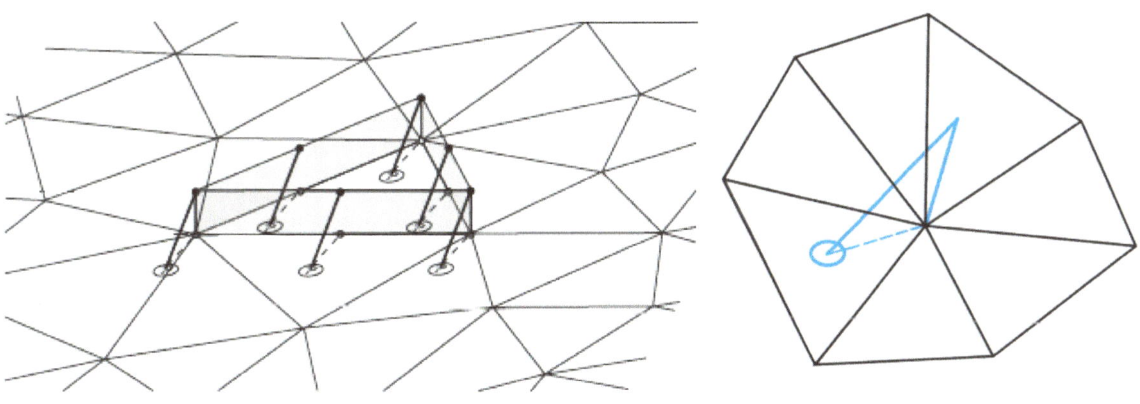

Figure 2.1. Characteristics extended down to the previous time level.

from the kth grid node and definitely will fall within a nearby triangle. The condition for a characteristic from the next time level to fall within nearby triangles is an essential condition for parallel computing.

2.2. Flowchart of parallel computing

Figure 2.2 displays a flowchart of parallel computing. The initialization step that comprises building a grid, dividing the grid into separate domains, and saving the domains in a file is performed on one core without the need for the Message Passing Interface (MPI) standard. The primary work cycle is arranged to run in parallel on a large number of cores. The work of each core begins with loading the grid prepared in the initialization step. Further work consists of successively integrating time steps. Each iteration, i.e., one time step, includes calculation of all grid points within a domain and exchange of information on border points with neighboring domains.

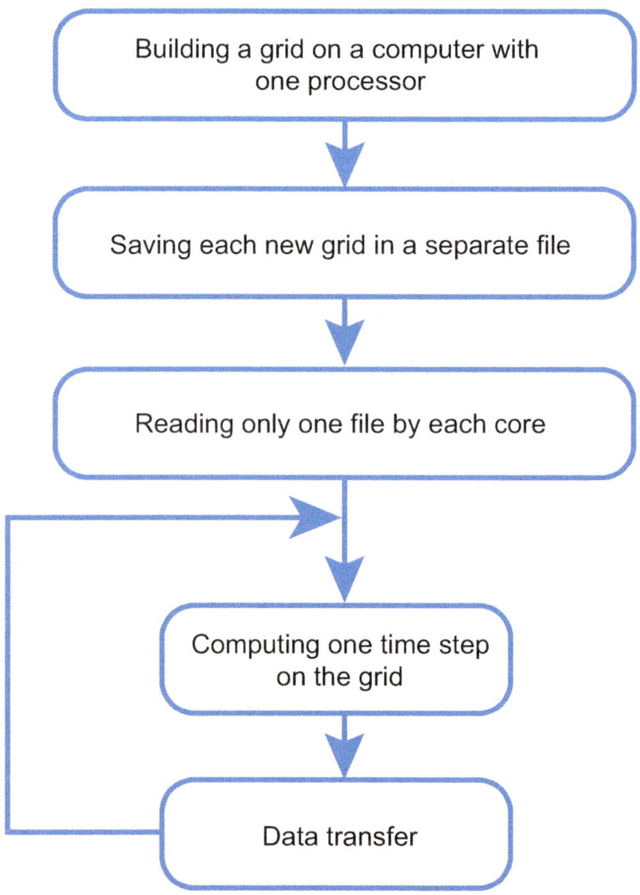

Figure 2.2. Flowchart of parallel computing.

2.3. Unstructured grids in two dimensions

To describe the state of a body in a numerical simulation, only a finite set of body points at which the values of variables are stored is defined. The solution at the other points can be approximated from the values stored at several nearby points of the set, so finding the required number of nearby points is important. A grid is understood to be an introduced set of points, in combination with additional information about their neighborhood. A numerical method is used to calculate the values at the grid points for the next moment of time from the previous values.

When using an irregular triangular grid, the fewest constraints are imposed on the shape and location of the nodes. For this reason, the numerical method must adaptively select an interpolation scheme in the neighborhood of each node. Let us consider an example in which a grid-characteristic method is used, and a transport equation is solved with first-order accuracy. The values at the grid nodes for the next time level are copied from the points determined by characteristics at the previous level. The values at these points are obtained using linear interpolation.

An essential condition is that the characteristic drawn from a point at the next time level should lie within a triangle adjacent to the vertex considered at the previous time level. This creates additional difficulties, especially for methods which are accurate at high orders. This disadvantage, however, is compensated by the fact that, in an irregular grid, we can provide any limits for the sizes of all nodes, regardless of the complexity of shapes of domains of integration.

We employ triangular grids because they will provide more accurate and flexible calculations than structured grids. This is important because, in our case, we need to model bodies with complex geometry. However, a significant disadvantage is that this type of grid can be used only for some body geometries, because the form of a body boundary and the presence of cavities impose restrictions on the possible positioning of grid nodes. As a rule, a structured grid can be built automatically only if we are examining a simply connected region. Furthermore, even if a regular grid can be constructed, it may be impossible to guarantee an identical density of nodes in all parts of the grid as well as a low curvature of grid lines. For instance, in a circular area, a deformed grid becomes both denser and more curved when approaching any four points at a boundary. These features lead to degradation of the solution accuracy of numerical grid-based methods. Moreover, changes in the optimal position of grid nodes due to deformation of a body often require a solution for the resource-intensive, multidimensional problem of minimizing a functional. Also, it is impossible to exclude arbitrarily some nodes from a regular grid, e.g., in order to reduce their density in areas of high concentration.

2.3.1. Grid structure

A grid consists structurally of four sets of elements, such as triangles, vertices, edges, and semi-edges. In a software implementation in C++ language, each of these elements is represented by the corresponding structure. For instance, a vertex of the grid is represented by a pair of numbers in a Cartesian coordinate system, and a triangle is represented by three indices that determine the location of its vertices. Such a description has these desired qualities:

1) Simplicity and minimal size. This, easy to interpret data structure occupies a small amount of RAM. Therefore, it is suitable for long-term storage of triangular grids.
2) Convenience. Some algorithms of computer graphics only need to go quickly through triangles, and to consider only the values of local vertices.
3) Wide applicability. Many applications working with a grid accept data in this form.

Moreover, for the other types of grid elements, which together form extended data structures, additional information is stored to enable navigation through the grid, i.e., for fast operation of iterators and, therefore, fast calculation of the stress tensor and velocity values at the points of a time level. This information represents the relationship between different structural elements of the grid.

The existence of these links significantly increases the speed of the following operations:

1) Grid modification, i.e., introduction of a new vertex and all edges incident to it.
2) Creating a single drawing of each grid edge; otherwise, if all of the edges of the triangles are displayed, each will be shown twice.
3) Going through all triangles adjacent to a given one.
4) Searching for a triangle in which a point with a fixed distance from a given node lies, and which is applied when re-interpolating to a new grid.

2.3.2. Restrictions on the grids used

First, we construct a triangular grid for the entire given area in which calculations will be conducted. Calculations performed to solve modern problems in a successive application with a single core use a grid containing about 4 million vertices. An initial grid of this size takes about 100 MB of RAM. Thus, with a 64-core cluster, high-accuracy calculations can be performed using a grid with 256 million vertices. The construction of the entire grid is conducted on one computing core. Therefore, the size of a 2D triangular grid in parallel computations, which are investigated in this study, is limited by three factors, namely:

1) the amount of memory in a computer used to build a grid for the entire area;
2) the amount of RAM provided by a cluster to a computing core; and
3) the amount of hard disk space in a cluster to store the results of calculations.

The first restriction is insignificant, because computers with RAM up to 500 GB are available. The second problem can be overcome by varying the initial size of a grid and selecting an appropriate method for dividing the grid into domains. There also are means to deal with the third restriction; for instance, if hard disk space is limited, we can store the investigated values at the nodes of another grid that is not coincident with the original, rather than the results obtained at the initial grid nodes. The new grid should be coarser to reduce the amount of data stored. The values at its nodes are calculated by interpolation from the original fine grid.

2.4. Grid division into domains

2.4.1. Division method

In our technology, a geometric model is implemented to divide a computation area into domains. The model is selected for test purposes to check the performance of this process as software packages for efficient topological division of unstructured grids, such as METIS, ParMETIS, and hMETIS, already exist. These packages provide algorithms to divide unstructured graphs into an arbitrary number of parts according to the number of cores used in a cluster. In a software implementation, a grid consists of four interrelated sets of elements, such as triangles, vertices, edges, and semi-edges. The grid is

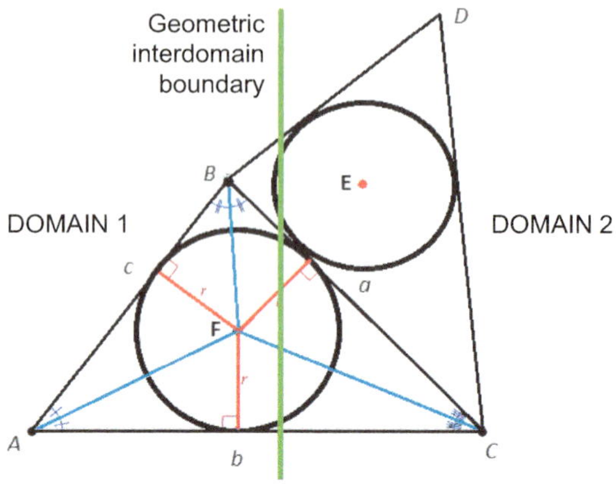

Figure 2.3. Algorithm to determine whether the triangle belongs to the domain.

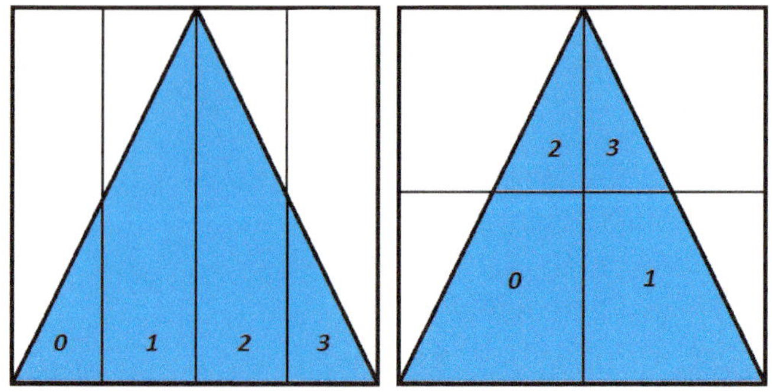

Figure 2.4. Examples of division of an area by only vertical lines (left) and by both vertical and horizontal lines (right).

divided geometrically into domains using triangles as defining elements. This is accomplished by assigning each triangle a special flag that indicates the domain to which the triangle belongs. The domain into which the triangle falls is determined by the position of the center of the inscribed circle. This occurs as follows (Figure 2.3).

For the triangle ABC with vertex coordinates (x_1,y_1), (x_2,y_2), and (x_3,y_3), respectively, the coordinates of the center F of the inscribed circle are

$$(x_F, y_F) = \left(\frac{ax_1 + bx_2 + cx_3}{a+b+c}, \frac{ay_1 + by_2 + cy_3}{a+b+c} \right). \quad (2.3)$$

Before division, the minimum-area rectangle comprising the area to be triangulated should be determined. Then, this rectangle is divided into the number of pieces specified by the calculator. For each triangle, the coordinates of point F are calculated. They are not fixed at any data structure. After that, the found point is checked to determine whether it belongs to the rectangular area geometrically allocated for a particular domain. Geometric division of two types, i.e., by parallel vertical lines and by both vertical and horizontal lines, is implemented (Figure 2.4).

A practical result of this standard division is illustrated in Figure 2.5. Here, additional elements of domains, obtained after further processing of the division results, are shown in grey.

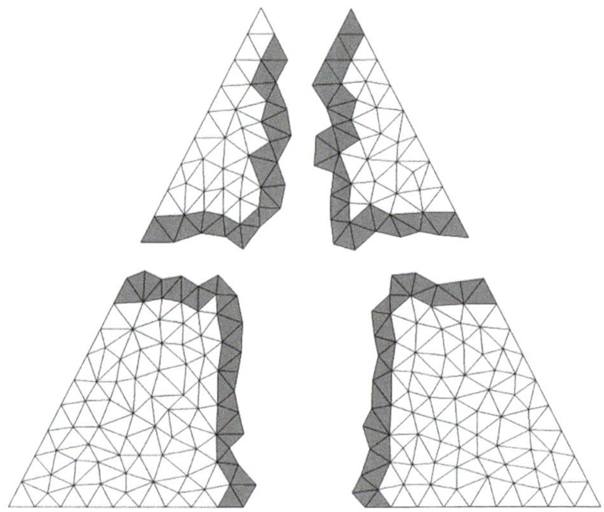

Figure 2.5. Division of a triangular area into four parts by vertical and horizontal lines.

2.4.2. Computation at points on the boundary of a domain

To compute values correctly at a point on a domain boundary, an additional layer of triangles is added to the domain. This allows us to neglect the fact that a characteristic may fall into triangles not belonging to the domain. The additional triangles are the other domains' triangles adjacent to the initial domain's boundary, which is obtained after geometric division (Figure 2.6). This means that, if we consider the set of vertices that lie on the geometric boundary of the domain after constructing the grid, then the desired triangles are those containing at least one of these vertices.

2.4.3. Method of interaction of calculators

The method of interaction of calculators is that, in one step in time, they receive data from and send data to each other. Each domain interacts with all adjacent domains. Because any domain has an additional layer of triangles, due to which it grows, any pair of domains has common triangles. It is the information contained in these triangles that the domains exchange with each other. The grid belonging to one calculator has two layers of

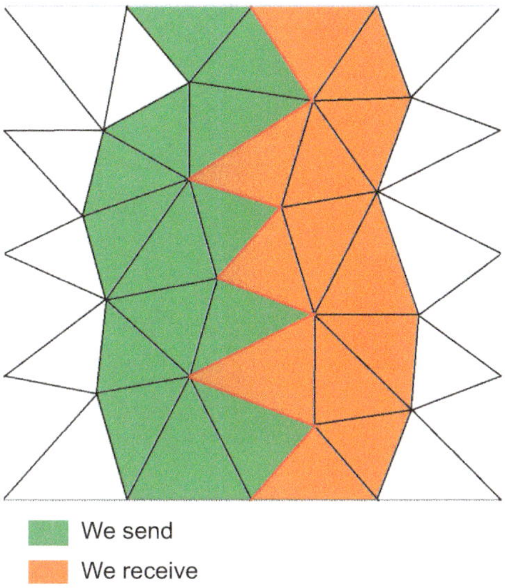

Figure 2.6. Zones to be sent and received at the boundary of a domain.

triangles, one of which is internal (to send data), and the other external (to receive data). A code starts running when all of the initial conditions in the zero time step are specified. To calculate the velocity values and the stress tensor components in the next step, the data are first exchanged, and then the calculation itself is performed, and so on in each step in time.

2.5. Additional elements of a domain

2.5.1. Detailed structure of the grid elements

The presence of additional data structures in an unstructured grid can significantly increase the speed of navigation through the grid and iteration over elements of the grid near a certain point. These data structures play an important role in parallelization of a code and data exchange between computing cores of MPI. The features of these elements essentially are used in algorithms for allocating additional elements of a domain, which are represented in the form of two boundary layers of triangles.

When navigating, the central auxiliary element of a grid is a semi-edge. It can be represented as a vector. Any semi-edge contains information about the starting vertex as well as the next semi-edge and triangle to the left. Note that any turn to the next semi-edge is always made in a counter-clockwise direction (Figure 2.7).

Thus, it allows going through all triangles containing a given vertex in constant time. Moreover, it allows iterating through all triangles adjacent to a given triangle in constant time.

In view of the additional data structures, a triangle is represented not only by a set of vertices and edges, but also by six semi-edges, because each edge can be divided into two semi-edges (Figure 2.8). This feature allows us to allocate contours in a body with a given direction of going around, and to denote the body boundaries. This is used when dividing a grid into domains. Then, the old boundary is adjusted and a missing part of the boundary of a new domain is created.

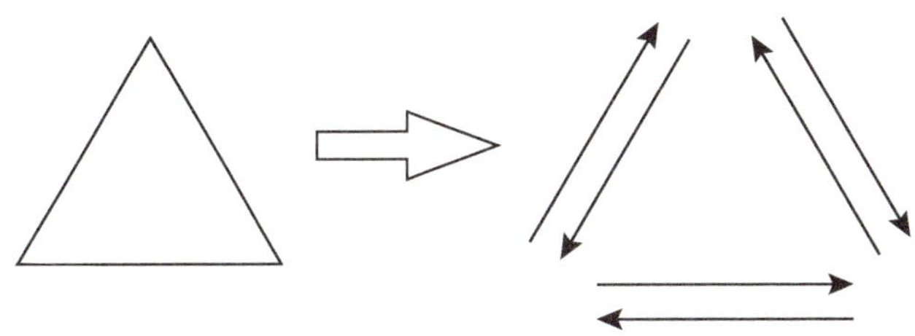

Figure 2.8. Presentation of a triangle as a list of its constituent semi-edges.

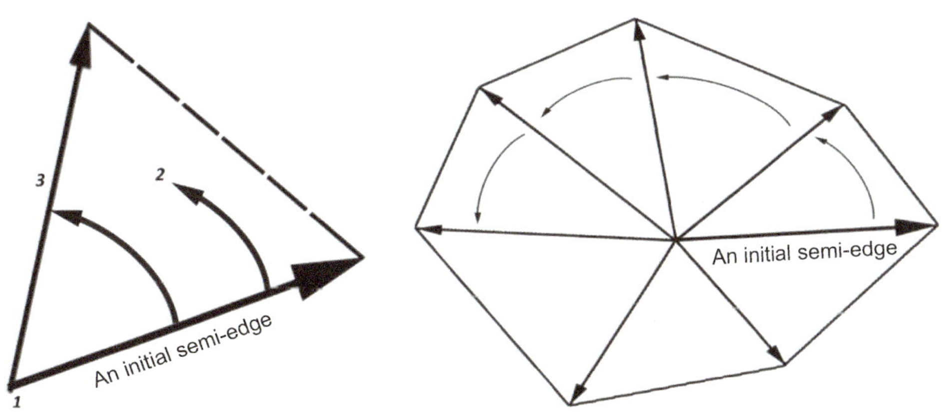

Figure 2.7. Numbering of initial semi-edges.

2.5.2. Determining additional triangles

When finding additional triangles, the domain's internal triangles adjacent to its boundary are determined first. These are the triangles in the neighborhood of which there are triangles that belong to another domain. In Figure 2.9, they are represented as the triangles labeled with the number 1. These are elements that will be sent from domain 0 to domain 1.

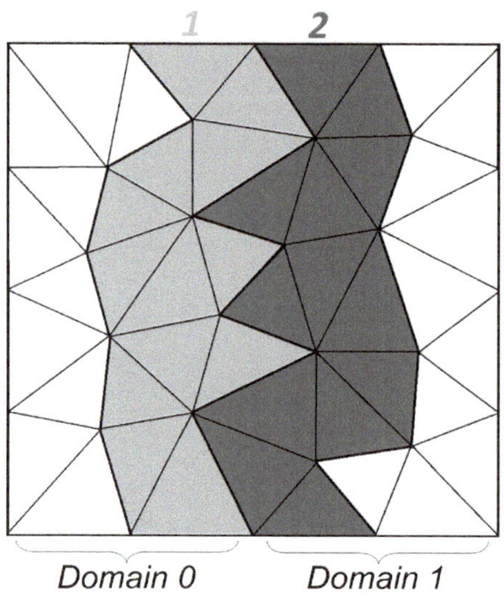

Figure 2.9. Determining the domain triangles adjacent to its boundary.

The next step is determining the received triangles, which are labeled with the number 1 in Figure 2.9. These triangles are external to the old domain when creating the entire graph of an unstructured grid, but are constituents of the new domain. They are found as the triangles from domain 1 that lie in the vicinity of the triangles from the internal boundary layer.

Because any domain can border many other domains, a container for additional triangles for each contact between domains is created in advance. This increases the speed of processing domain boundaries and data transfer. Also, this allows RAM to be saved.

The triangles of a distributed domain can be divided into three groups:

1) the external layer of triangles representing the received information,
2) the internal layer of triangles representing the sent information, and
3) all other triangles of the domain.

2.5.3. Determining additional edges

All of the additional edges belong to the external layer of triangles. All of the edges of a new distributed domain can be divided into four groups (Figure 2.10).

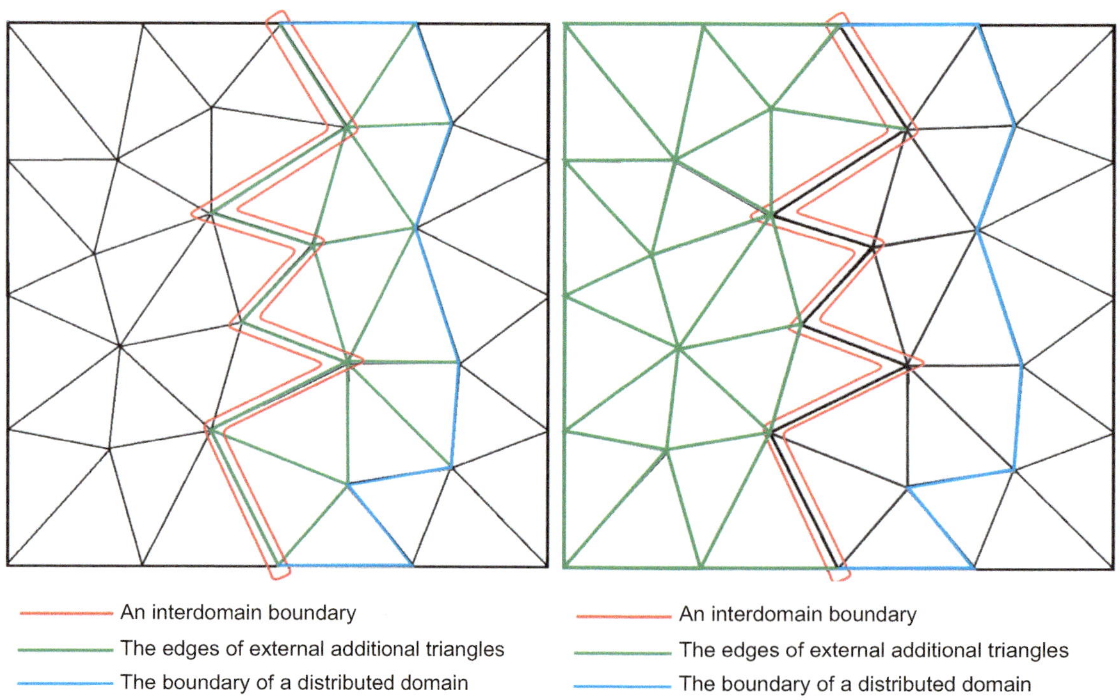

Figure 2.10. Locating the edges adjacent to a boundary.

1) The internal edges of the domain that belong to all domain triangles obtained immediately after building the original grid.
2) The edges at the boundary of the distributed domain that form the interdomain boundary and belong to the external layer of triangles.
3) The edges of all external additional triangles except for the edges referred to in point 2.
4) The edges of the external layer of triangles that form the boundary between the sets of triangles belonging to different contact boundaries of the domain. (Note that the number of adjacent domains coincides with that of contact boundaries.)

The search for additional edges is performed using the data obtained from the search for additional triangles.

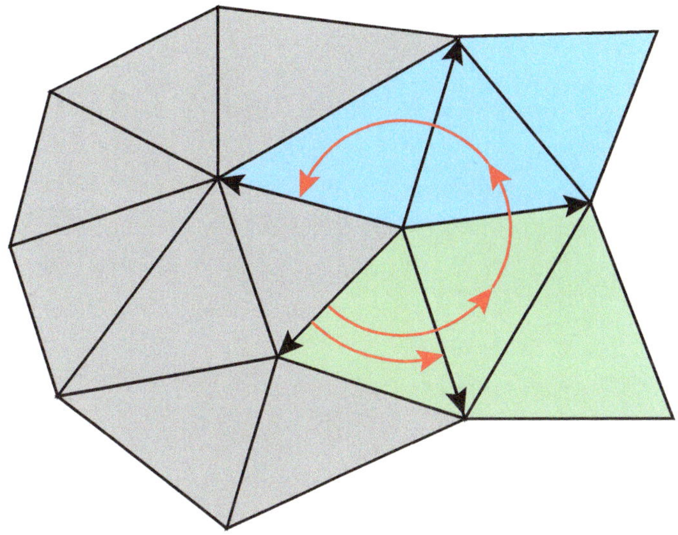

Figure 2.11. Updating information in boundary half-edges.

2.5.4. Determining additional semi-edges

The edges and semi-edges are stored in different containers and have a one-to-one correspondence. Each edge corresponds to two semi-edges, direct and inverse. Methods for finding the indices of the two corresponding semi-edges from the index of an edge are implemented in the software. In this connection, the classification of semi-edges coincides with the above classification of edges.

When creating a new domain, the values stored at the additional data structures are adjusted. The strongest changes are those that affect the semi-edges because each of them stores information about the nearest semi-edge and triangle. When building a grid in a new domain, some edges are located at the boundary with another domain. The corresponding semi-edges should be amended so as to indicate correctly the new domain boundary. This is because the elements referred to by a semi-edge at a boundary with the other domains now may not belong to the domain under consideration (Figure 2.11). In addition, the semi-edges adjacent either to the junction of two boundary sets of triangles or the junction of the boundary between two domains and the boundary of the whole body require additional processing to mark which domain they belong to, or which boundary condition will need to be applied.

2.5.5. Determining additional vertices

The search for additional vertices is performed using additional triangles and additional edges. In this connection, the classification of vertices coincides with the above classification of edges.

2.6. The issue of reindexing domain elements

In our code, the triangles, vertices, edges, and semi-edges are stored in a container that implements an advanced array. The elements are numbered starting with 0. Calculations are performed, provided that all of the elements are strictly ordered and all of the relationships between them are maintained. When allocating a certain piece of a grid as a new part of full value to be suitable for computing, the new elements chosen should be rearranged so that they are successive in the container and their new relationships with each other are maintained. During calculation, there is no need to save the entire grid on a cluster because it occupies too much space on a hard disk drive. It is more convenient to introduce a coarser grid and save interpolated values at its nodes. Therefore, it is necessary to have a one-to-one correspondence between old and new indices for only the sent and received data, i.e., for the two layers of triangles, internal and external, adjacent to the boundary. This mapping is stored in a separate additional data structure.

2.7. Grid division examples

The following are examples of division of a grid into domains generated by a computing code. Figure 2.12 shows the results of division of a simple rectangular area into domains by only vertical lines and by both vertical and horizontal lines. In this case, such an approach pro-

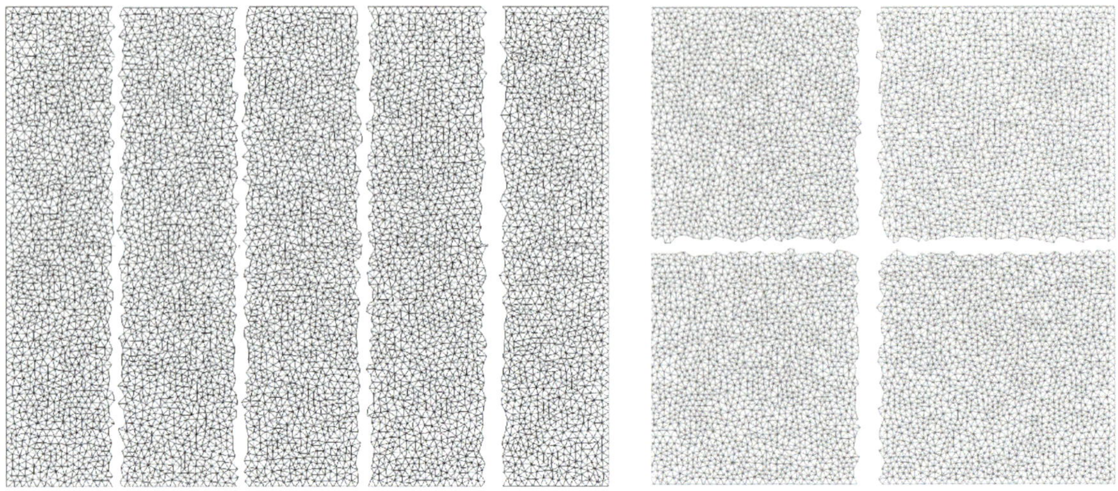

Figure 2.12. Division of a rectangular area into domains by only vertical lines (left) and by both vertical and horizontal lines (right).

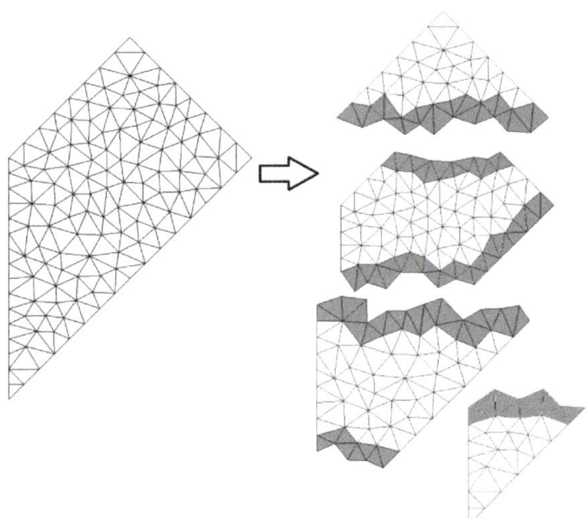

vides decomposition into domains with an equal number of elements, while minimizing the amount of data transferred between the domains.

Figure 2.13 shows an example of a more complex area divided into domains by vertical and horizontal lines. It is evident that, in this case, the number of elements in the domains is substantially different. This decomposition leads to inefficient use of the computational resources of those cores to which a domain with a small number of elements belongs. In numerical simulations of seismic problems, the majority of computational grids have a simple rectangular shape, so the approach with vertical and horizontal decomposition is justified.

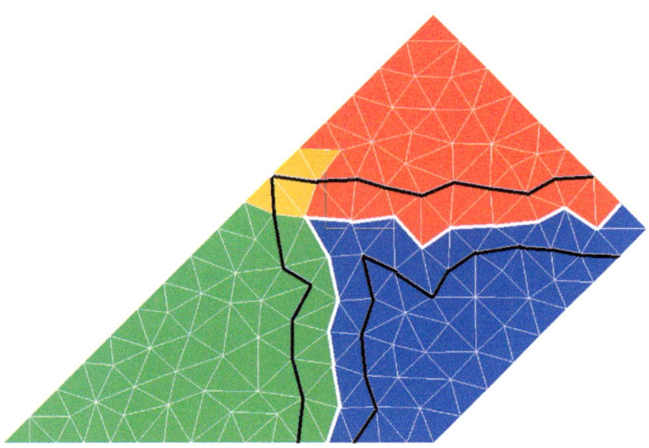

Figure 2.13. Division of a complex area by vertical and horizontal lines.

2.8. Conclusions from Chapter 2

This chapter has considered the parallel implementation of the grid-characteristic method on unstructured 2D (triangular) grids. The computational method itself is the same within all domains, whereas the primary challenge is to distribute a grid between domains and arrange data transfer between computing cores. Both issues have been addressed in detail in this chapter.

The described algorithm has been implemented as a software package in C++ and used for numerical simulations of most of the tasks that will be presented in subsequent chapters.

Chapter 3

The Grid-characteristic Method in 3D

To solve system (1.9) numerically, the grid-characteristic method with tetrahedral grids is used. It allows us to build numerical algorithms for calculating boundary points and points at the interface between two media with different Lamé parameters and/or different densities.

3.1. Interpolation in a tetrahedron

To determine a polynomial field of degree N dependent on x, y, and z, knowledge of the values at $\frac{(N+1)(N+2)(N+3)}{6}$ reference points is required.

The following method to set reference points is proposed. In the tetrahedron $ABCD$, we draw planes that are parallel to its faces and divide each of its edges into N equal parts. The reference points for $N=3$ are numbered as shown in Figure 3.1.

The planes divide the tetrahedron into similar tetrahedra, as well as into octahedra. When $N=2$, we have four small tetrahedra and one octahedron, which are shown in Figure 3.2.

The Courant number is calculated using the minimum height of all small tetrahedra.

Let the vectors of the tetrahedron vertices be designated as $\vec{r}_A$, $\vec{r}_B$, $\vec{r}_C$, and $\vec{r}_D$. For each given N, their respective formulae to calculate the weights $w_{abcd}(\vec{r})$ of the reference points are used. The value of the polynomial at point $\vec{r}$ is defined by the formulae

$$v(\vec{r}) = \sum_{a,b,c,d} w_{abcd}(\vec{r}) v_{abcd}, \qquad (3.1)$$

where $v_{abcd} = v(\vec{r}_{abcd})$ denotes the value of the interpolated function at reference point $\vec{r}_{abcd}$.

Let us describe an algorithm for constructing an interpolant with a limiter which is intended to limit spurious oscillations (thus the name "limiter") on tetrahedral grids that is based on interpolation by a polynomial of order N.

1) Using polynomial interpolation of order N, we determine the value of a test function at the given point $\vec{r}$. Let it be $v_N(\vec{r})$.
2) We determine into which small element, a tetrahedron or octahedron, point $\vec{r}$ falls. If the point falls into an octahedron, then an axis is drawn in this octahedron in one of three possible ways. The axis divides the octahedron into four tetrahedra with equal volumes but not similar to each other. Then, we

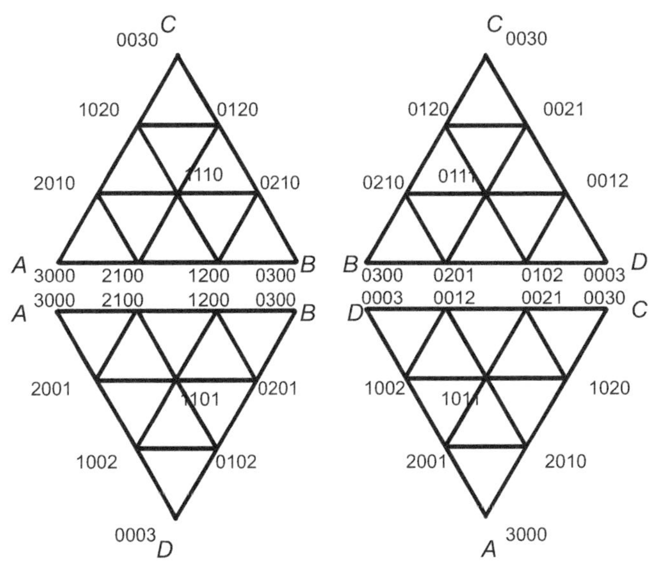

Figure 3.1. Reference points in a tetrahedron for cubic interpolation.

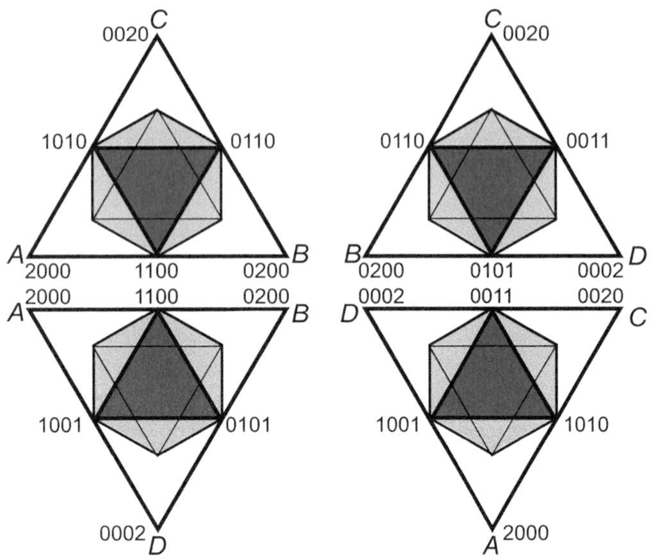

Figure 3.2. Reference points in a tetrahedron for parabolic interpolation.

determine into which of the four tetrahedra the point falls.

3) We compare $v_N(\vec{r})$ with the minimum, m, and maximum, M, values at the vertices of the tetrahedron.
   - If $m \leq v_N(\vec{r}) \leq M$, the interpolant value at point $\vec{r}$ is $v_N(\vec{r})$.
   - If $v_N(\vec{r}) < m$, the interpolant value at point $\vec{r}$ is m.
   - If $v_N(\vec{r}) > M$, the interpolant value at point $\vec{r}$ is M.

The use of interpolation with the limiter allows us to eliminate spurious oscillations in the polynomials arising from the presence of discontinuities in the interpolated polynomial functions.

In addition to this limiter, monotone interpolation is implemented, which is a hybrid of linear and parabolic methods. This interpolation method is based on a property of second-order polynomials in a tetrahedron, i.e., if the polynomials have extrema, they will be at the tetrahedron edges.

Let us successively designate all of the segments shown in Figure 3.2 as AB. The value of v_M is obtained via parabolic interpolation at the midpoint of AB. Let v_A and v_B be the values at the reference points A and B, respectively. If v_M is not between the values v_A and v_B, it is replaced by $\frac{v_A + v_B}{2}$. This operation is performed for each of the 31 segments. Then, we conduct piecewise parabolic interpolation using the real reference points and the additional ones, v_M. The latter are the midpoints of the segments, where the values are calculated according to the described procedure. If v_M does not lie between the values v_A and v_B for all edges of a small tetrahedron, piecewise linear interpolation is conducted in it. In either case, the resulting interpolating function will not have extrema except, perhaps, the reference points.

There are two ways to increase the accuracy of calculations when the spatial resolution is fixed. The first is to increase the order of approximation. The second is to divide a tetrahedron into smaller ones inside the interpolator. We have implemented piecewise parabolic interpolation that provides a greater accuracy when the sampling rate of an initial tetrahedral grid is fixed. Also, this type of interpolation reduces the consumption of resources for the construction and storage of a grid.

3.2. Hierarchic grids with a varying time step

Next, we explain how to use hierarchical grids with a varying time step utilizing the system of two transport equations,

$$u_t + u_x = 0 \tag{3.2}$$

and

$$v_t - v_x = 0. \tag{3.3}$$

Let's consider a grid with a time step factor of two. For the solution of these transport equations, the following expressions are valid:

$$u(t + \tau, x_{bnry}) = u(t, x_{bnry} - \tau), \tag{3.4}$$

$$v(t + \tau, x_{bnry}) = v(t, x_{bnry} + \tau), \tag{3.5}$$

$$u(t + 2\tau, x_{bnry}) = u(t + \tau, x_{bnry} - \tau), \tag{3.6}$$

and

$$v(t + 2\tau, x_{bnry}) = v(t, x_{bnry} + 2\tau). \tag{3.7}$$

In expressions (3.4) through (3.7), x_{bnry} represents the coordinate to the left of which the grid is small and to the right the grid is large (see Figure 3.3), t denotes a current time, and τ is the step of integration over time in the area with a fine grid. In the area with a larger grid, the step is 2τ, respectively.

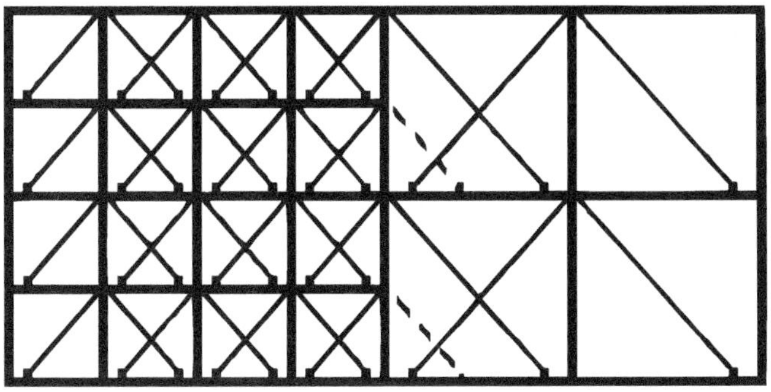

Figure 3.3. A 1D slice of a hierarchical grid with a time step factor of two.

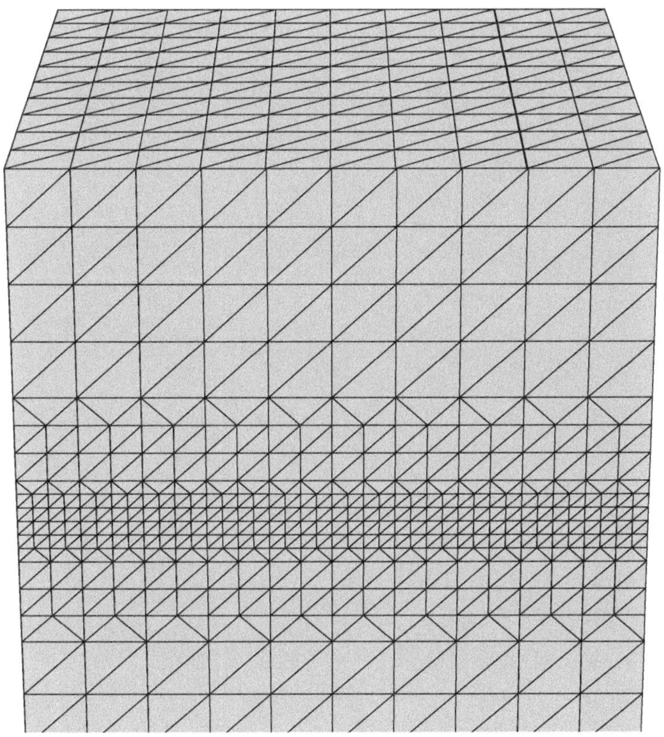

Figure 3.4. An example of hierarchical unstructured grids of multiplicity two.

In solving the system of equations for the state of a linear elastic medium, two methods to calculate the boundary points with the coordinate x_{bnry} can be used. The first is to calculate and sum the values of the characteristics from different points that depend on both the time step and the grid type of the areas into which the characteristics can be divided. The second method uses the contact corrector at a rigid boundary. Mathematical and numerical studies show that the second approach is better generalized to the corner points and does not produce spurious oscillations at these points. Figure 3.4 presents an example of hierarchical unstructured grids of multiplicity two.

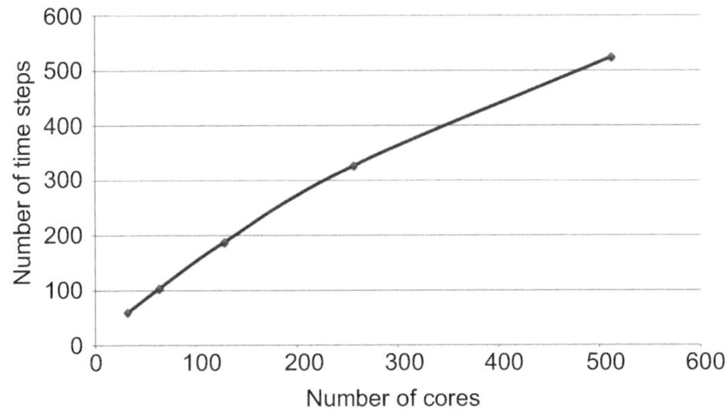

Figure 3.5. The dependence of the number of time steps made in 10-minute calculation on the number of cores used.

3.3. Conclusions from Chapter 3

A 3D implementation of the grid-characteristic method basically repeats a 2D implementation, but it has a number of significant features. One of the features of a 3D grid is the presence of an octahedron when partitioning a tetrahedron into smaller tetrahedra. Another feature is much heavier demands on computational resources, which requires the introduction of a hierarchical grid with varying time step.

The described algorithm has been implemented as a software package in C++. We have tested our software package to address spatial dynamic problems in complex heterogeneous environments using the grid-characteristic method with tetrahedral grids. Figure 3.5 shows the dependence of the number of time steps made in 10-minute calculation on the number of cores used.

The implemented method demonstrates a stable counting efficiency with up to 512 computing cores.

Part Two

The Investigation of Seismic Responses from Multiscale Single Fractures and Fracture Systems

Introduction

Over the last decade, research into fracturing of hydrocarbon reservoirs typically has followed one of two approaches:

1) estimation of the parameters of a fractured medium through the determination of the anisotropic characteristics of reflected waves passed through the medium, using wide-azimuth surveys, or
2) the use of wavefields scattered by fractures.

These approaches include special processing methods of 3D data, such as focusing transformation (FT) technology, migration imaging of scattering features (MISF), and the common scattering point method, with maps and volumes of the scattered energy as an outcome. The special data acquisition technique using the side-view seismic locator (SVSL) accomplishes the same purpose.

The first approach entails indirect estimation of fracture parameters from the properties of waves generated by deep-seated reflectors. Although wide-azimuth 3D surveys now are widely applied, estimates of the fractured subsurface properties they produce are not always confirmed by subsequent hydrodynamic simulation studies, which provide estimates of hydroconductivity, permeability, etc.

The primary challenge of the second approach is the inherent non-uniqueness of data interpretation. Confidence in the identification of wave energy anomalies associated with fractured zones may be low, and thus geologic interpretation of scattered wave anomalies to identify or predict the location of fractured reservoirs is ambiguous. In this situation, it seems appropriate to assess the possibility of applying methods for the direct use of seismic responses from single fractures and fractured zones for their characterization. This should not be limited only to the energy estimate of the response on the conventional vertical *Z*-component, but also should use the characteristics of the wave response on the horizontal *X*- and *Y*-components.

The first step in this approach is defining the response characteristics of the basic types of multiscale fractures and fractured zones. Because practically no analytical solution to this problem exists, the wavefields, i.e., the responses from fractures, have to be computed by numerical simulation. In essence, it is necessary to solve the direct seismic problem for fractured media. To address the problem, we employ the grid-characteristic method described in Part One.

Terminology

To facilitate the subsequent discussion, it will be helpful to clarify our definition of the following terms.

Fracture. A fracture is a break or dislocation in a rock mass that (unlike a fault) does not display substantial displacement of the faces relative to the long axis of the fracture. MacBeth (1995) and Kozlov (2006) define several categories of fracture, including:

Microfracture. This category of fractures is of a microscopic scale, typically with apertures (openings) of tens of microns and lengths of a few centimeters.

Mesofracture. These fractures have apertures of hundreds of microns and lengths of a few meters. Typically, they are observed in core samples and can be identified during well-log surveys using electric scanners, e.g., formation microimager (FMI), and acoustic tools, e.g., cross dipole sonic (XMAC).

Macrofracture. In this category, apertures range from a few fractions of a millimeter to as much as a few millimeters. Their lengths (or heights) range from a few tens to hundreds of meters (as a rule within the bed thickness). Thus, their size is comparable to or greater than seismic wavelengths (λ).

Micro-, meso-, and macrofractures are partially or completely filled by fluid. At relatively shallow depths, they may be filled by gas. A generally uniform orientation of these fractures, caused by tectonic stress, creates anisotropy in physical properties of fractured rocks.

Megafracture. These are the largest fractures, with heights and lengths similar to or greater than those of macrofractures and thicknesses measured in meters. In contrast to macrofractures, megafractures are filled by fragmented material with a high fluid permeability. These are often called ***fracture corridors*** or ***fluid-conductive channels***, and they may intersect more than one bed.

Brief overview of earlier investigations of rock fracture responses using numerical simulation

Over the last two decades, the finite-difference method has been used widely to simulate numerically the propagation of elastic waves in fractured media. The method is based on a linear-slip fracture model (Schoenberg, 1980) and uses the finite-difference algorithm to approximate a wave equation, as proposed by Coates and Schoenberg (1995). The fracture model assumes stress to be continuous, allows for discontinuity in displacement, and assumes a linear relationship between the slip across a fracture and stress. The latter condition implies the use of an equivalent medium, with faults specified in finite-difference grid cells. With the fractures discretely specified in the models, the method ensures a more complete wave response pattern and, in particular, makes it possible to record scattered (diffracted) waves. For these reasons, the method has been adopted widely.

Haugen and Schoenberg (2000) use the finite-difference approach to analyze the response from a single fracture of large planar extent and show that the fracture's reflectivity depends on its inclination and the frequency of the seismic wavefield. As the inclination increases, the amplitudes of PP- and SS-reflections decrease to a certain degree, and the extraction of fracture information from the scattered waves is more successful.

Wu et al. (2005) simulate the response from a single vertical fracture using two models: (1) a thin fluid-filled bed embedded in an isotropic background medium, using a variable grid – the TLM model (Groenenboom and Fokkema, 1998); and (2) a linear-slip interface embedded in an anisotropic background medium with a constant grid cell size – the LSM model (Schoenberg, 1980). In addition to direct and head P-waves, the responses (identical for both models) contain fracture-induced P- and converted diffracted waves. On the *X*-component, both additional waves exhibit a polarity reversal at the vertical-scatter point.

As discussed by Nakagawa et al. (2002), Willis et al. (2004), and Bakulin et al. (2000), the numerical simulation of responses from clusters of near-vertical unidirectional fractures reveals the following effects: (1) transmitted waves attenuate due to scattering; (2) scattered component amplitudes increase as the incidence angle increases relative to the vertical; and (3) a "coda" (i.e., a multi-phase sequence) of waves appears as a result of multiple reflections between fracture planes. These studies are limited to the use of only the vertical component.

In more detail, the features of the wave response from a single near-vertical fracture of finite length comparable to a seismic wavelength (macrofracture) are reported by Fang et al. (2010), through a finite-difference numerical simulation of Schoenberg's linear-slip fracture model (Hsu and Schoenberg, 1993). A frequency-dependent solution is used in which the fracture response is dependent on several factors (e.g., velocity, ratio of linear size of a fracture height to wavelength), with the dominant factor being the ratio of the normal (*Zn*) and tangential (*Zm*) compliances. Single fracture computations have established that, for the P–P source–receiver system, a significant amount of scattered energy is lost downward. For the P–S system, however, although part of the energy goes downward, a considerable portion goes upward, that is, back-scattering of energy occurs. For near-vertical fractures, and also for fracture clusters, diffraction from the fracture tips is observed. The usefulness of the results is degraded by the use of empirical coefficients such as fracture compliances; furthermore, the response is considered as scattered energy, rather than as specific waves recorded on the earth's surface.

Sayers (2009) characterizes the seismic response from multiple fracture sets of different orientations in a productive reservoir. He too uses the linear-slip fracture model and compliance coefficients, and employs offset-and-azimuth-dependent variations in reflection coefficients in order to determine the components of the second-rank fracture compliance tensor. The tensor component directions might be used to guide directional drilling.

The advent of powerful computers has made it possible to investigate wave responses from fractures in three dimensions. In one of the earliest works, Xu et al. (2010), using Coates and Schoenberg's finite-difference algorithm, investigate the impact of fracture spacing on P- and S-waves of the seismic response in 3D models. Snapshots of P- and S-wavefronts in the *XZ* and *YZ* planes show that, as fracture spacing decreases, concentric P-wavefronts gradually change to elliptical forms with increasing axial ratio, which indicates an increase in anisotropy. Here the energy of scattered waves attenuates.

In recent years, numerical simulation has been used to investigate secondary waves that form when P-waves encounter fluid-filled fractures. Frehner and Schmalholz (2010) employ the finite-element method to examine how Stoneley waves reflect and scatter at the tips of fluid-filled fractures. For this mechanism, only a portion of the Stoneley wave energy is spent; the remainder propagates inside fractures and forms resonance nodes with higher but rapidly attenuating amplitudes. This makes it challenging to detect the Stoneley waves.

Krauklis waves have been examined by Korneev (2008, 2011) and Frehner (2014). These are waves initiated by P- and S-wavefronts traveling through fluid-filled fractured rocks. P- or S-waves each form two Krauklis waves at each fracture tip when they reflect from and are scattered by the fractures. They propagate along fluid-filled fractures and may travel within the fracture tip to tip, back and forth many times, imitating resonant oscillations with the dominant frequency of the seismic signal. These waves may initiate a seismic source inside the fracture. The wave amplitude strength, however, is very dependent on the orientations of fractures relative to the incident P-wave. They are weakly initiated at low angles and in the 40° to 65° range, but are strongest when the P-wavefront and fracture plane are parallel. S-waves commonly initiate much higher Krauklis wave amplitudes than do P-waves and these amplitudes are strongest at angles close to 50°. Here, all of the Krauklis wave energy remains inside the fracture. These waves, however, are of importance in the exploration of fractured reservoirs.

Much attention has been given in the geophysical literature in the recent years to the simulation of fracturing.

The most complete analysis, based on the finite-difference method in numerical simulation of seismic responses from fractures of various types and from fracture clusters, generalizes previous experience and the researchers' own experience in the study of fracture properties and is presented in Hall and Wang (2012). They consider the impact of fracture properties, such as density, degree of variation in orientation, sizes, infill property, and clusterization (see Chapter 9), on the generalized wavefield of responses from sets that contain several tens to several hundreds of fractures. However, the correspondence of the study results to the real properties of fractured reservoirs is not obvious because of the inadequacy of model assumptions. This drawback is due to the use of the finite-difference method and the theory of equivalent media. To be adequately representative of real fractured reservoirs, the models should have discrete positions of tens of thousands of fractures with an aspect ratio on the order of 1/1000 to 1/100,000. The applicability of the analysis is degraded by its consideration of a generalized wavefield rather than individual waves of the response from the fractured media.

Lisitsa et al. (2013) discuss the numerical simulation of scattered waves from fractured reservoirs by applying the finite-difference method and grid downscaling. The researchers use fractured reservoir models with a minimal fracture aperture (opening) of 10 cm (grid cell size), which is about 100 times greater than the average real aperture. Such simulation can provide only a rough estimate for the total energy field of scattered waves from the fractured reservoir. The use of this parameter not infrequently results in the ambiguous identification of zones of its abnormally high values, as usually is the case with fractured prospects.

Liu and Martinez (2012) set forth the basic details of seismic anisotropy, the equivalent modeling method for fractured rocks, methods to estimate fracturing parameters by analyzing azimuthal dependence of compressional waves, and application of multicomponent data to characterize the fracturing. The use of an effective model of the medium, however, fails to ensure adequate representation of real fractured reservoirs, and hence high reliability of the prediction.

Chapter 4

General Properties of P-wave Responses from Fractures when Specifying Boundary Conditions on Fracture Surfaces

4.1. Character of seismic responses from fractures as a function of fracture length, inclination, and infill

The investigations highlighted in this and subsequent chapters have been made using only incident compressional P-waves.

For this study of seismic responses from fractures using the grid-characteristic method with specified boundary conditions on the fracture surfaces, it is appropriate to begin with establishing the effects of the principal fracture parameters on the seismic responses. In our computer model, the subsurface parameters are selected to represent a sedimentary section that has these values: $Vp = 2000$ m/s, $Vs = 1400$ m/s, and $\rho = 2.0$ g/cm^3, with a macrofracture of finite length and infinitely small thickness. For simplicity, a plane horizontal wavefront with a 30 Hz Berlage wavelet (Aldridge, 1990) is selected to be the seismic energy source.

Figure 4.1 compares the absolute value of the wave response from a macrofracture of 20° dip depending on its length (50, 100, 200, and 400 m) when the fracture is dry (filled with gas) and filled with water.

Similarly, Figure 4.2 shows how the macrofracture response varies with changing the angle of macrofracture from the vertical (20°, 35°, 55°, and 70°), while the fracture length remains fixed ($L = 200$ m). As well as in Figure 4.1, the computations are performed for the gas- and water-filled fractures.

Analysis of the resulting wavefields allows the following to be noted: reflections from the macrofracture (compressional PP and converted PS) are much stronger than diffractions but are localized in a space, their dimensions slightly exceeding the fracture length. Diffractions (*Dpp* and *Dps*) are generated at the macrofracture tips and form two groups of concentric wave pairs, with the circumference centers at the fracture tips. Straight-line reflection wavefronts (PP and PS) tangentially link concentric wavefronts of the associated diffractions (Dpp_u–Dpp_l and Dps_u–Dps_l, where subscripts u and l denote upper and lower, respectively).

The effect of varying the length of the macrofracture is clear. When the macrofracture length (height) L is decreased from four or five seismic signal lengths δ[1] to less than one signal length ($L < \delta$), the straight-line reflection wavefront shrinks and turns into a strong element of the concentric diffraction wavefront. As the macrofracture length is decreased, the concentric diffraction wavefronts become closer to each other, both in circumference size and separation distance. So, when the fracture length is smallest, as a result of the merger of the identical types of waves emanating from different fracture tips, the response is represented practically by only two diffractions, compressional *Dpp* and converted *Dps*. In this situation, the fracture starts to operate as a point diffractor (see Figure 4.1).

The effects of macrofracture inclination on the components of the seismic response recorded on the earth's surface are different for reflections and diffractions. Reflected waves, although they are formed in almost the entire range of angles, can be recorded on the surface only when macrofracture dip angles do not exceed the limiting values at which the reflected wave propagates along the earth's surface. For compressional reflections, according to geometric ray theory, the critical angle is 45° (for zero-

[1] δ: Seismic signal length is the product of signal duration in time τ and velocity V of signal propagation.

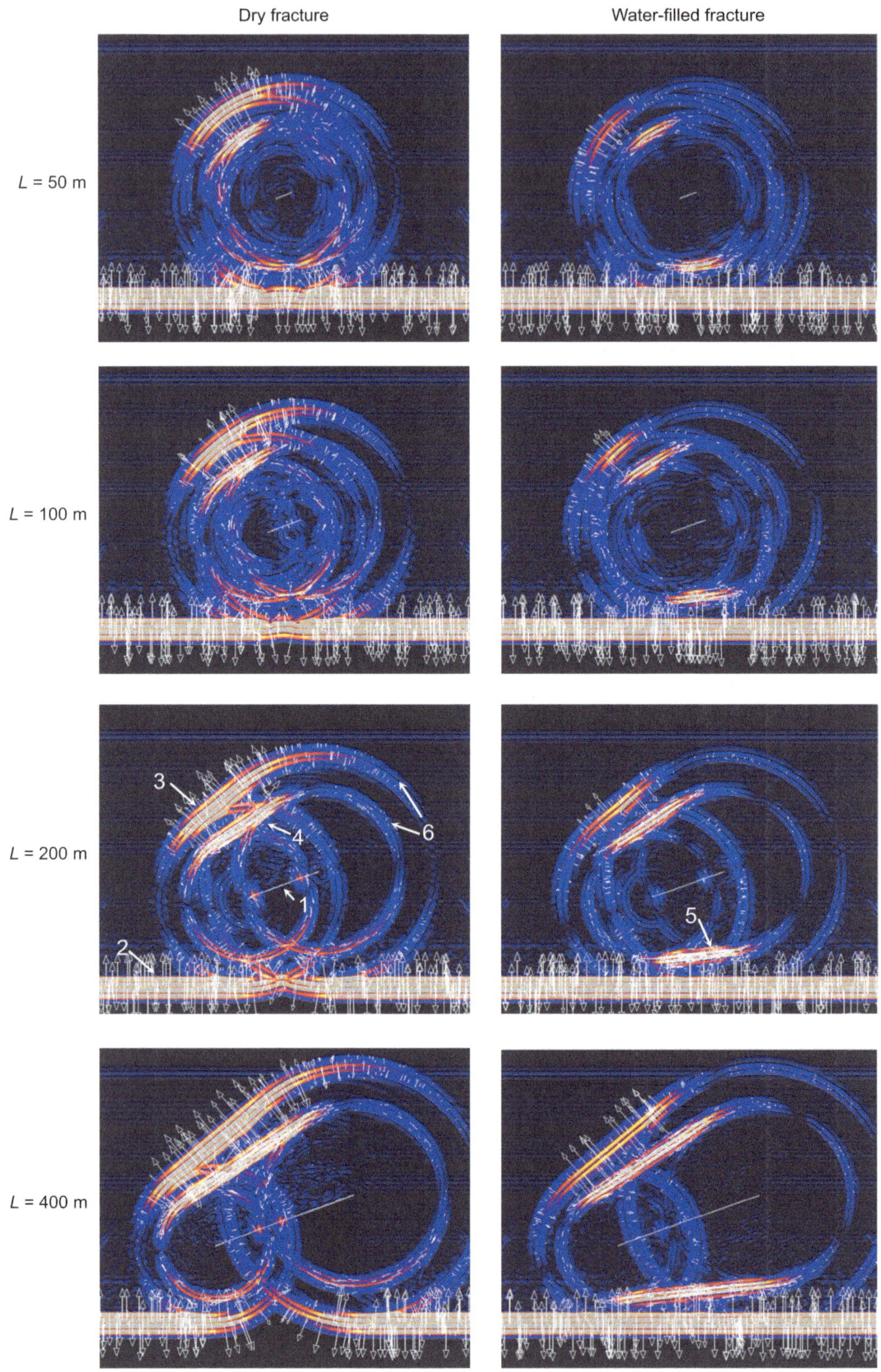

Figure 4.1. Dependence of the absolute value of the wave response from a dry or water-filled macrofracture of 20° dip on its length *L*, where 1 denotes the macrofracture, 2 incident compressional wavefront, 3 compressional PP reflection, 4 converted PS reflection, 5 transmitted shear wave, and 6 diffracted compressional and diffracted converted waves.

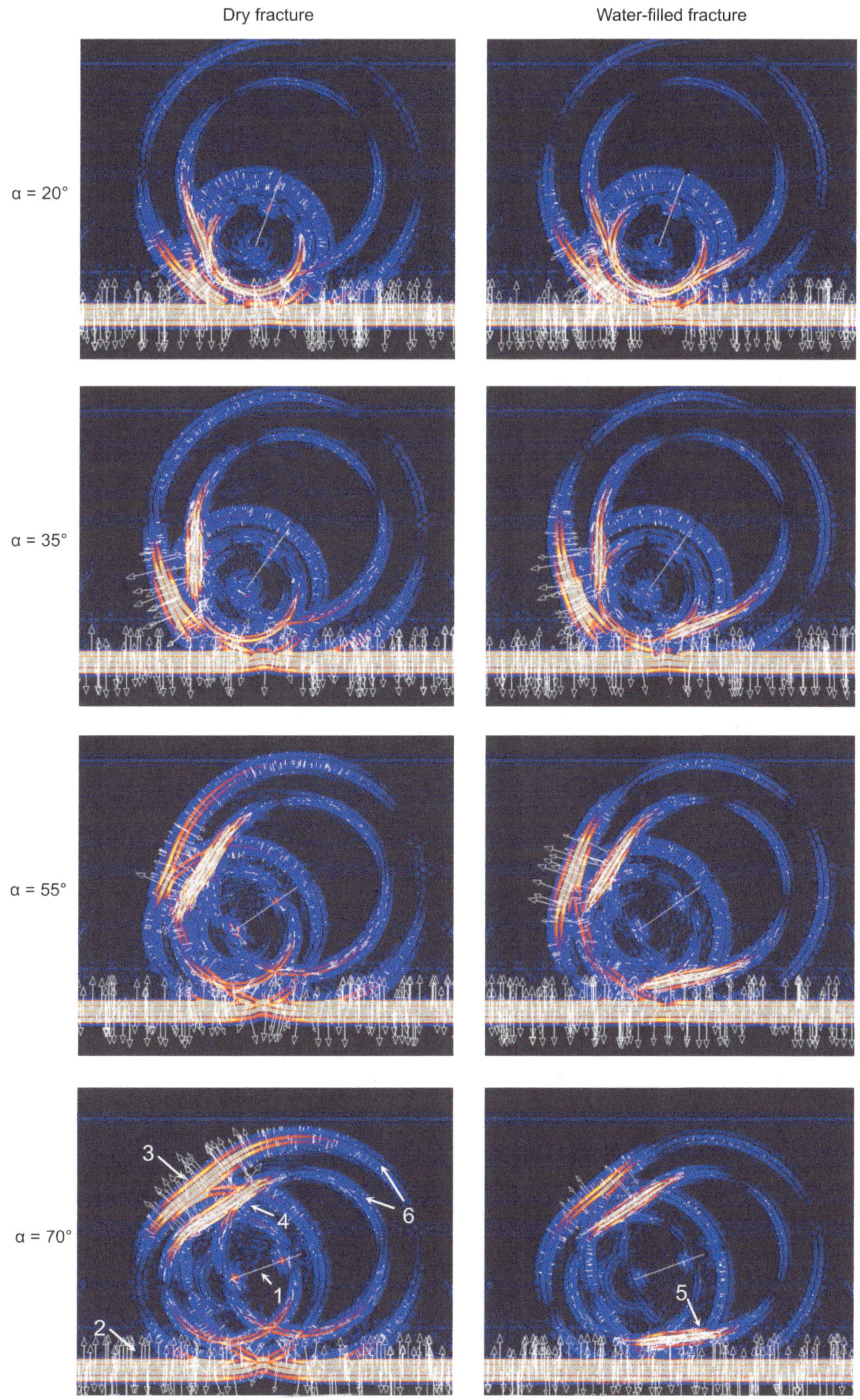

Figure 4.2. Dependence of the absolute value of the wave response, from a 200 m long dry or water-filled macrofracture, on the deviation α of its plane from the vertical, where 1 denotes the macrofracture, 2 incident compressional wavefront, 3 compressional PP reflection, 4 converted PS reflection, 5 transmitted shear wave, and 6 diffracted compressional and diffracted converted waves.

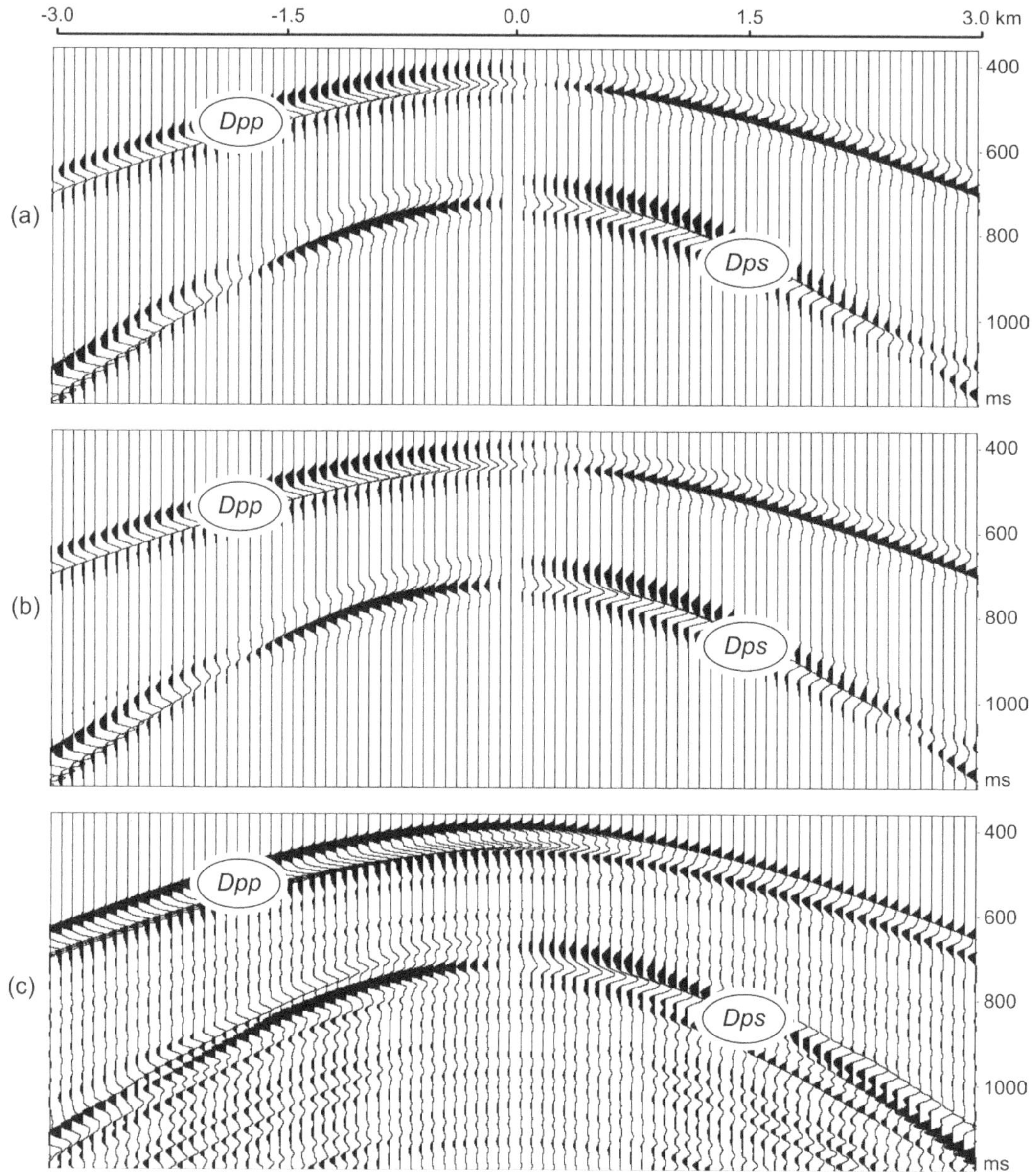

Figure 4.6. Comparison between the *Z*-components of the responses from a single fracture with different aspect ratios: (a) infinitely thin, (b) AR = 1/1000, and (c) AR = 1/100. No gain is applied.

Table 4.2. Estimates of the amplitude maxima of the main diffractions from a single fracture and the behavior of their polarity.

Model type	Compressional wave						Converted wave				
	***X*-component**			***Z*-component**			***X*-component**		***Z*-component**		
	Amplitude		**Polarity reversal**	**Amplitude**		**Polarity reversal**	**Amplitude**	**Polarity reversal**	**Amplitude**		**Polarity reversal**
AR = 1/100	5	2	Yes	6		No	12	No	4	3	Yes
AR = 1/1000	3		No	3	3	Yes	11	No	3	3	Yes
ITF model	3		No	3	3	Yes	11	No	3	3	Yes

A fracture having an aspect ratio of 1/100 manifests itself in a compressional wave as a normal reflector, whereas a converted wave does not practically respond to the fracture opening, and behaves differently than a reflection from a long boundary. This indicates that the converted wave does not respond to the fracture opening.

When the aspect ratio is 1/100, the amplitudes of compressional waves increase twofold, whereas the amplitudes of converted waves increase by about 10%.

The behavior of the polarity of the converted *Dps* diffractions for an ITF is identical to that for a fracture with an aspect ratio of 1/100.

For both models, the *Z*-component undergoes a polarity reversal, whereas the *X*-component does not. Note that there is a difference in the behavior of the compressional *Dpp* wave. Specifically, on the *X*-component, the wave related to the ITF model does not undergo a polarity reversal while the wave related to the fracture with an aspect ratio of 1/100 does. On the *Z*-component, the opposite is true, i.e., the wave related to the ITF model undergoes a polarity reversal while the wave related to the fracture with an aspect ratio of 1/100 does not. This is consistent with the known polarization property of any wave: if on one component a polarity reversal occurs at a point where the traveltime slope changes its sign, it does not occur on the other component.

The comparison of the *X*- and *Z*-components of the wave responses from ITFs and fractures with a finite opening allows us to draw the following conclusions.

1) The high similarity of the responses from fractures with an aspect ratio of 1/1000 and from infinitely thin fractures justifies the applicability of the ITF model for mesofractures with aspect ratios of up to 1/1000.
2) An almost complete similarity of converted diffractions and a considerable difference in compressional diffractions are shown on the *X*- and *Z*-components for all of the single fractures in this comparison of ITF and finite-opening fractures with aspect ratios of 1/1000 and 1/100.
3) On the *X*-component, converted diffractions dominate.

4.2.4. Analysis of polarity reversals on the horizontal component of converted diffractions

We can learn much about the character and structure of fracturing in productive reservoirs from the relatively strong converted diffracted (scattered) wave and the absence of polarity reversals on its *X*-component.

Our conclusion on the absence of any polarity reversal in the converted diffracted wave from a single fluid-filled fracture contradicts the results reported by several authors, e.g., Wu et al. (2005), who indicate the presence of such a phenomenon in this wave. Naturally, it is necessary to explain this contradiction. Note that these authors consider vertical fractures, whereas all of the computations we have described address fractures deviating from the vertical at a small (5°) angle. To verify that the difference in polarity behavior is due to the fracture being nonvertical, we employ the discontinuous Galerkin method (Miriyakha et al., 2015). With this method, *X*-component seismograms are computed for a nonvertical ITF, nonvertical fractures with aspects ratios of 1/1000 and 1/100, and a vertical fracture with an aspect ratio of 1/100. In Figure 4.7, for the first three fracture models, the converted wave is continuous, whereas it exhibits a polarity reversal for the vertical fracture model.

The good quality of the computations is evidenced by a multicyclicity inherent to all responses both from vertical and nonvertical fractures with aspect ratios noticeably exceeding 1/1000, for which the ITF model is inappropriate. The multicyclicity, as noted, is related to a series of successive reflections from the fracture tips caused by the wave (guided wave) traveling inside the fracture (Frehner and Schmalholz, 2010).

A polarity reversal of the converted wave from a vertical fracture also is observed in Figure 4.8b. This computation is conducted on a model used in Wu et al. (2005). The resulting wave pattern is identical to that in Wu et al. (2005) and also exhibits a polarity reversal on the *X*-component of the converted diffracted wave. The wave pattern computed for the same model but with a 5° fracture deviation from the vertical, which is shown in Figure 4.8a, has no polarity reversal.

Comparing the infinitely thin fracture (ITF) and finite-opening fracture results allows us to establish the following.

1) Application of the ITF model, used in the grid-characteristic method, is appropriate for the two major types of fluid-filled hydrocarbon reservoir fractures, macrofractures and mesofractures, with aspect ratios of 1/100,000 and 1/10,000 through 1/1000, respectively.
2) *X*- and *Z*-component seismograms from ITF and finite-opening fractures with aspect ratios of 1/100 and 1/1000 show a similarity in both amplitude

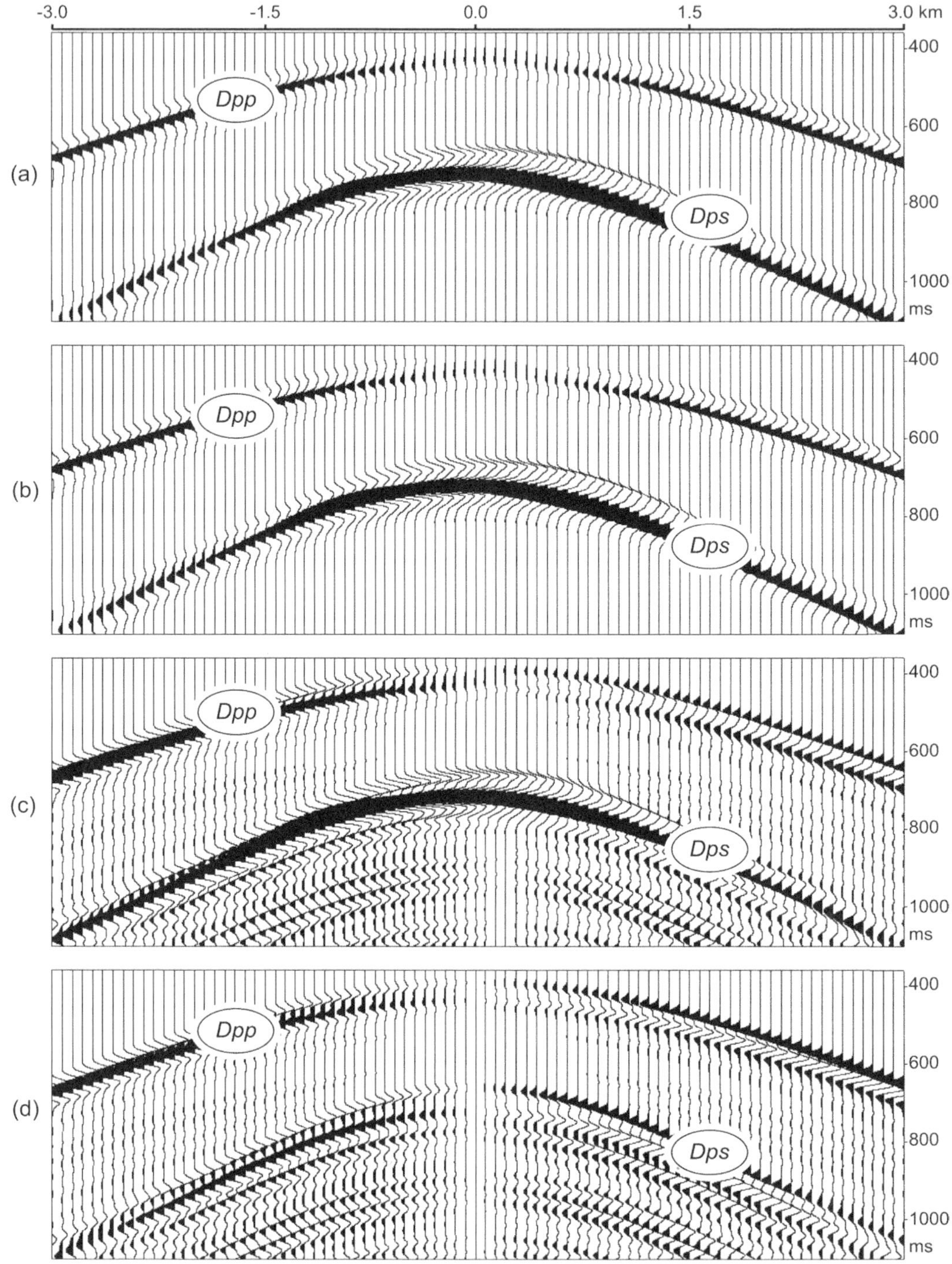

Figure 4.7. Comparison between the *X*-components of the responses from a single fracture with different aspect ratios and different deviations from the vertical: (a) ITF with a 5° deviation, (b) fracture with 5° deviation and 1/1000 aspect ratio, (c) fracture with 5° deviation and 1/100 aspect ratio, and (d) vertical fracture with 1/100 aspect ratio. No gain is applied.

strength and polarity behavior of converted diffractions as well as a substantial difference in both amplitude strength and polarity behavior of compressional diffractions.

3) Even a minor (5°) fracture deviation from the vertical is sufficient to cause the absence of a polarity reversal on the *X*-component of the converted diffraction, which greatly enhances the chance of its identification and its value as a diagnostic tool for evaluating fractures.
4) A polarity reversal on the *X*-component of the converted diffraction from a vertical fracture occurs when the fracture opening is appreciable, i.e., its aspect ratio is about 1/100 or greater.

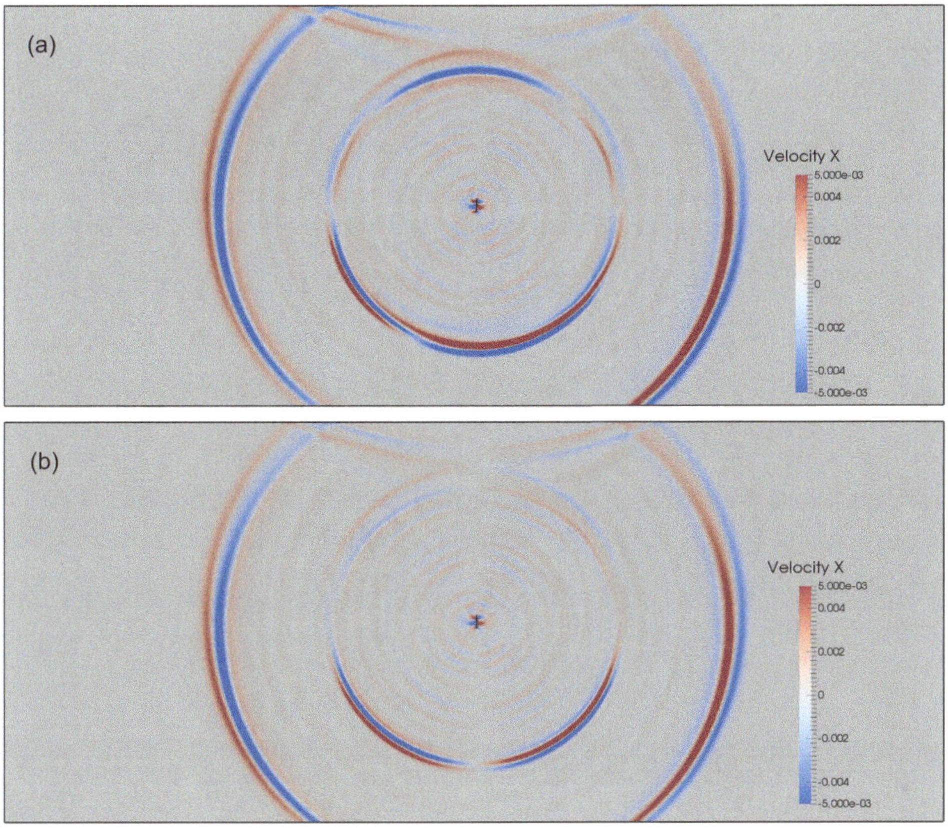

Figure 4.8. Comparison between the *X*-component snapshots at 0.52 s from (a) a nonvertical and (b) a vertical single fracture with an aspect ratio of 1/100.

4.3. Features of responses from vertical and near-vertical fluid-filled fractures

Exploration targets in oil and gas fields typically lie at depths of a few kilometers, where a high vertical stress caused by the overburden results in a vertical and near-vertical orientation of fractures. Inclined and horizontal fractures normally close under this stress.

The response from a vertical fracture is governed by two factors, such as the angle of incidence and the degree of fracture opening (aspect ratio). Three situations are possible.

1) When a fracture has a small opening (AR < 1/1000) and an energy source is located at the projection of the fracture onto the surface, the incident wave propagates vertically downward and tangentially to the fracture, thus producing almost no reflection (Figure 4.9).
2) When a fracture has a small opening (AR < 1/1000) and the incident wave propagates from an energy source located away from the fracture projection onto the surface, the seismic response is recorded as diffracted compressional and diffracted converted waves with minimal arrival times at the projection point.
3) When a fracture is wider (AR > 1/1000 to 1/100) and the incident wave propagates vertically, both compressional diffractions and converted diffractions of the response have polarity reversals on the *X*-component (see Figures 4.7 and 4.8).

Near-vertical fluid-filled fractures with any small aspect ratio, up to an infinitely thin fracture, form compressional and converted diffractions in their responses.

It is unlikely that any fracture would be strictly vertical. Fractures with a deviation from the vertical in the range of 5° to 15° seem to be most common.

Both for the upgoing compressional diffraction and the upgoing converted diffraction from the tip of a near-vertical fracture, the amplitude strength comprises 2% to 3% of that of the incident P-wave (see Table 4.1). A significantly greater portion of the diffraction energy generated by the fracture propagates downward. This is well illustrated by the snapshot in Figure 4.2, which is obtained by numerical simulation for a fracture with a 20° deviation from the vertical.

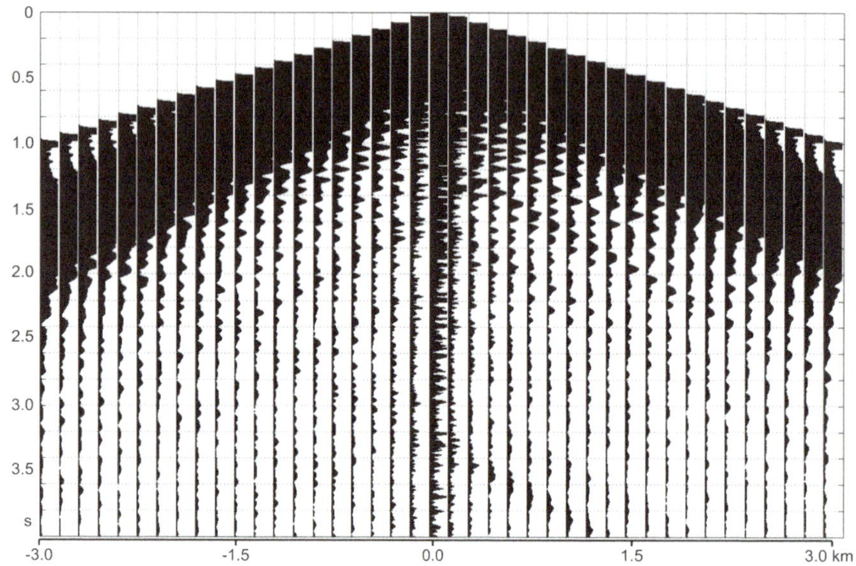

Figure 4.9. Absolute values of the response from a vertical fracture located at a depth of 2000 m, with a point energy source coinciding with the projection of the fracture onto the surface. No reflection from the fracture occurs.

It is important to note that, for near-vertical fluid-filled fractures with an aspect ratio of less than 1/1000, which are typical of hydrocarbon reservoirs, converted diffractions have much higher amplitudes than those of compressional diffractions and have no polarity reversal on the X-component.

Numerical simulation of the wave responses from near vertical fractures illustrates one more of their properties, asymmetry of the diffractions, i.e., a difference between their left and right parts, even at a small (5°) deviation from the vertical. Increasing the deviation angles (from 0° to 15°) enhances this difference (Figure 4.10).

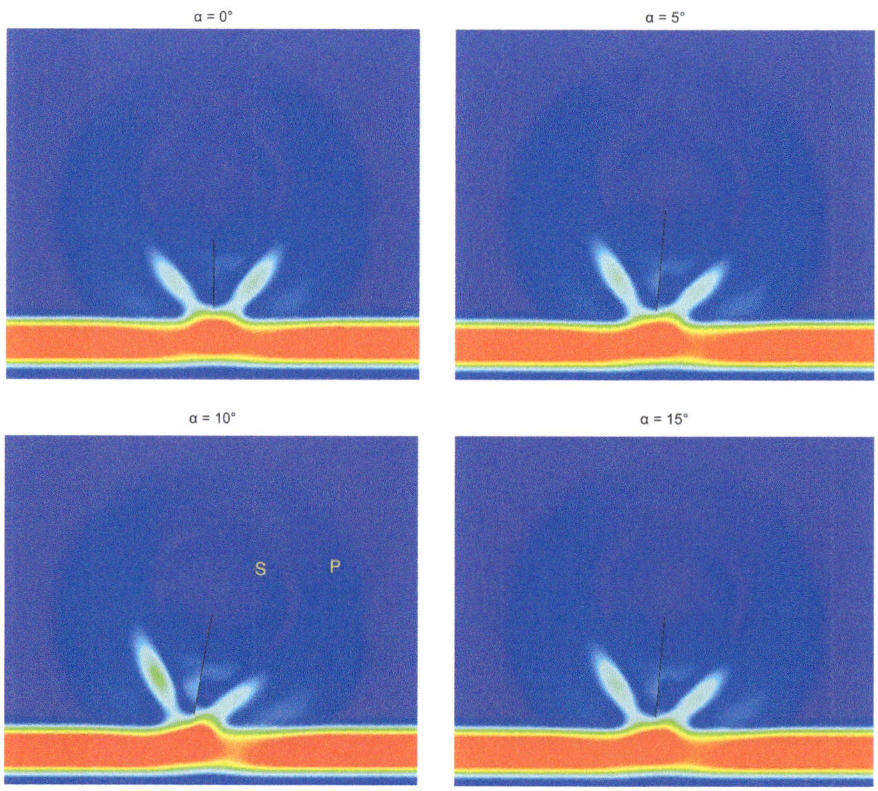

Figure 4.10. The asymmetry of the response from almost vertical single macrofractures increases as the fracture deviation angle increases.

4.4. Effects of fracture-surface asperity on wave-response character

Outcrops, cores, thin slices, and scans of well walls all evidence the variability of fractures in asperity, thickness, and, most importantly, areas of contact of fracture faces. Such data have been presented in investigations of fractured reservoirs by Bagrintseva (1999), Bertotti et al. (2003), Nekrasov (2006), and others. It is important to evaluate and understand how much direct contact may reduce or distort a response from a fracture, if the response is to be used as a diagnostic tool.

Next we present the result of a series of numerical experiments designed to evaluate the effects of partial contact of fracture walls on the wave response.

4.4.1. Computation technique

We exploit different fracture models that account for partially closed fractures. Depending on the type of contact, at some points of these models, we impose one of the following three types of boundary conditions, i.e., free sliding for fluid-filled fractures,

$$\vec{v}_a \cdot \vec{p} = \vec{v}_b \cdot \vec{p} = V_p, \quad f_p^a = -f_p^b, \quad \vec{f}_\tau^a = \vec{f}_\tau^b, \quad (4.1)$$

free boundary for dry (gas-filled) fractures,

$$\vec{f}_a = \vec{f}_b = 0, \quad (4.2)$$

full contact rigidly closed fractures,

$$\vec{v}_a = \vec{v}_b = \vec{V}, \quad \vec{f}_a = -\vec{f}_b, \quad (4.3)$$

where $\vec{v}$ are the velocities of touching boundary points, $\vec{f}_p$ and $\vec{f}_\tau$ are, respectively, the normal and tangential components of the external force $\vec{f}$ affecting the boundary, and subscripts a and b denote, respectively, the left and right sides of a fracture.

The contact conditions along a fracture are specified in a random manner, with the fraction of contact fracture area varying from 0% to 30%. In these computations, the grid-characteristic method employs a second-order scheme on an unstructured triangular grid. For fluid-filled and gas-filled fractures, the closure factor g, representing the ratio of the contact area of a fracture to its total area, is varied from 0% to 30%, with an increment of 5%.

4.4.2. Model specifications

In this experiment, the physical parameters of the host medium are chosen to be $Vp = 2300$ m/s, $Vs = 1300$ m/s, and $\rho = 1200$ kg/m^3. In this medium, 11 fractures of 100 m height, 5° deviation from the vertical, and 50 m spacing are imbedded at a depth of 400 m.

For fluid-filled fractures, the amplitude of the *Dps*-wave is found to be linearly dependent on factor g. This amplitude gradually decreases as the percentage of the contact surface increases (Table 4.3). The percentage decrease in the amplitude of the converted *Dps* wave is shown to correlate well with g (Figure 4.11).

For gas-filled fractures, we have considered the dependence of both *Dpp*- and *Dps*-waves on g (Table 4.4).

The *Dpp*-wave amplitudes also decrease linearly with g. However, the linearity of the relationship between the *Dps*-wave amplitude and g is no longer maintained when the fracture face contact exceeds 10%, because the response amplitude in this case becomes comparable to the computation error (Figure 4.12).

Thus, the following should be noted.

1) When the contact area (the portion of the fracture faces that are in contact) is moderate (up to 30%), the converted diffraction on the *X*-component decreases by approximately the same percentage as g and retains the ability to carry information about the fracture.

Table 4.3. Variation of *Dpp* and *Dps* amplitudes with changing contact surface percentage for fluid-filled fractures.

Wave \ g (%)		0	5	10	15	20	25	30
Dps	Vx	144	135	128	123	117	111	109
	Vz	21	16	15	16	18	12	11
Dpp	Vx	31	31	29	28	28	29	28
	Vz	33	27	25	25	24	21	19

Table 4.4. Variation of *Dpp* and *Dps* amplitudes with changing contact surface percentage for gas-filled fractures.

Wave \ g (%)		0	5	10	15	20	25	30
Dps	Vx	211	122	57	43	36	33	31
	Vz	11	9	9	9	9	9	9
Dpp	Vx	167	119	92	81	71	69	64
	Vz	381	324	311	285	282	264	248

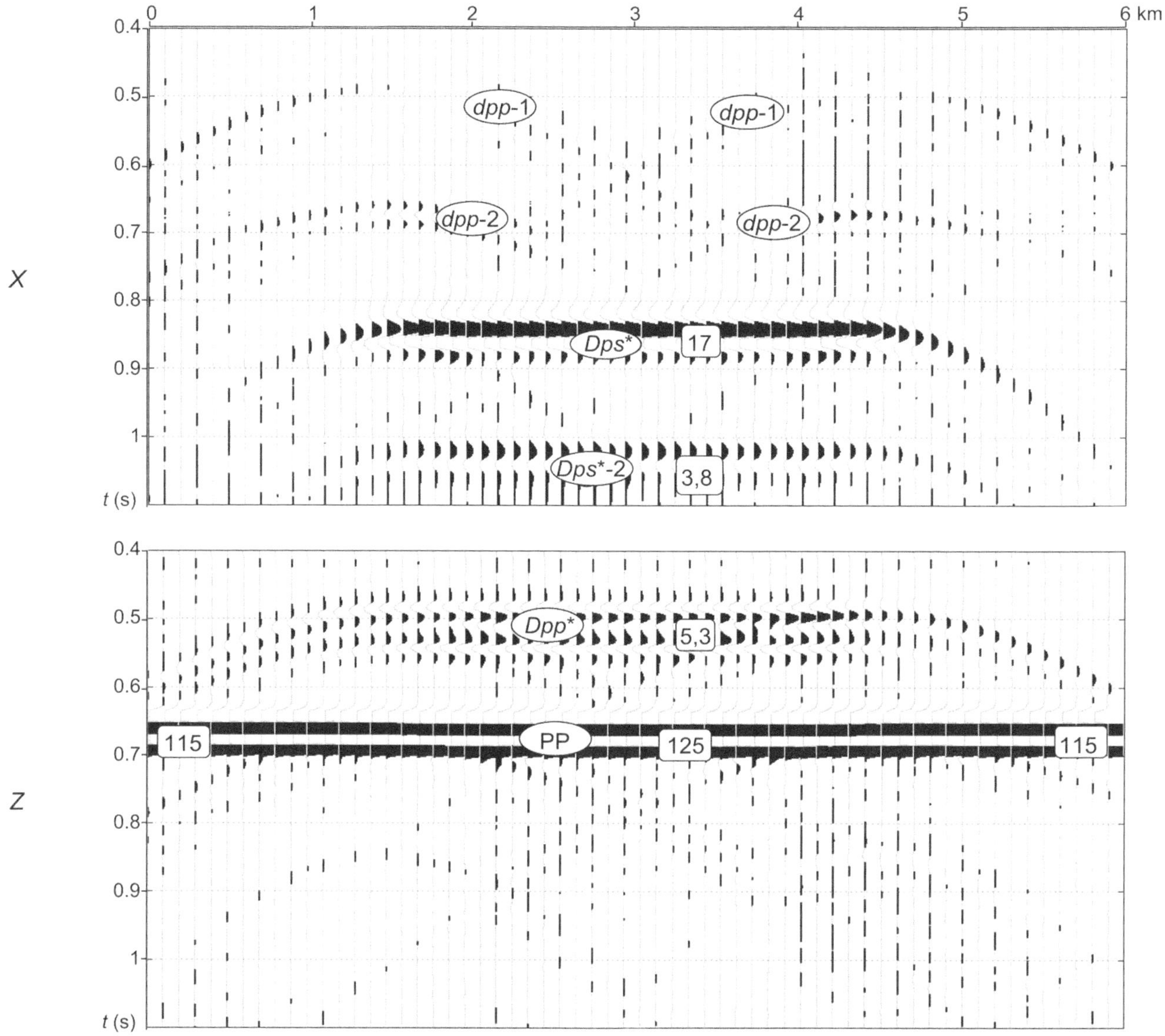

Figure 5.3. Horizontal (*X,* top) and vertical (*Z,* bottom) components of the response from a bed-bounded fluid-filled mesofracture zone generated by a plane wavefront: *Dps** = scattered converted wavefront; *Dpp** = scattered (diffracted) compressional wavefront; and PP = compressional reflection from the lower horizontal boundary. Values in boxes indicate amplitudes in relative units[1] (r.u.).

horizontal fronts of compressional waves (3), converted waves (4), and scattered waves (6), as well as the concentric wavefronts from the edges of the mesofracture zone, develop in time and space. At a time of 350 ms, the scattered compressional wavefront (3) is clearly visible as two separate waves, from the top and base. On most of the snapshots (at 150, 200, and 600 ms), the scattered converted wavefront (4) remains an unchanged single wave formed by the top rows of the fractures.

The primary elements of the wave response from the mesofracture zone shown in Figure 5.2 also are recorded on the *X*- and *Z*-components at the earth's surface, as shown in Figure 5.3.

These results demonstrate the following.

The primary elements of the wave response, scattered converted (*Dps**) and scattered compressional (*Dpp**) wavefronts, and reflections from the bed boundaries (PP) are linear for the horizontal boundaries.

The principal scattered wavefronts (*Dps** and *Dpp**) are formed by the superimposition of individual converted and compressional diffractions, respectively, from the mesofracture tips of the first rows of fractures in the fractured bed.

[1] Presentation of wave amplitudes in relative units (r.u.) ensures their true ratio within the analyzed combination.

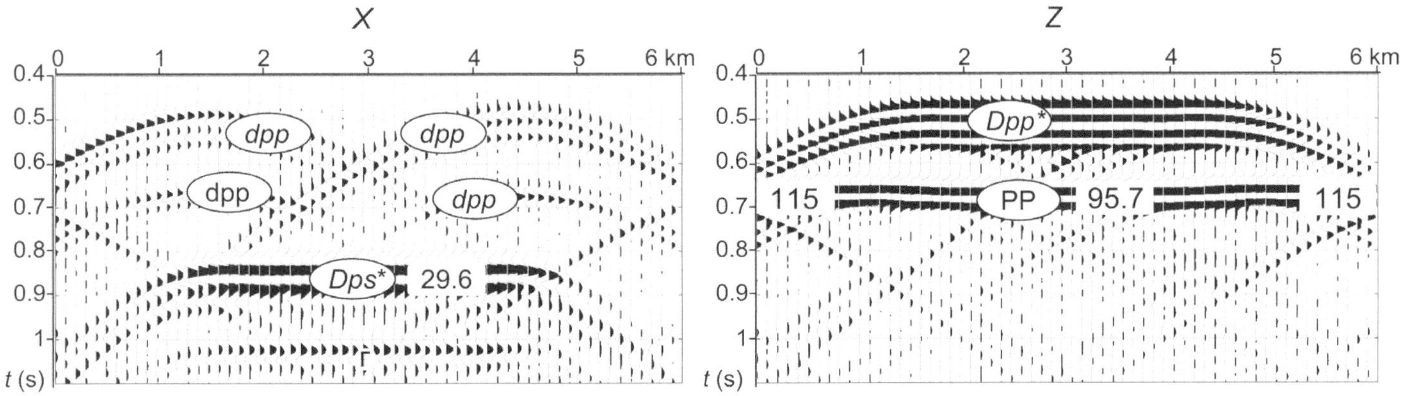

Figure 5.4. Horizontal (*X*, left) and vertical (*Z*, right) components of the response from a bed-bounded gas-filled mesofracture zone generated by a plane wavefront. The wave type labelling is the same as that in Figure 5.3.

Substantially weaker response elements are related to the edge hyperbolas (*Dps* and *Dpp* in Figure 5.3) and to the scattered wavefronts formed by the reflection coming from beneath (Dps^*_2 in Figure 5.3).

Comparing the scattered converted and scattered compressional wavefronts generated by the plane-wavefront energy source, we observe:

- the converted wavefront (*Dps**) on the *X*-component is substantially stronger (by several times) than the compressional wavefront (*Dpp**) on the *Z*-component, approximately as with a point energy source;
- the signal of the scattered converted (*Dps**) wavefront formed by the first rows of mesofractures at the bed top is lower-frequency one-pulse signal shape, and
- a complex two-pulse signal of scattered compressional wavefront (*Dpp**) occurs, which is apparently related to the generation of diffractions from mesofractures of both top and base rows.

The response from the base model with gas-filled mesofractures is illustrated in Figure 5.4. The wave composition of the response remains the same. The response, however, is stronger in total and differs in the ratio of the amplitudes of scattered converted and scattered compressional waves on both vertical and horizontal components. This distinction is largely related to complete reflections from the surfaces of gas-filled fractures, a condition which is incorporated in the method algorithm.

The analysis performed to this point is based on data obtained with a plane-wavefront energy source and a model of mesofracture bed with no difference in the physical properties at the top and base boundaries, when no reflections are generated at these boundaries. The goal is to obtain a wave response from the mesofractured bed uninfluenced by any other objects.

The characteristics of the wave response from the mesofractured bed-bounded zone are as follows:

1) The principal response elements are scattered (diffracted) converted and compressional wavefronts.
2) For fluid-filled mesofracture zones, which are of most interest for hydrocarbon exploration, the scattered converted wavefront is considerably stronger than the scattered compressional wavefront.
3) The scattered converted wavefront signal is one-pulse and low frequency, whereas the scattered compressional wavefront signal is two-pulse due to the interference of signals formed at the top and base of the bed.
4) The diffraction from mesofracture tips (edges of the fractured zone) on the *X*-component does not change its sign, whereas on the *Z*-component the sign does change.

5.1.2. Investigating the mechanism of P- and S-wave propagation in a mesofractured bed

Computation of propagation of converted and compressional waves inside a fractured bed provides an explanation of how these scattered wavefronts form and what distinguishes them. The base model of a fractured bed embedded in a homogeneous half-space and a plane-wavefront energy source of compressional waves are used in the computation.

To improve the response resolution, the incident wavefront frequency is increased to 60 Hz. The results are presented in Figure 5.5 as snapshots on the *XZ*-plane. As distinct from those shown in Figure 5.2, these data show greater detail, such as individual mesofractures in

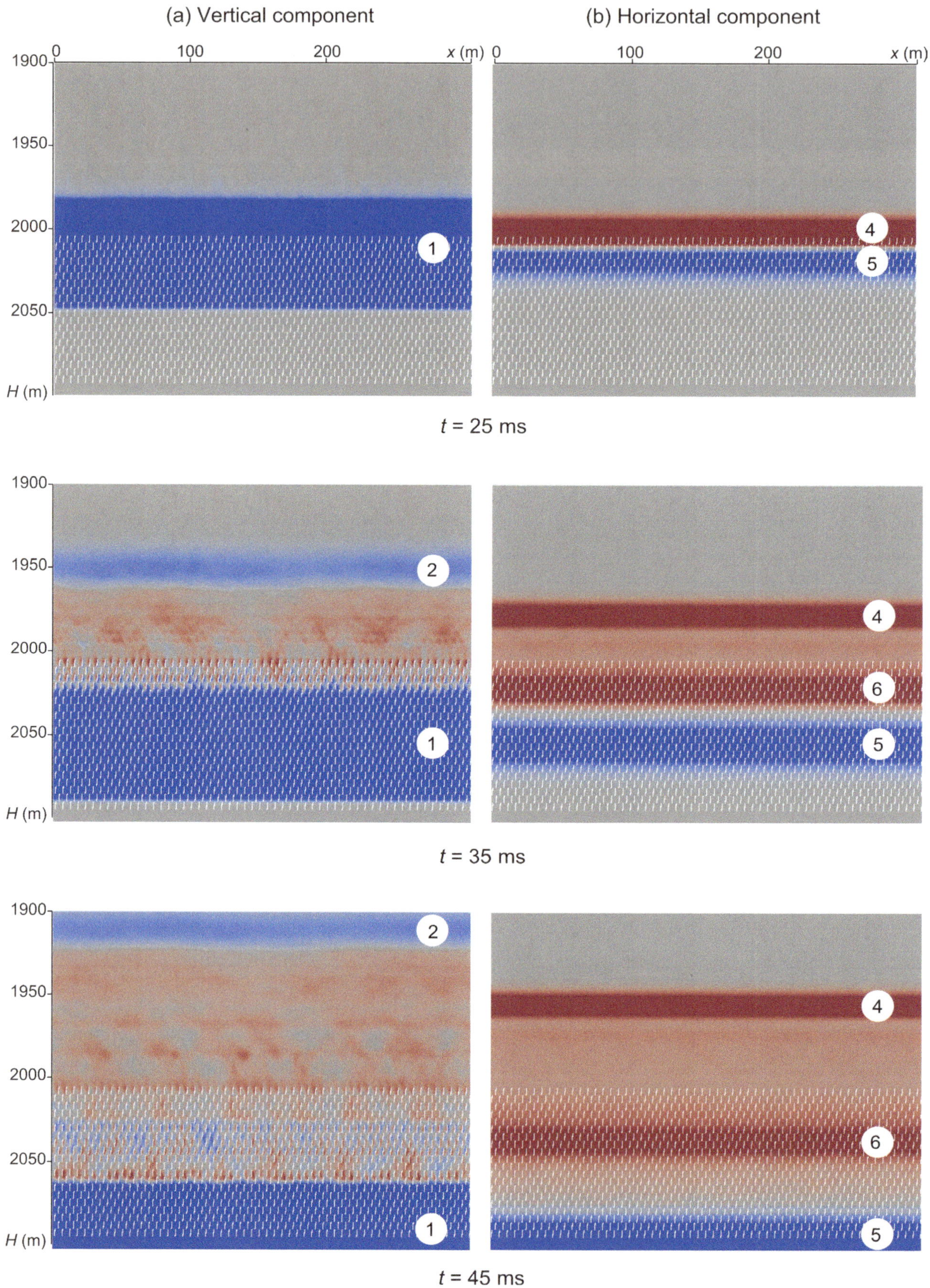

Figure 5.5. (*Continued*)

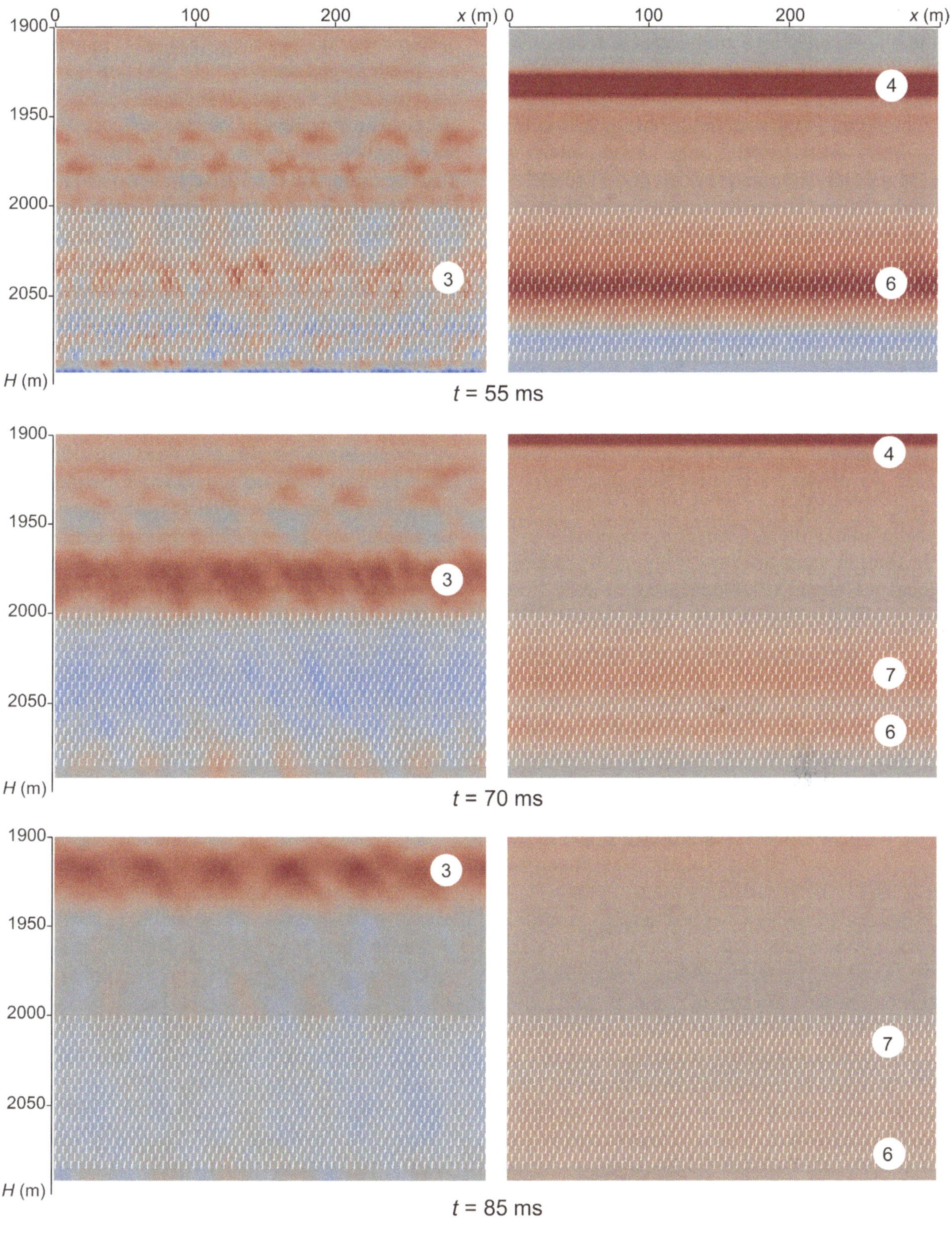

Figure 5.5. ***(Continued)*** Wave propagation through a fractured bed: 1 = incident plane compressional wave; 2 = front of scattered (diffracted) compressional waves propagating upward from the top rows of fractures; 3 = front of scattered compressional waves, of negative amplitudes, propagating upward from the base rows of fractures; 4 = front of scattered (diffracted) shear waves, of negative amplitudes, propagating upward from the top rows of fractures; 5 = horizontal component of compressional wavefront (1); 6 = front of scattered shear waves, of negative amplitudes, propagating downward from the top rows of fractures; and 7 = front of scattered (diffracted) shear waves propagating upward from the base rows of fractures.

Table 5.1. Parameters used to assess the effect of their variation on the seismic response from a zone of mesofractures.

Parameter to analyze / Parameter analysis conditions	Density d (m)	Height h (m)	Deviation from the vertical α (degrees)	Frequency f (Hz)	Bed thickness (m)
Density	1, 2, 4	2	5	30	100
Height	2	2, 4, 8	5	30	100
Deviation from the vertical	2	2	5, 10, 15	30	100
Frequency	2	2	5	15, 30, 60	100
Bed thickness	2	2	5	30	2, 4, 10, 25, 50, 100, 200

Note: Each parameter being analyzed has a corresponding row of values of the other parameters defining the models of a particular group.

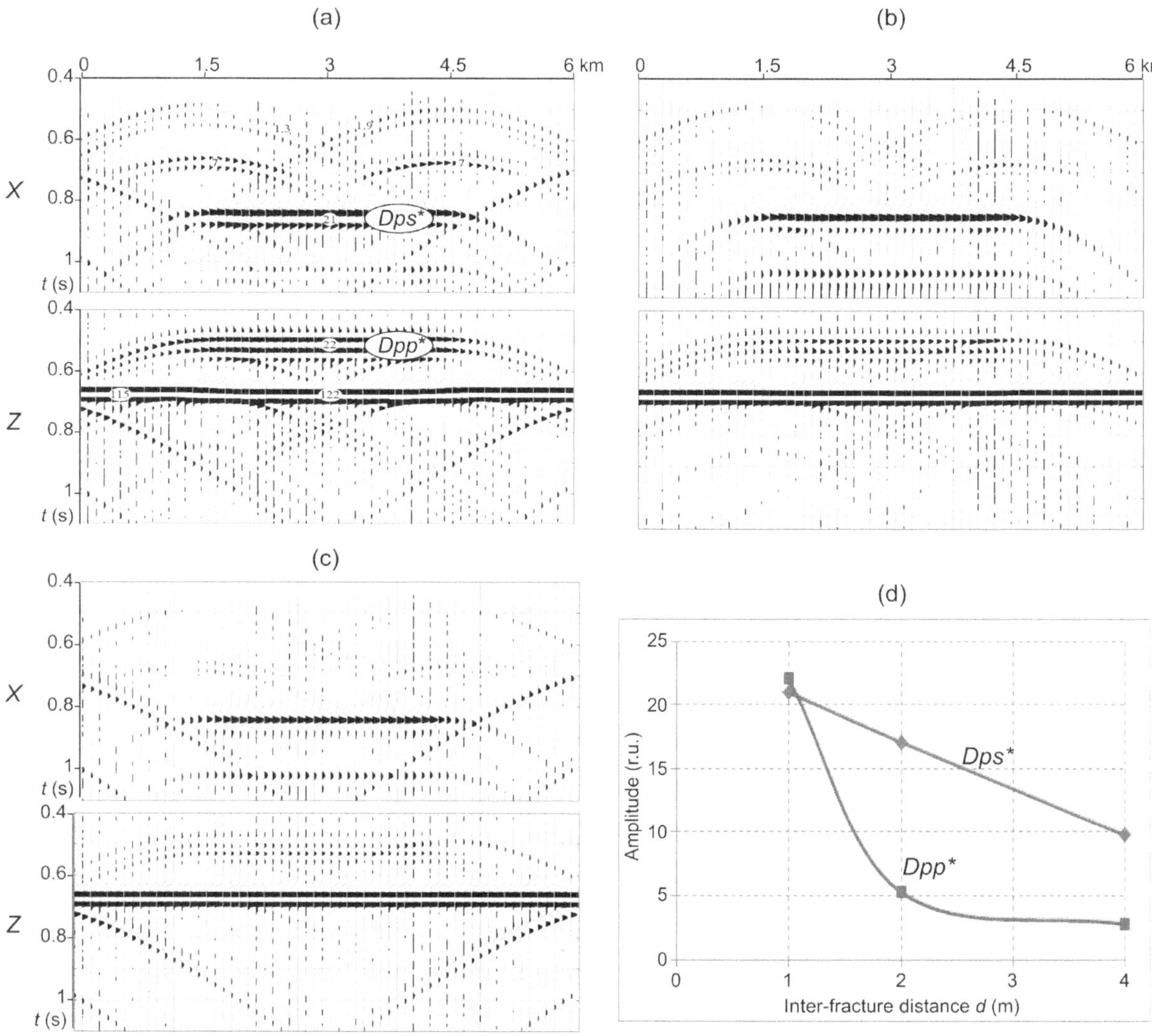

Figure 5.7. Common-source gathers for different values of mesofracture spacing: (a) 1 m, (b) 2 m, (c) 4 m, and (d) plots showing the dependence of the amplitude strengths of scattered converted (*Dps**) and compressional (*Dpp**) wavefronts on fracture spacing. In (a)–(c), waves similar to those shown in Figure 5.3 are displayed. The spurious inclined events are formed by the side walls of the model used.

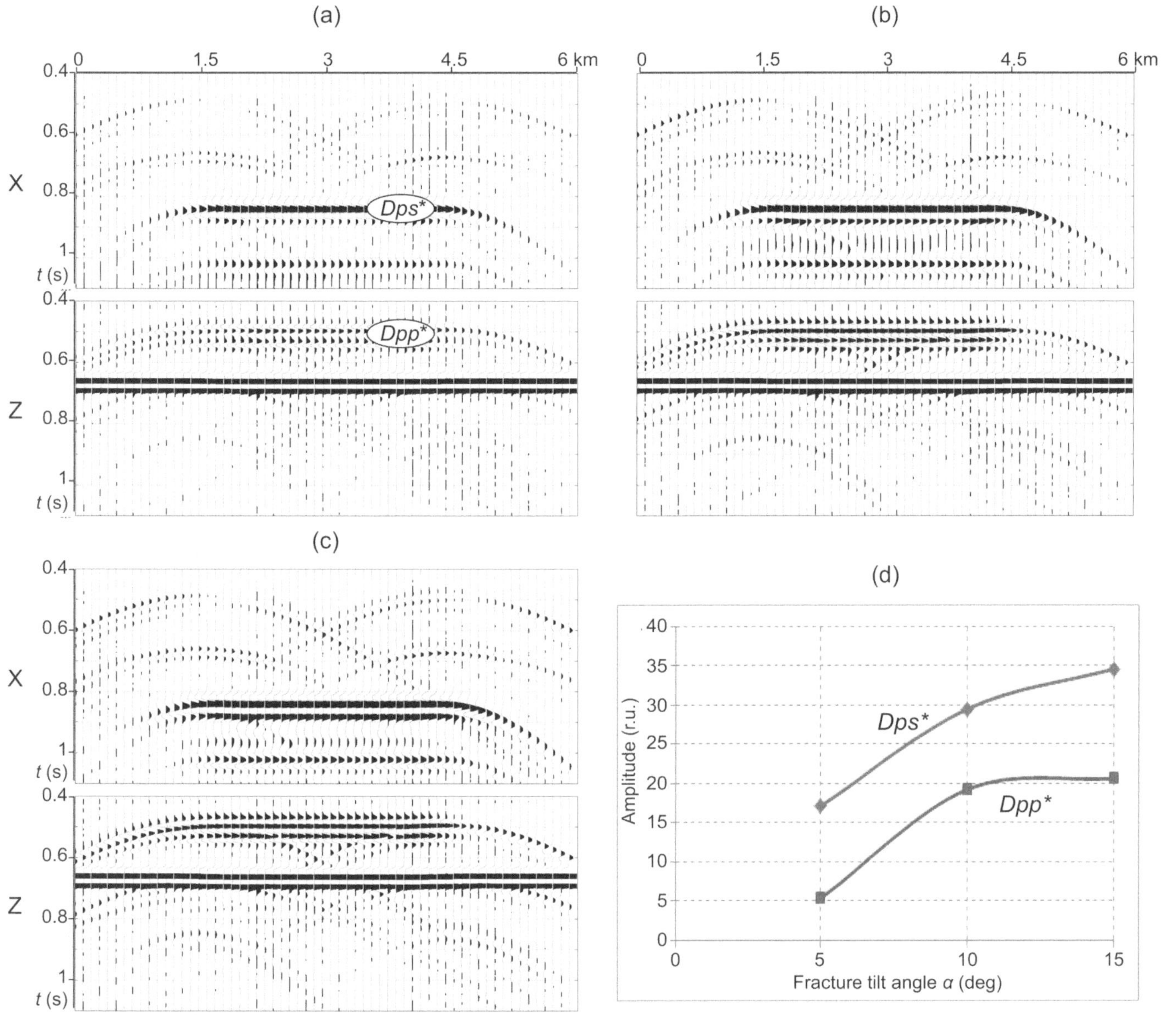

Figure 5.8. Common-source gathers for different mesofracture deviations from the vertical: (a) 5°, (b) 10°, (c) 15°, and (d) plots showing the dependence of the amplitude strengths of scattered converted (*Dps**) and compressional (*Dpp**) wavefronts on the fracture deviation. In (a)–(c), waves similar to those shown in Figure 5.3 are displayed.

Fracture deviation from the vertical α.—The effect of varying the deviation from the vertical is illustrated in Figure 5.8. At a triple increase in the deviation angle from 5° to 15°, the amplitude strength of scattered wavefronts is shown to increase substantially, i.e., double for the converted wavefront and fourfold for the compressional wavefront. The double amplitude increase of the scattered converted wavefront also is observed at the maximum fracture deviation from the vertical (Figure 5.8d).

Fractured bed thickness H.—Figure 5.9 shows the effects of varying the thickness of the fractured zone, within a wide range of thickness values: 2, 5, 10, 50, 100, 200 m (dots on the plots). We find the following changes in the wave response: a hundredfold increase in this parameter does not significantly change the shape and amplitude of the scattered converted wavefront, with the minor exception of a 20% amplitude decrease compared with the overall amplitude level for $H = 2$ m (Figure 5.9). Unlike this wave, the scattered compressional wavefront substantially changes its form and amplitude strength in the 2 to 50 m thickness range of the fractured bed. As the thickness is increased further from 50 to 200 m, the signal amplitude remains constant, while the signal itself splits more and more into two wavelets. Such changes in the wave response from the mesofracture zone apparent-

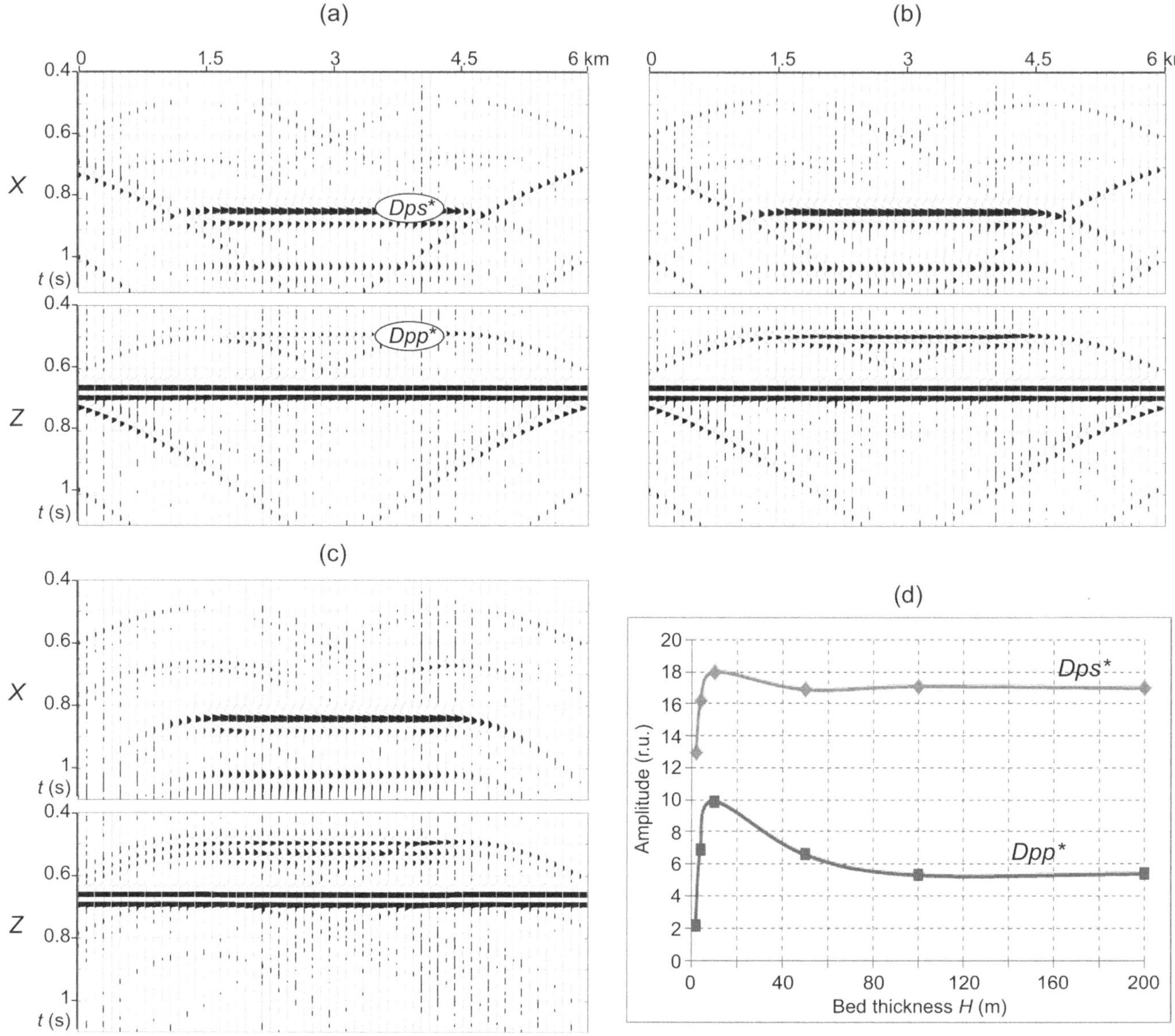

Figure 5.9. Common-source gathers for different values of the bed thickness: (a) 2 m, (b) 10 m, (c) 100 m, and (d) plots showing the dependence of the amplitude strengths of scattered converted (*Dps**) and compressional (*Dpp**) wavefronts on bed thickness. In (a)–(c), waves similar to those shown in Figure 5.3 are displayed.

ly can be attributed to the formation of the scattered converted wavefront by only the first top rows of mesofractures.

Diffractions from the bed base rows attenuate when they propagate upward in the fractured bed, as shown in Figure 5.5.

The scattered compressional wavefront, which is obviously dependent on the fractured bed thickness, is formed by the first top rows and the last base rows of mesofractures. The shape of this interference signal depends on the difference between the arrival times from the top and base of the fractured bed, which is determined by bed thickness H.

Signal frequency f.—A fourfold decrease in the frequency of an incident wave from 60 to 15 Hz considerably changes the relationship between the amplitudes of scattered wavefronts (Figure 5.10). It substantially increases the amplitude of the converted wavefront in comparison to a negligible increase in the amplitude of the compressional wavefront. Therefore, this effect may be used for more confident detection of mesofracturing zones.

Mesofracture height h.—Common-source gathers for mesofractures with different heights differ little from one another. A fourfold change in fracture height from 2 to 8 m changes negligibly the amplitudes and frequen-

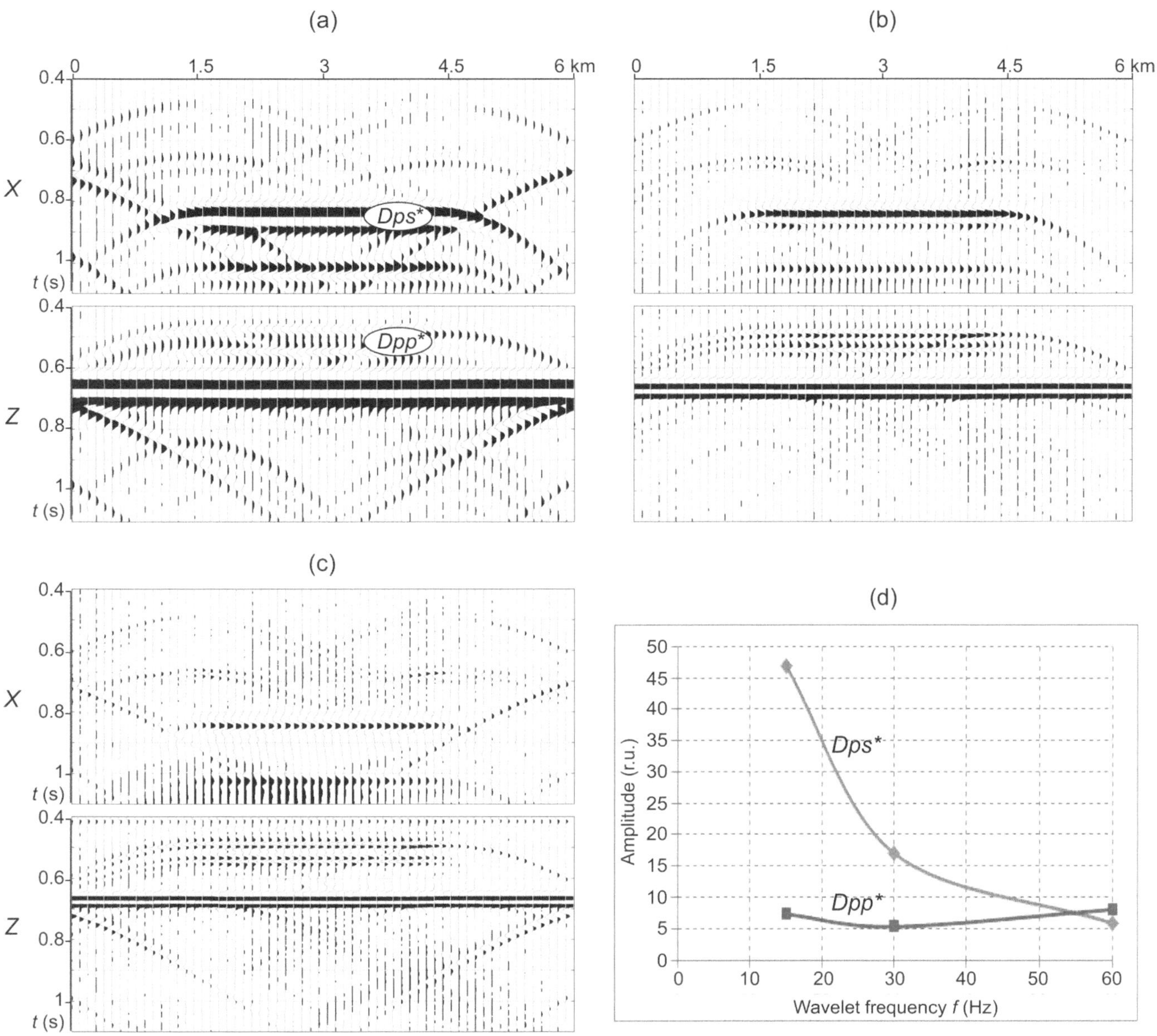

Figure 5.10. Common-source gathers for different values of signal frequency: (a) 15 Hz, (b) 30 Hz, (c) 60 Hz, and (d) plots showing the dependence of the amplitude strengths of scattered converted (*Dps**) and compressional (*Dpp**) wavefronts on signal frequency. In (a)–(c), waves similar to those shown in Figure 5.3 are displayed.

cies of the major response elements, scattered wavefronts, and retains the triple amplitude predominance of the converted wavefront over the compressional one.

The following conclusions can be drawn from the preceding discussion of the fluid-filled fractures.

1) Changes in fracture spacing, deviation from the vertical, and signal frequency substantially influence the amplitude strength of scattered wavefronts.
2) Changes in fracture height and thickness of the fractured bed only marginally influence the response waves except for the signal amplitude of the compressional wavefront when H is small.
3) When the parameters of the mesofracture zone are varied, the scattered converted wavefront is much more stable than the scattered compressional wavefront.
4) The amplitude of the scattered converted wavefront is nearly always greater, in many cases by many times, than that of the scattered compressional wavefront. This amplitude can be made still higher by shifting the spectrum to lower frequencies.
5) The scattered converted wavefront, which may be a prospective indicator of mesofracture zones, possibly can be formed by very thin fractured beds, up to a few meters thick.

5.1.4. Assessment of how real factors may affect the identification of responses from bed-bounded zones of mesofractures

Layering of the target interval.—In most cases, seismic surveys are conducted over near-horizontal layered sedimentary formations. For this reason, we use a five-layer model to assess the possibility of detecting mesofracturing zones using the wave responses emanating from such zones. Computations are performed for a medium imitating a carbonate formation with the parameters listed in Table 5.2.

The fractured bed is located in layer 4 and has the parameters of the base model. The *Z*- and *X*-component common-source gathers for this layered model (Figure 5.11) allow us to see the following. On the vertical *Z*-component, reflections from the horizontal boundaries dominate. The scattered compressional wavefront (3) is many times weaker in amplitude than the reflection events (1), which originate from the boundaries with small reflectivities (about 10% and less). The scattered compressional wavefront is comparable to the observed multiples (4).

Table 5.2. Parameters of a five-layer carbonate formation model containing a mesofracture zone in Layer 4.

Layer	*Vp* (m/s)	*Vs* (m/s)	ρ (g/cm^3)	*H* (m)	*R*
1	4130	2150	2.50	800	–
2	4540	2420	2.55	400	0.109
3	4800	2540	2.60	600	0.072
4	5150	2650	2.65	800	0.085
5	5660	2800	2.70	400	0.108

On the *X*-component common-source gathers, the situation is different. Here, only the scattered converted wave (2) dominates. A complete absence of converted reflections is caused by a normal incidence of the compressional plane wavefront onto the horizontal boundaries. When reflectors are non-horizontal and point sources of seismic energy are used, these waves may become a major obstacle to identifying scattered converted wavefronts.

Variability of mesofracture spacing along a bed.—In most cases, zones of mesofractures in reservoirs do not have sharp boundaries. Transitions from areas with a small fracture spacing to areas with a large fracture spacing are commonly diffusive. Figure 5.12 schematically shows scattered converted wavefronts for a pair of such zones. Wave responses in this figure differ from those of the base model (see Figure 5.3) in that diffractions from the zone edges are reduced substantially. This is attribut-

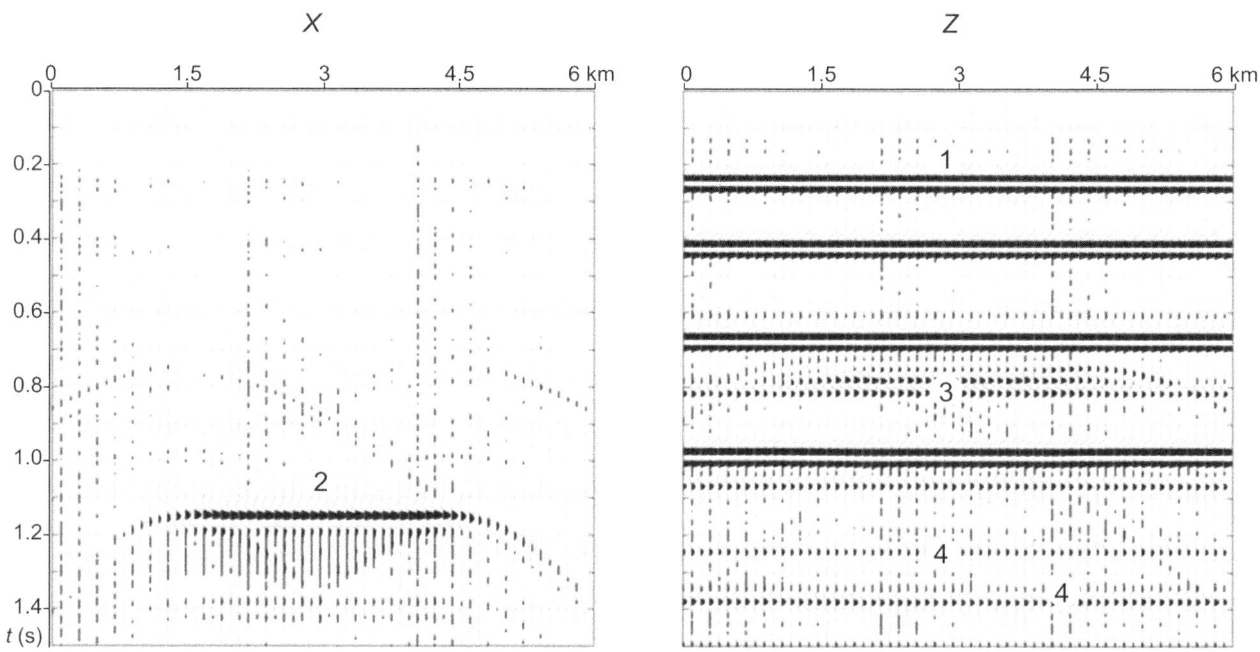

Figure 5.11. *X*-component (left) and *Z*-component (right) of a synthetic common-source gather for a five-layer model containing a zone of mesofractures: 1 = reflections from horizontal boundaries of the subsurface; 2 = scattered converted wavefront; 3 = scattered compressional wavefront; and 4 = weak multiples.

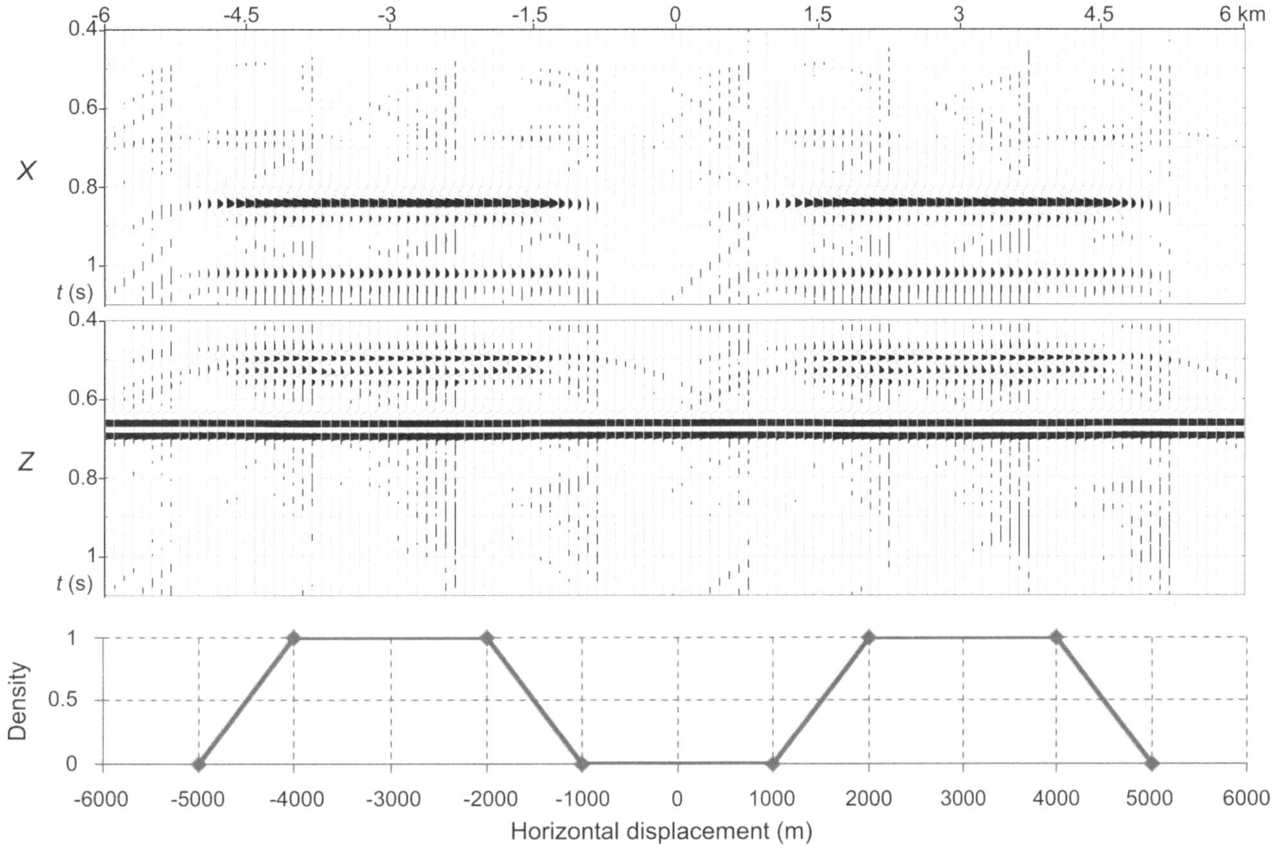

Figure 5.12. *X*- and *Z*-components of a common-source gather (top) for a model of mesofracture density gradually varying in space (bottom).

ed to a gradual variation in the mesofracture spacing, as distinct from the sharp zone edges in the base model.

Dispersion of the mesofracturing parameters.—All of these computations use idealized models containing fractures with identical sizes, inclinations, orientations, spacings, and distributions, which is rarely feasible in real conditions. These parameters clearly have certain variability levels, or dispersion. In order to assess the effects of dispersion on the possibility of using the scattered converted wavefront as an indicator of mesofracture zones, we perform computations using a technique previously employed for the similar purpose for macrofractures (Leviant et al., 2013).

Figure 5.13b shows the *X*- and *Z*-components of the response from a mesofracture zone with a 20% variability (dispersion) of both fracture spacing and angle of fracture deviation from the vertical, specified by a random number generator. Only barely noticeable changes in the scattered converted wavefront are observable by visually comparing this common-source gather with the wave response from the base model, which has idealized fracture parameters (Figure 5.13a). Quantitative estimates of the wave amplitude strength are shown in Figure 5.13c. They indicate a minor (10%) decrease in the converted wavefront amplitude and a double increase in the scattered compressional wavefront amplitude at the 20% dispersion of spacing and angle.

Non-horizontal reflectors.—The absence of converted reflections on the *X*-component in Figure 5.11 is attributed to the presence of strictly horizontal reflectors only. In real conditions, small deviations from the horizontal may occur, although most reservoirs are found in typical platform environments at low angles not exceeding a few degrees. In layered media even with such low dips, identification of a target scattered wave may be hindered by the *X*-components of compressional and converted reflections.

In order to assess the impact of this factor, acoustic boundaries of a fractured bed with 1° dip are included in the base model. Figure 5.14 illustrates how the horizontal component of compressional reflections manifests itself on the *X*-component even in the case of such a small dip. Note that this result is obtained using a plane-wavefront energy source. The almost fourfold predominance of the scattered converted *Dps** wave over the *X*-component of the PP-reflection, which is shown clearly

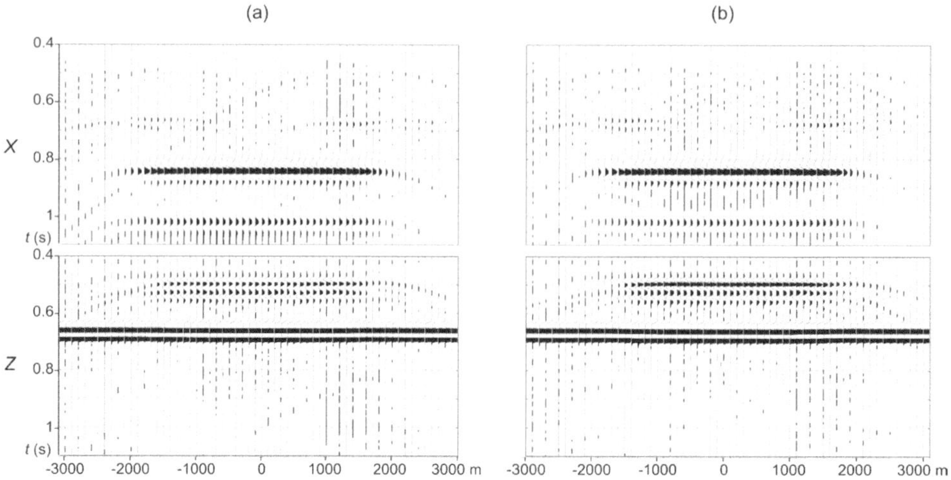

(c) Amplitude estimates

Recorded Component	**Vx**	Vx	Vx	Vx	Vx	Vx	**Vz**	Vz	Vz
Estimate Coordinate	**0**	0	−500	500	−1500	1500	**0**	0	−2500
Amplitude Estimate Time	**0,8–0,9**	1,0–1,1	0,5–0,6	0,5–0,6	0,6–0,7	0,6–0,7	**0,45–0,55**	0,6–0,75	0,6–0,75
Regular Distribution – A	**17,4**	4,3	0,4	0,5	1,9	1,8	**5,6**	124,9	115,7
Irregular Distribution – B	**15,7**	2,7	0,5	0,5	2,1	2,1	**12,5**	120,1	115,7

Figure 5.13. *X*- and *Z*-components of the responses (a) from a model with idealized fracture parameters and (b) a model with a 20% dispersion of both fracture spacing and angle of fracture deviation from the vertical, and (c) amplitude estimates for both models.

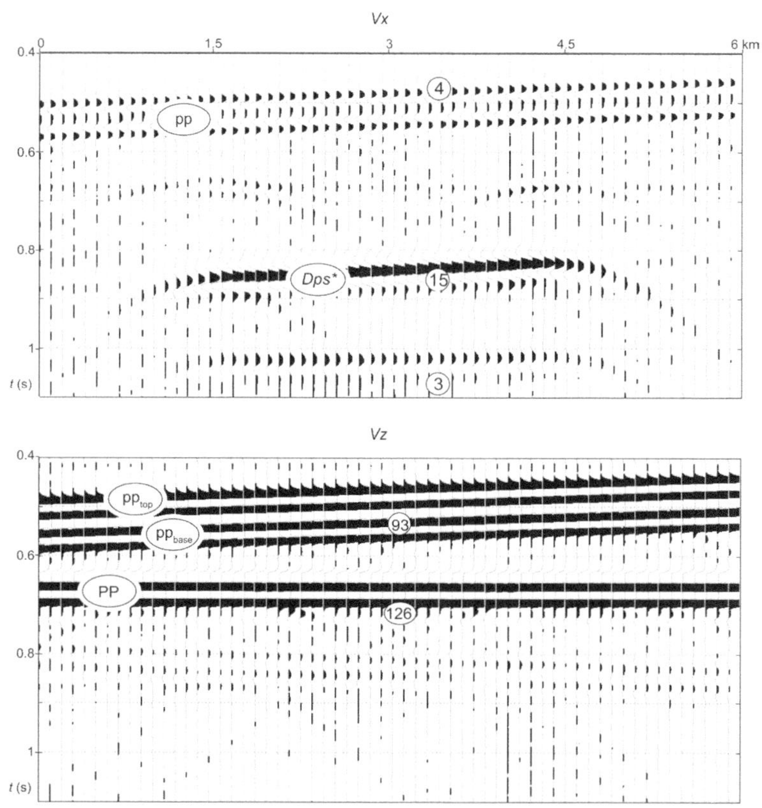

Figure 5.14. *X*-component (top) and *Z*-component (bottom) of the response from a bed-bounded mesofracture zone of a 1° dip.

in Figure 5.14, may substantially decrease in real conditions, and additional measures may be needed to identify this type of scattered waves.

5.1.5. Conclusions from Section 5.1

1) Using a plane-wavefront source has helped us establish that the principal elements of the seismic response from a bed-bounded zone of fluid-filled mesofractures are the scattered converted and scattered compressional wavefronts. The converted wavefront dominates on the *X*-component.
2) It is established that shear waves suffer abnormally high scattering (absorption) when propagating parallel to the fracture planes through the fractured medium. This mechanism severely reduces (to zero sometimes) the amplitude of scattered converted waves at the bed base.
3) The scattered converted wave is primarily a simple one-pulse wavelet of signal formed by interference of diffractions from mesofractures of the top rows.
4) For a scattered compressional wave with low absorption in a fractured medium, the wavelet of signal is typically formed by interference of two pulses from the top and base of a fractured bed.
5) Changes in spacing and inclination of fractures and in signal frequency substantially affect the amplitude strength of the scattered converted wave, whereas changes in the height of fractures and bed thickness have little effect.

5.2. Characteristics of wave responses at long offsets from bed-bounded zones of mesofractures for point sources

Modern seismic reflection methods use long offsets (up to twice the target depth) and point energy sources, and commonly record only the vertical component of wavefields. It is obvious that, in real conditions, the character of scattered compressional and scattered converted waves from fractured zones will differ from what has been described for plane-wavefront energy sources and rather short offsets. Considering the current relevance of detecting fractured reservoir zones using scattered seismic waves, the discussion that follows will present the results of numerical simulations of wave responses from these features in typical geologic settings. This analysis is performed separately for vertical and horizontal components.

5.2.1. General characteristics of wave responses from mesofracture zones on the vertical component

Fractured reservoirs most often are encountered in tight rocks (e.g., limestones, dolomites, cemented sandstones) at depths of 2.5 to 4 km.

A geologic setting to be considered in this context exhibits low-velocity contrasts. It has a fairly high acoustic impedance but with little positive and negative variations in the impedance and *Vp/Vs* ratio (Poisson's ratio) at all inner boundaries. A typical crossplot shown in Figure 5.18a substantiates this statement by an almost horizontal alignment of *Vp/Vs* values in the 12,000 to 14,000 impedance range. Consequently, reflection coefficients at the boundaries equal only a few percent. This type of geologic setting most frequently is encountered in East Siberia.

Another type of geologic setting is characterized by significant velocity contrasts. In this case, smooth acoustic boundaries have reflection coefficients typically equal to or more than 10%. Such conditions are common in the petroliferous basins of the Volga-Ural and Timan-Pechora provinces. The acoustic characteristics of fractured beds in those basins differ, as a rule, from those of the host rocks.

These factors are considered when constructing numerical models of fractured beds and bed-bounded fractured zones to study wave responses from such geologic features.

5.2.1.1. Model characteristics

The base models have physical parameters typical of carbonate rocks: $Vp = 4500$ m/s, $Vs = 2500$ m/s, and $\rho = 2500$ kg/m^3. The fractured bed, 100 m thick, is specified as lying at a depth of 2000 m. The size of mesofractures is 5 m and their deviation from the vertical is 5°. The fracture spacing is 5 m. The mesofractures are parallel to each other and filled with fluid. At a depth of 2400 m, there is a reflector that is the top boundary of the underlying half-space having the parameters $Vp = 5000$ m/s, $Vs = 2800$ m/s, and $\rho = 2.6$ kg/m^3.

An 8×3.5 km integration domain is chosen to characterize waves generated by the fractured bed and a 6×3 km domain to assess the impact of the most significant factors. The waves are generated by a point energy source on the earth's surface with a 30 Hz Berlage wavelet. The base model designed for use in the subsequent investigations is shown as four modifications in Figure 5.15, where the fracturing structure is depicted in the circular fragments. In models A and B of low-contrast media, the physical parameters of both the fractured bed and the fractured zone (3 km long) are equal to the parameters of the host medium. The reflection from the top boundary of the second bed is rather small because of a reflection coefficient of 0.081. This type of geologic setting is typical of the most part of East Siberia.

The base model C in Figure 5.15 contains a fractured bed in which the acoustic characteristics $Vp = 4800$ m/s, $Vs = 2550$ m/s, and $\rho = 2550$ kg/m^3 differ from those of the host medium. So, this bed has reflecting boundaries at the top and base with a reflection coefficient of 0.08. Parameters of the underlying half-space ($Vp = 5500$ m/s, $Vs = 3000$ m/s, and $\rho = 2600$ kg/m^3) cause a considerable reflection coefficient of 0.22 at its boundary. This type of geologic setting is typical of the largest portions of the Volga-Ural and Timan-Pechora petroliferous provinces. The base model D containing a bed-bounded fractured zone 3 km long is designed for this type of geologic setting.

5.2.1.2. Seismic characteristics of scattered compressional waves from a fractured bed

Figure 5.16 shows the *X*- and *Z*-components of a common-source gather recorded on the earth's surface for model A. Three waves with traveltimes close to hyperbolas are shown on both components. The first to arrive (at 0.9 s) is a scattered compressional wave *Dpp** that is formed by summation of all diffractions from the top tips of mesofractures in the fractured bed. The second wave, with its minimal arrival time of 1.08 s, is a compressional PP-reflection from the horizontal boundary. The last to arrive, with its minimal arrival time of 1.27 s, is a scattered converted *Dps** wave.

In Figure 5.16a, the *X*-components of all of the waves undergo a polarity reversal in the vicinity of the seismic energy source. In Figure 5.16b, the *Z*-components of all of the waves are continuous.

In accordance with Huygens' principle, the traveltimes of the scattered compressional *Dpp** wave should correspond strictly to those of the compressional reflection from a hypothetical boundary passing through the top tips of fractures. Table 5.3 provides comparative arrival times of the scattered compressional wave from the fractured bed top and the specular reflection from a hypothetical boundary coincident with this top, both at a depth of 2000 m.

We note that the traveltimes of the scattered compressional *Dpp** diffraction and the PP-reflection from a boundary coincident with the top of the fractured bed ('PPtop-reflection') are practically matching in form but have a systematic 4 ± 1 ms shift, which corresponds to a 1/8 wave period. The probable cause of the delay of

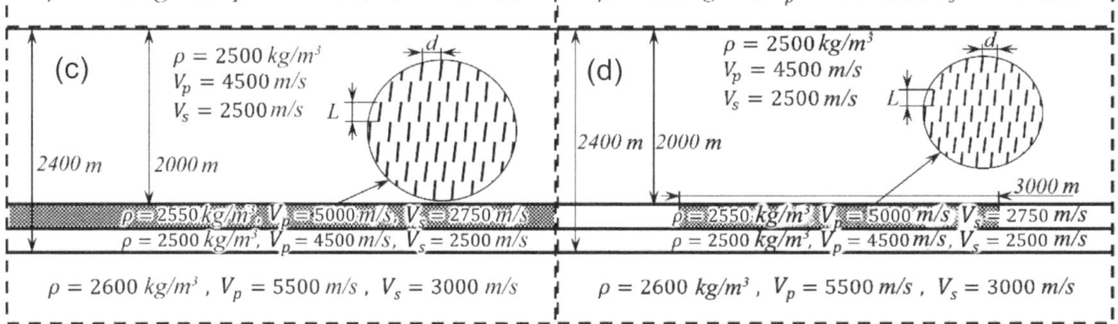

Figure 5.15. Base model modifications: (a) physical parameters of the fractured bed are equal to the parameters of the host medium; (b) physical parameters of the fractured zone 3 km long are equal to the parameters of the host medium; (c) physical parameters of the fractured bed are different from the parameters of the host medium; and (d) physical parameters of the bed containing a fractured zone 3 km long are different from the parameters of the host medium.

*Dpp** is that this signal is formed as a combination of responses from several near-top rows of mesofractures, which leads to a delay in the maximum amplitude of the interference signal. Because the arrival times of both waves are practically equal, the kinematic parameters of the scattered compressional wave from the top of a fractured bed and the compressional reflection from the same boundary can be regarded as matching. This is important for data processing.

Comparison between the left- and right-side arrival times of the scattered *Dpp** wave in Table 5.3 confirms that they are symmetric.

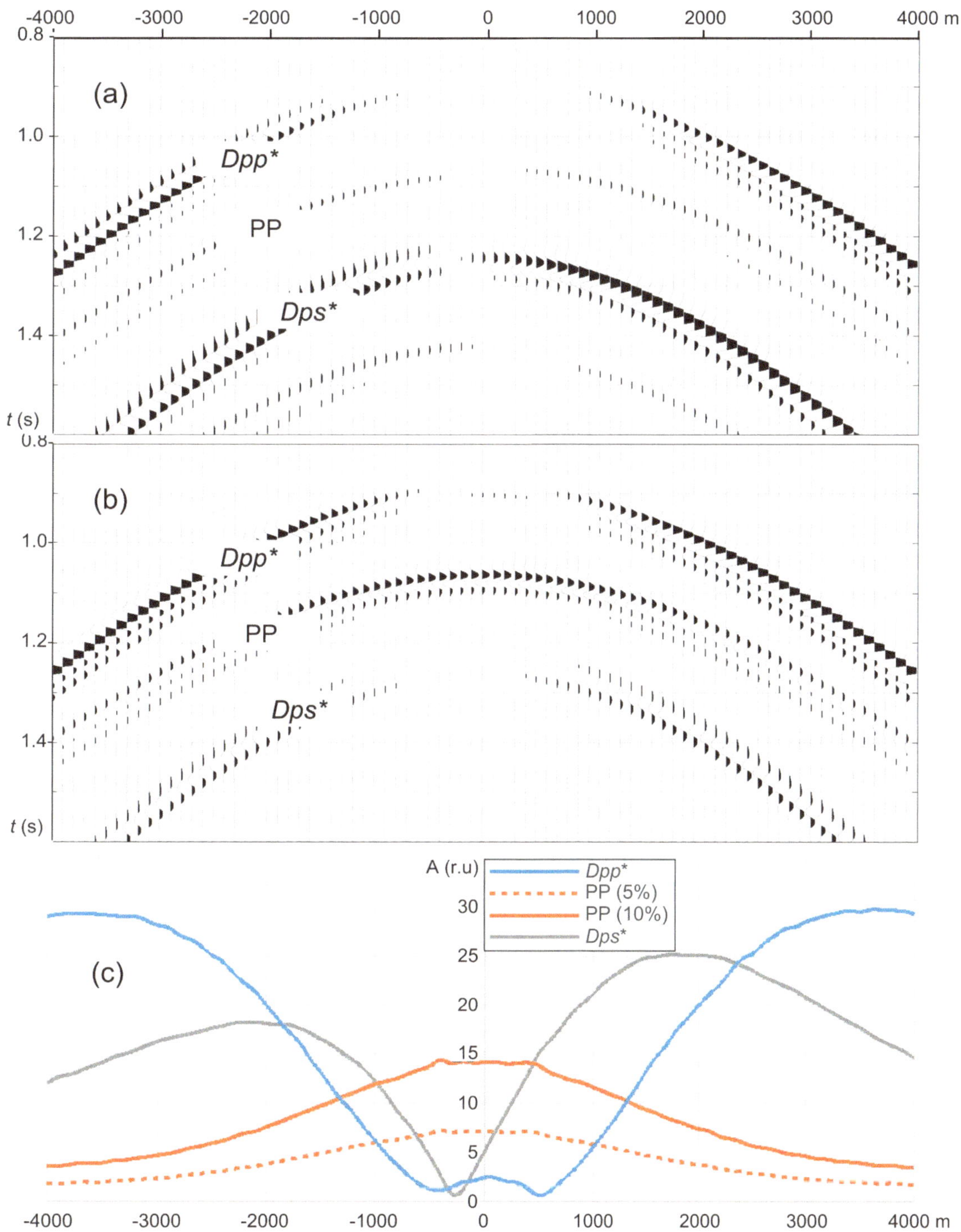

Figure 5.16. (a) *X*-component and (b) *Z*-component of a common-source gather; (c) amplitude profiles of the *Z*-component of *Dpp**, the *Z*-component of PP, and the *X*-component of *Dps** for model A.

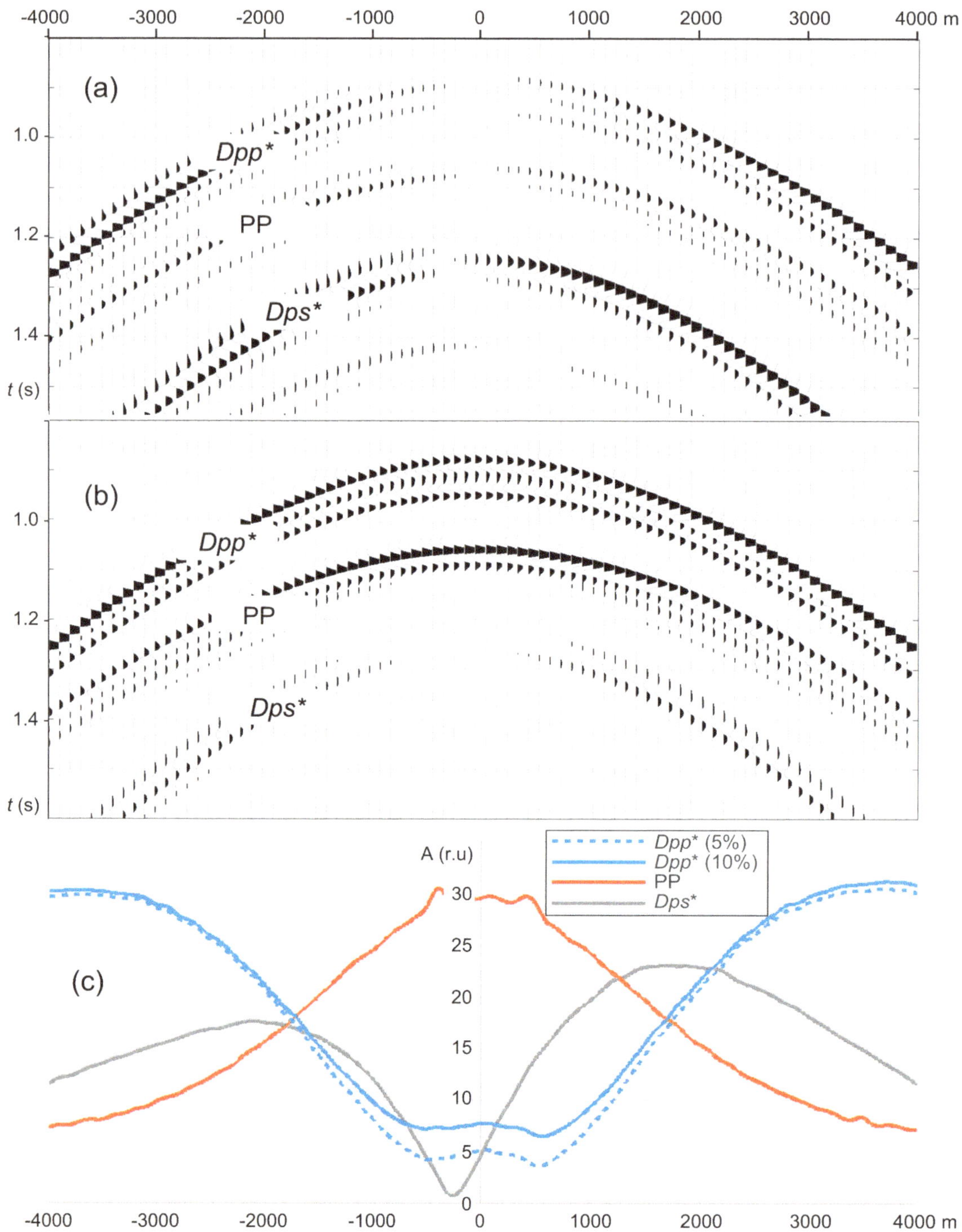

Figure 5.17. (a) *X*-component and (b) *Z*-component of a common-source gather; (c) amplitude profiles of the *Z*-component of *Dpp**, the *Z*-component of PP, and the *X*-component of *Dps** for model C.

A change in the deviation of mesofractures from the vertical by 5° to 15° greatly affects the scattered compressional *Dpp** wave, while the amplitude profile of the PP-reflection from the horizontal boundary remains practically unchanged (Figure 5.20). This increase from 5° to 15° in fracture deviation from the vertical noticeably flattens the amplitude profile of the *Dpp** diffraction by decreasing the amplitudes at long offsets and increasing the amplitudes at short offsets. The difference between the summed amplitudes at short and long offsets increases by a factor of 3.5, while the width of the zone of minimal amplitudes de-

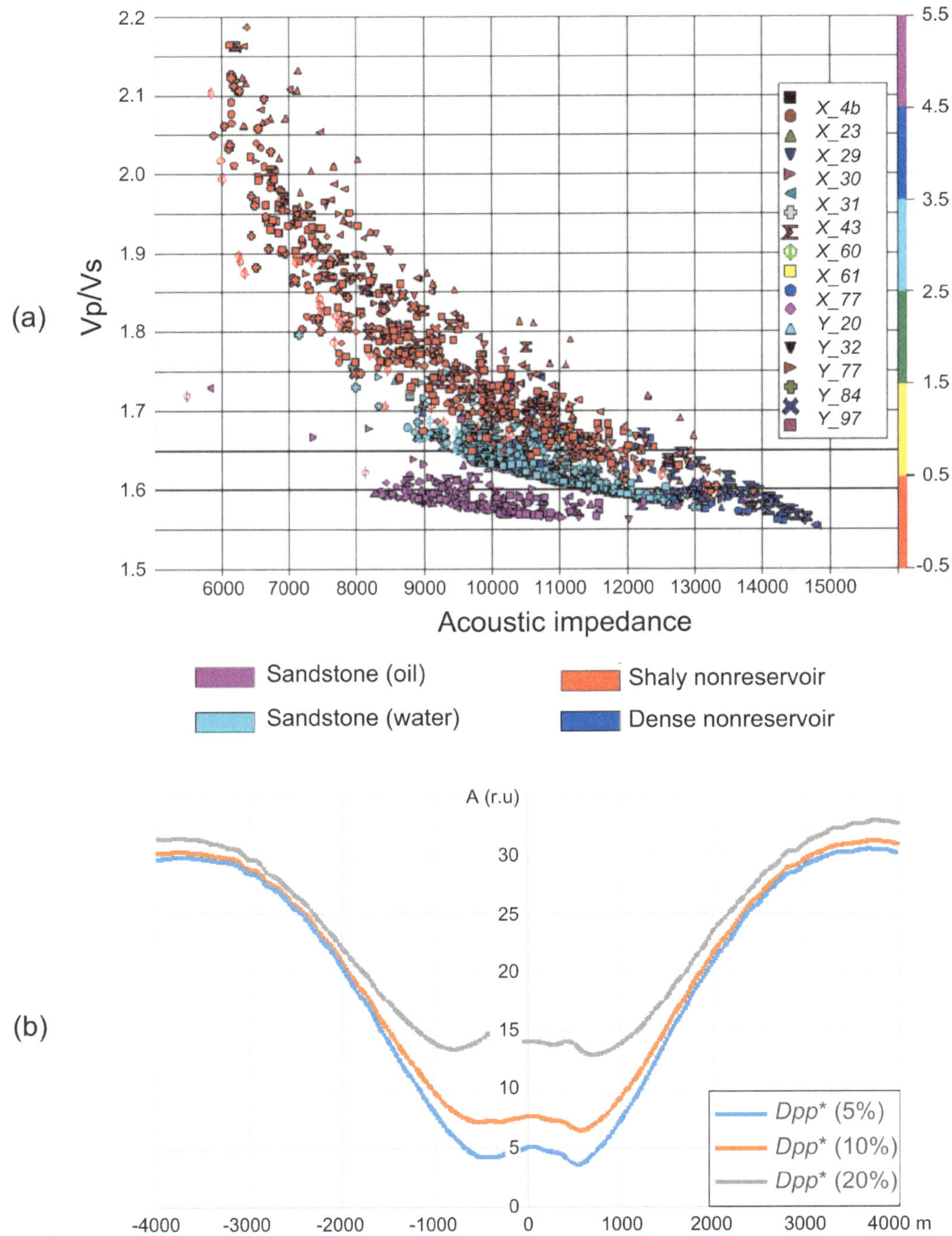

Figure 5.18. (a) Typical crossplot of *Vp/Vs* ratio vs. impedance for different lithotypes (after CGG); (b) dependence of the amplitude of the combined response containing both *Dpp** and PPtop waves on the reflection coefficient at the top of a fractured bed.

fined at a 10% level of the maximum amplitude grows double.

A reduction in signal frequency from 60 to 30 Hz (Figure 5.21) has almost no impact on the amplitude profiles of the scattered *Dpp** and reflected PP waves. A further frequency reduction to 15 Hz increases the amplitudes of both scattered and reflected waves at far offsets, while the *Dpp** wave amplitudes at short offsets grow a little. Overall, this factor only slightly influences the ratio of the summed amplitudes of the *Dpp** wave at short and long offsets and the width of the zone of small amplitudes of this wave (Figure 5.21c and 5.21d).

A 15 to 100 m change in the fractured bed thickness also only slightly impacts the major characteristics of the

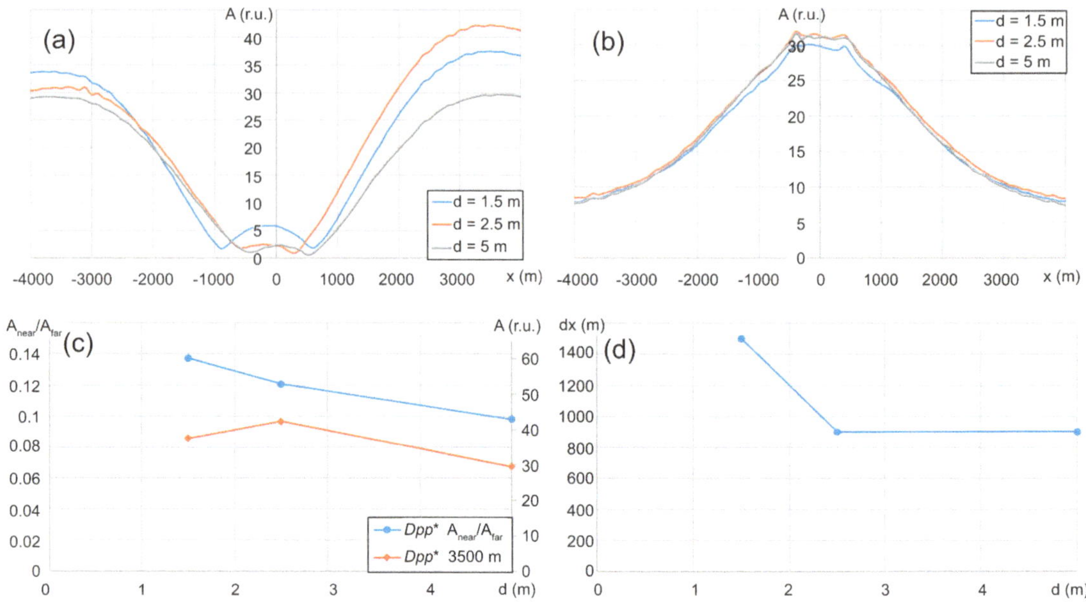

Figure 5.19. Amplitude profiles of (a) the *Dpp** wave and (b) the PP wave; (c) the ratio of the summed amplitudes of the *Dpp** wave at short (ranging from –1000 to 1000 m) and long (ranging from –3000 to –2000 m and from 2000 to 3000 m) offsets and the amplitude of the *Dpp** wave at the 3500 m offset; and (d) the width of the zone of minimal amplitudes of the *Dpp** wave for different values of fracture spacing *d*.

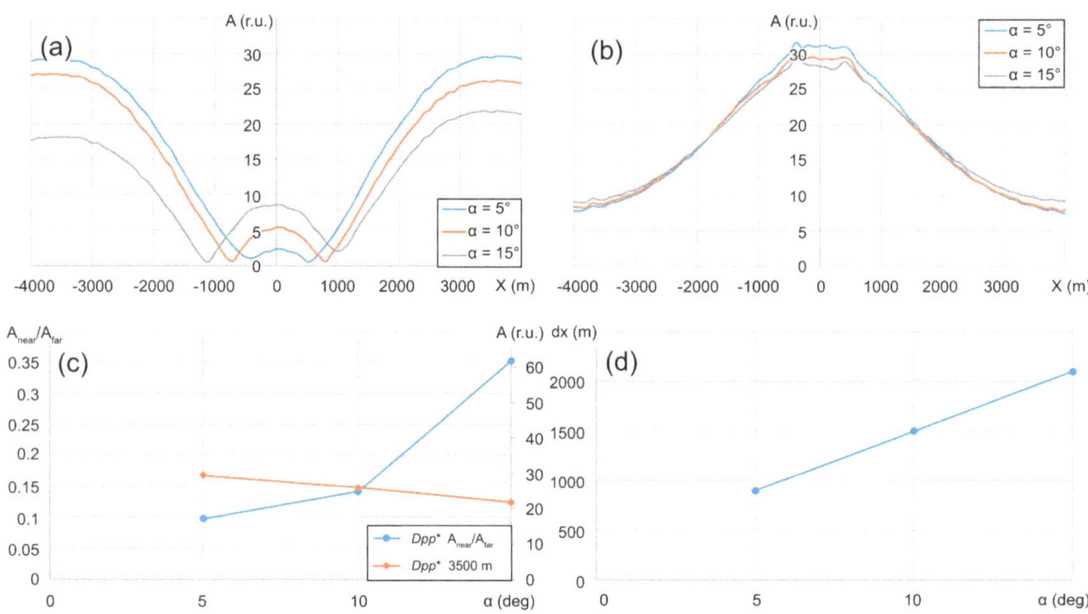

Figure 5.20. Amplitude profiles of (a) the *Dpp** wave and (b) the PP wave; (c) the ratio of the summed amplitudes of the *Dpp** wave at short (ranging from 0 to 1000 m) and long (ranging from –3000 to –2000 m and from 2000 to 3000 m) offsets and the amplitude of the *Dpp** wave at 3500 m offset; and (d) the width of the zone of minimal amplitudes of the *Dpp** wave for different values of fracture deviation from the vertical α.

scattered compressional *Dpp** wave (Figure 5.22c and 5.22d). When bed thicknesses are rather small (5 to 15 m), the *Dpp** wave amplitudes at the 3500 m offset decrease, while the PP-reflection amplitudes at far offsets increase.

So, when the fractured bed thicknesses are small, the differences in the waves being compared decrease and, therefore, the possibility of recognizing them diminishes. The width of the zone of minimal amplitudes defined at a 10% level of the maximum amplitude remains almost un-

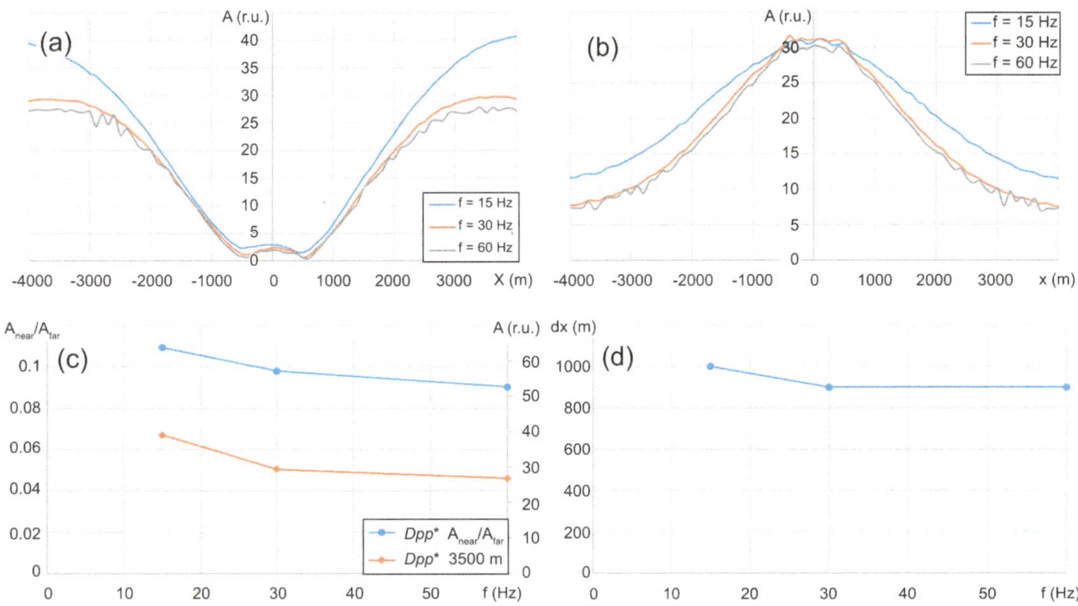

Figure 5.21. Amplitude profiles of (a) the *Dpp** wave and (b) the PP wave; (c) the ratio of the summed amplitudes of the *Dpp** wave at short (ranging from –1000 to 1000 m) and long (ranging from –3000 to –2000 m and from 2000 to 3000 m) offsets and the amplitude of the *Dpp** wave at 3500 m offset; and (d) the width of the zone of minimal amplitudes of the *Dpp** wave for different values of signal frequency *f*.

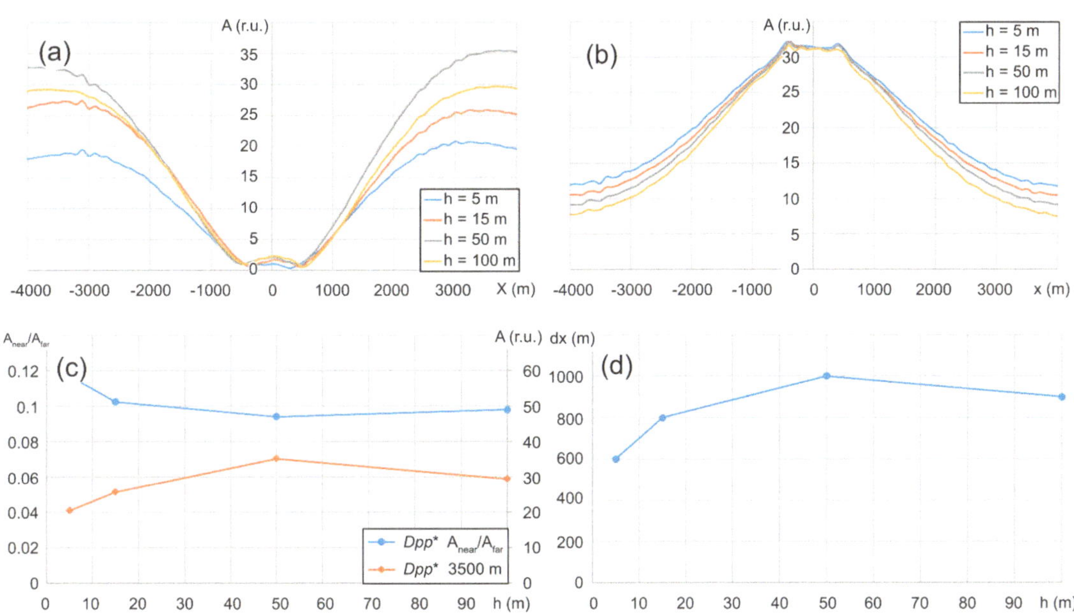

Figure 5.22. Amplitude profiles of (a) the *Dpp** wave and (b) the PP wave; (c) the ratio of the summed amplitudes of the *Dpp** wave at short (ranging from –1000 to 1000 m) and long (ranging from –3000 to –2000 m and from 2000 to 3000 m) offsets and the amplitude of the *Dpp** wave at 3500 m offset; and (d) the width of the zone of minimal amplitudes of the *Dpp** wave for different values of bed thickness *h*.

varied in the 10 to 100 m thickness variation range (Figure 5.22d).

The results of computations presented in Figures 5.19 through 5.22 for the four factors under discussion show the following.

- Despite the variety of fractured bed parameters and signal frequencies used, the overall character of the amplitude profile of the scattered compressional wave, with its wide and deep amplitude minimum in the vicinity of the source location and sharply pronounced maxima at far offsets exceeding the fractured bed depth, remains generally consistent.

- Greater stability is observed with the compressional reflection from a near-horizontal boundary, whose amplitude profile has a maximum in the vicinity of the source location.
- The ratio of the summed amplitudes at short and long offsets is sufficiently constant for all of the factors examined except fracture inclination. This is significant for wave type identification.
- The maximum amplitude value of the scattered compressional wave at the 3.5 km offset is stable enough. It changes by 25% to 30% when the parameter being analyzed is varied three- to fourfold.
- The width of the zone of minimal amplitudes of the scattered compressional wave defined at a 10% level of the maximum amplitude remains similar in all of the computations.

5.2.1.4. Conclusions from Subsection 5.2.1

1) The scattered compressional waves generated by fractured beds have a wide zone of relatively small amplitudes in the vicinity of the source location. The amplitudes substantially increase away from the source and are highest at maximum offsets, i.e., their AVO curve has a large positive gradient. The ratio of amplitudes at short and long offsets far exceeds unity.
2) In the case of a tight geologic formation with low-velocity contrasts as well as very minor changes in *Vp/Vs* ratio and impedance, in accordance with the formulae by Aki and Richards (1983) for reflection coefficients and with typical relations between *Vp/Vs* ratios and impedances, the amplitude profiles of reflections from the boundaries have a gentle character regardless of sign and, thus, an insignificant AVO gradient.
3) In tight geologic formations with high-velocity contrasts, AVO curves with a maximum in the vicinity of the source location and a negative gradient are typical of smooth acoustic boundaries with a positive change in impedance and an insignificant change in Poisson's ratio (or *Vp/Vs* ratio). The ratio of amplitudes at short and long offsets is less than unity.
4) The scattered compressional waves are shown to have rather high amplitudes, and the strongest amplitudes of this type of waves are comparable to those of reflections with $R = 25\%$ and substantially exceed those of reflections with reflection coefficient $R = 5\%$ to 10%.
5) The high amplitudes of the scattered compressional wavefront are attributed to the response formed by fluid-filled mesofractures deviated from the vertical and to constructive interference of individual diffractions that have no polarity reversal.
6) The kinematic characteristic (normal moveout – NMO) of the scattered compressional wave emanated from the fractured bed does not differ from that of the reflection from the hypothetical boundary coincident with the bed top, except for a 4 ms shift.
7) The acoustic boundaries present at the top and base of the fractured bed reduce the difference between the maximum and minimum amplitudes of the wave response from the fractured bed. They, however, have no effect on the almost linear character of the positive and considerable gradient of the AVO curve, at midrange offsets ranging from 1.0 to 2.5 km.
8) The effects of fracture spacing, signal frequency, and fractured bed thickness on the AVO curve and characteristics of the scattered wave, such as the difference between maximum and minimum amplitudes and the width of the zone of minimum amplitudes, are found to be insignificant.

5.2.2. Characteristics of the *X*-component of wave responses at midrange offsets

The preceding subsection shows that the scattered compressional *Dpp** wave manifests itself best at long offsets ranging from 3 to 4 km. The use of long offsets entails the risks of masking the effects of specular reflections and other types of noise on the *Dpp** wave. Furthermore, we cannot confidently recognize *Dpp** waves in high-contrast media, when reflections from boundaries with an appreciable negative change in impedance have a positive AVO gradient. Both of these factors make us assess the possibility of using scattered converted *Dps** waves to detect naturally fractured zones.

In Figures 5.16c and 5.17c, the maximum amplitudes of *Dps** waves are observed at offsets of approximately 1000 to 2500 m, not exceeding the target depth in the base model. Common-source gathers to aid in analyzing the properties of the *X*-component of these waves are computed utilizing the base model of Figure 5.15 in these four versions:

- a low-velocity-contrast tight ("massive") geologic medium with a bed and a bed-bounded zone, which is typical of East Siberia (models A and B);
- a high-velocity-contrast tight geologic medium with a bed and a bed-bounded zone both having reflectors

at the top and base, which is typical of the Volga-Ural and Timan-Pechora basins (models C and D).

The common-source gathers are generated with a 100 m receiver spacing along a 6 km line with a source located at the spread center.

Figure 5.23 presents the *X*-components of common-source gathers and AVO curves for models A and B. A scattered compressional *Dpp** wave (at 0.88 s), scattered converted *Dps** wave (at 1.24 s), and compressional PP-reflection (at 1.07 s) are visible.

From these waves, the following can be observed.

- The scattered converted *Dps** wave dominates along most of the spread, at offsets of 0 to 2500 m, with a predominance of 1.5 to 2 times or more at offsets with maximum wave amplitudes;
- Asymmetry of the wavefield is caused by a deviation of the mesofractures from the vertical. The difference between maximum amplitudes on the split-spread sides is 30% to 34% of the maximum amplitude.

Figure 5.24, which contains common-source gathers for fractured zones, exhibits the same types of waves as those shown in Figure 5.23.

With respect to character, only scattered waves are distinguishable, i.e., the converted *Dps**, whose maximum becomes more sharply pronounced, and the compressional *Dpp**, whose amplitude decreases double at maximum offsets. The *Dps** exceeds both PP and *Dpp** by 3 to 5 times. The scattered converted *Dps** wave is at its maximum amplitude at offsets ranging from 0.7 to 2 km. The relatively high amplitudes of the scattered compressional wavefront are attributed to the response formed by fluid-filled mesofractures deviated from the vertical and constructive interference of individual diffractions that have no polarity reversal from each single mesofracture.

The *Dps** traveltime hyperbola is symmetric and coincident (except for a small time delay) with the arrival times of the converted reflection from the top boundary of the fractured bed (for model D) or with the converted reflection from a hypothetical boundary running through the upper tips of fractures (for model B) (Table 5.4). This validates the applicability of standard processing procedures for specular converted reflections to scattered converted waves.

On the *X*-component, the scattered converted *Dps** wave, like all other waves, has a polarity reversal at the source location.

Asymmetry of the response manifests itself even when the megafracture's deviation from the vertical is minor (5°). The difference between the maximum amplitudes of the scattered converted wave on the left and right split-spread sides is 30% of the maximum amplitude for model B (Figure 5.23c and Table 5.5) and 27% for model C (Figure 5.24c and Table 5.5). This property is due to asymmetry of individual diffractions from near-vertical fractures (Figure 4.5), which form the total scattered wavefield.

In addition to asymmetry of the scattered converted wave, the fracture deviation from the vertical entails a horizontal shift of the zero value of the AVO curve, which occurs at the polarity reversal point.

The *X*-component of the scattered converted *Dps** wave typically has a one-pulse signal, due to the response being formed by top fracture rows only. No response arises, as a rule, from the fracture tips in base rows, which is due to the strong absorption of shear waves propagating from the bed base.

Figure 5.24 presents the *X*-components of common-source gathers for models of a fractured bed-bounded zone with and without reflecting boundaries. The scattered converted wave is shown to be many (up to 10) times stronger than the converted reflection from the same boundary.

It is, however, not only amplitude but also phase relationships that are important in assessing the impact of the converted reflection on the *Dps** diffraction, as modeled in Figure 5.25. This figure compares a *Dps** diffraction with PS-waves produced by fracture-free beds with opposite signs of the reflection coefficient and with total responses produced by the same but fractured beds. It also depicts corresponding wave profiles. Table 5.4 provides arrival times of the scattered converted *Dps** wave and PS-reflections.

Figure 5.25 allows us to note the following.

a) Converted PS-reflections from the top boundary of a fractured bed with opposite signs of the reflectivity arrive concurrently but with opposite polarities, as shown clearly in Figures 5.25b and 5.25c.
b) The scattered converted *Dps** wave is delayed by 5 ± 1 ms relative to the converted PS-reflection (Table 5.4). The probable cause of this systematic delay, which corresponds to a 1/8 wave period, is that the scattered wave is formed as a combination of responses from several near-top rows of mesofractures, which leads to a delay in the maximum amplitude of the interference signal.
c) The total response of the (*Dps** + PS) waves is formed in accordance with their phase relationship,

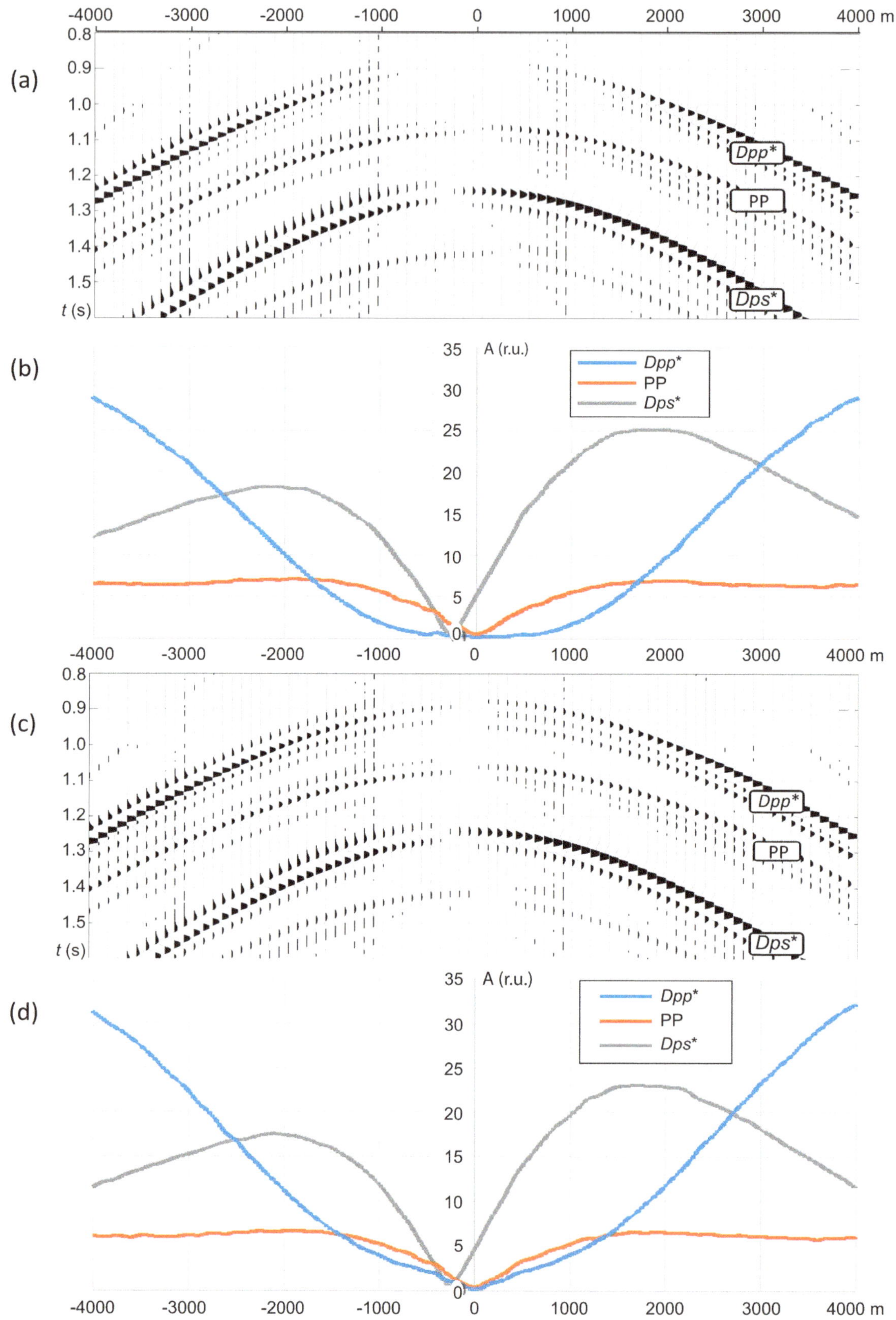

Figure 5.23. (a) *X*-component of a common-source gather and (b) AVO curves for a low-contrast medium with a fluid-filled fractured bed (model A); (c) *X*-component of a common-source gather and (d) AVO curves for a high-contrast medium with a fluid-filled fractured bed (model C).

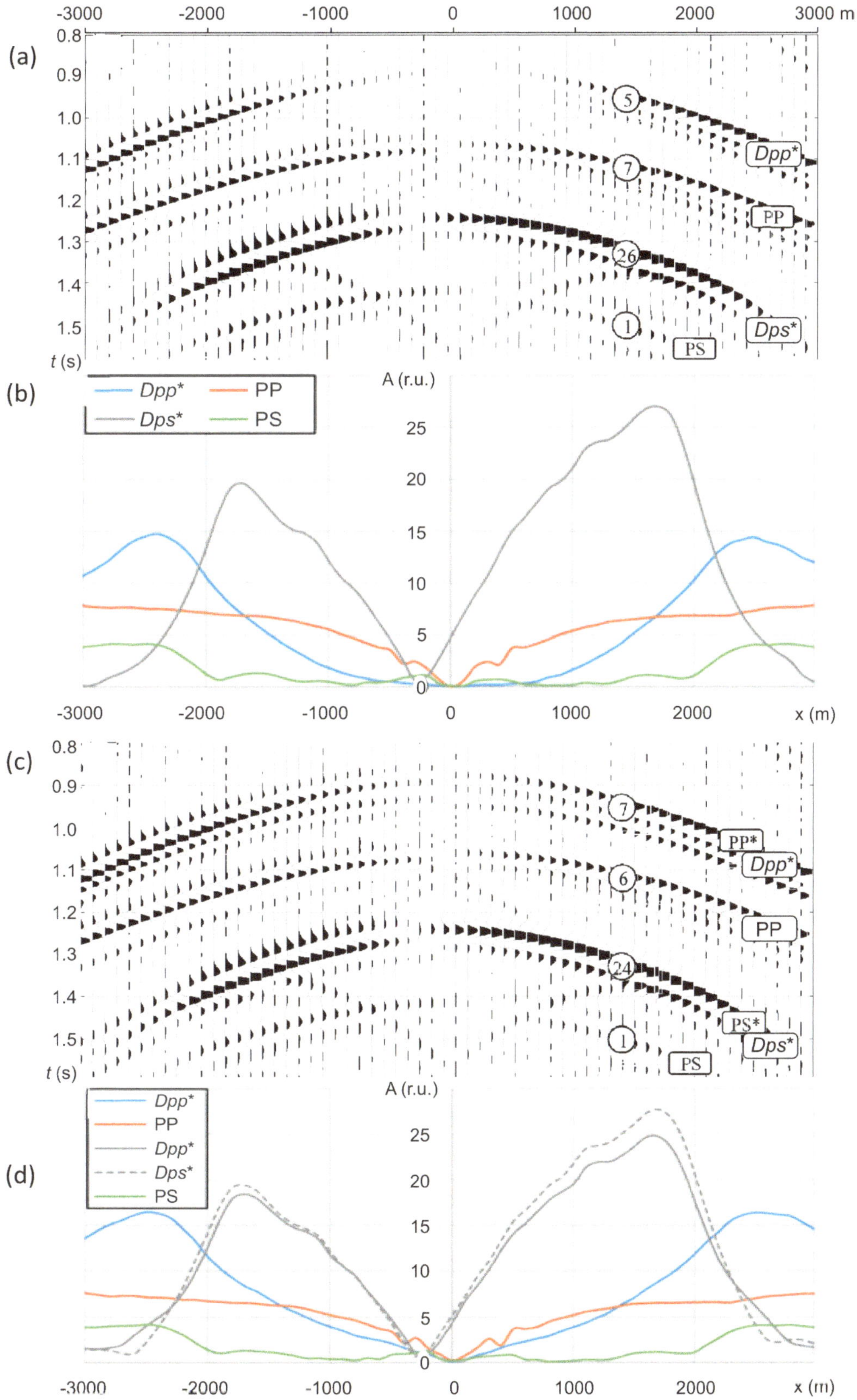

Figure 5.24. (a) *X*-component of a common-source gather and (b) AVO curves for a low-contrast medium with a fluid-filled bed-bounded fracture zone (model B); (c) *X*-component of a common-source gather and (d) AVO curves for a high-contrast medium with a fluid-filled bed-bounded fracture zone (model D).

verted reflection from the bed top is added to the scattered converted wave from this bed or zone. The converted reflection is, as a rule, weaker than the scattered wave, but how much they differ is dependent on the reflection coefficient of converted waves at the boundary. The impact of the PS-wave on the *Dps** wave depends on, first, a small systematic time delay of the *Dps* wave relative to the PS reflection arrival time and, second, the amplitude and polarity of the PS-wave.

5.2.3. Conclusions from Section 5.2

These investigations enable us to note the following.

1) The scattered compressional waves generated by fractured beds or bed-bounded zones have been discovered to have a property previously unknown to geophysicists. This is the anomalous feature of their AVO curves in showing a considerable positive gradient due to minimum at the zero offset and maximum at long offsets.
2) The AVO curves of reflections have, as a rule, a different character. For tight rocks with small changes in both the *Vp/Vs* ratio and the impedance, AVO gradients are insignificant. Furthermore, when these low-contrast conditions are met and the impedance change is positive, the AVO gradient is negative.
3) It has been established that the amplitude strength of scattered compressional waves is relatively high.
4) It has been found that the shapes of AVO curves of scattered compressional waves from fractured reservoirs are sufficiently insensitive to changes in fracturing parameters.
5) The *X*-component of scattered converted waves from fractured beds or bed-bounded zones has been found to have potential as a diagnostic tool. These waves have maximum amplitudes at midrange offsets of 700 to 2500 m, which corresponds to 0.5 to 1.0 of target depth, and dominate in amplitudes over all other waves.
6) Scattered converted waves have been discovered to possess such important properties as wavefield asymmetry at the opposite sides of the receiver spread and a horizontal shift of the zero value of AVO curves relative to the source position. The latter may be caused even by a minor deviation of mesofractures from the vertical.
7) A match between the kinematic characteristics of both scattered compressional and scattered converted waves and the respective characteristics of compressional and converted reflections from the top of a fractured bed confirms the scope for applying standard processing to the scattered types of waves.
8) Where acoustic boundaries are present at the top and base of a fractured bed or bed-bounded zone, the scattered compressional and scattered converted waves are summed with their respective reflections, with due regard for the amplitude and sign properties that are determined by the physical parameters at these boundaries.

5.3. Assessing the feasibility of direct detection of mesofractured reservoirs using scattered waves

The results that have been presented about the character of seismic wave responses from beds and zones of mesofractured reservoirs, specifically of scattered waves, allow us to proceed to validating methods to detect these subsurface targets. With this purpose in mind, we test them by numerical simulations on common midpoint (CMP) lines with typical seismic exploration parameters.

5.3.1. Examples of applying methods to recognize scattered compressional waves from fractured zones using *Z*-component data

The properties, discussed above, of scattered compressional *Dpp** waves generated by fractured beds provide us with a basis for detecting and distinguishing these waves on seismic data from the many other reflections generated at acoustic boundaries.

In tight rocks (limestones, dolomites, and cemented sandstones) containing fractured beds at depths common for oil exploration, rather small positive and negative changes in the *Vp/Vs* ratio (see the crossplot in Figure 5.18) and the impedance are typical. In these conditions, the AVO curve is gentle with quite low gradients. In these circumstances, it is worthwhile to use a method based on analyzing the AVO gradient for recognizing scattered waves.

For a wave scattered from a fractured bed, the AVO gradient is supposed to be positive and considerable. For reflections, the gradient may be either positive or negative but much smaller in magnitude. The recognition method proposed for scattered waves is also efficient for media with a series of positive changes in impedance and low variations of the *Vp/Vs* ratio. These conditions are similar to those used in the models (Figure 5.15), for which the AVO gradients of reflections are negative.

The method analyzes both the magnitude and sign of a gradient. Next, the difference between amplitudes at offsets of 1000 and 2500 m divided by 1.5 km is used as a gradient estimate for specific waves.

- A scattered compressional *Dpp* wave from a fluid-filled fractured bed has a considerable and positive AVO gradient. In Figures 5.16c and 5.17c, the gradient is +13.3 r.u./km.
- A reflection from a boundary in a low-velocity-contrast medium (first type) with minor changes in both impedance and *Vp/Vs* ratio exhibits an insignificant positive or negative gradient. In Figure 5.16c, the gradients of the PP-waves for $R = 10\%$ and 5% are –4.7 r.u./km and –2.0 r.u./km respectively.
- A reflection from a boundary with a large positive change in impedance and an insignificant change in the *Vp/Vs* ratio in the high-velocity-contrast medium (second type) has a significant negative gradient. In Figure 5.17c, the gradient of the PP-wave for $R = 22\%$ is –9.3 r.u./km.

In these cases, the AVO gradients are computed for midrange offsets, where the amplitude curves have a linear character.

The second method for recognizing scattered compressional waves from fractured targets uses common midpoint (CMP) stacks. It is applicable to geologic conditions characterized by increasing impedance with depth (positive abrupt impedance changes at the boundaries). Figure 5.26 illustrates this second method.

As to its kinematic (traveltime) characteristics, discussed in Subsection 5.2.1, a scattered compressional wave from a fractured bed top does not differ significantly from a specular reflection from a smooth hypothetical boundary coincident with this fractured bed top. Consequently, standard CMP processing procedures can be applied readily to such scattered compressional waves. When long offsets are used, e.g., for attenuation of multiples, events corresponding to the scattered waves from fractured beds will not differ on standard CMP images from specular reflections. In order to distinguish between these two types of waves by the character of amplitude variation away from the source location, partial stacks for near and far offsets should be used.

Figure 5.26(1) shows schematic amplitude profiles of the two types of waves discussed. They have opposite AVO gradients. Full stacking of all offsets (Figure 5.26(2)) is insensitive to this difference, so it is impossible to distinguish between the two types waves in such stacks (Figure 5.26(3)).

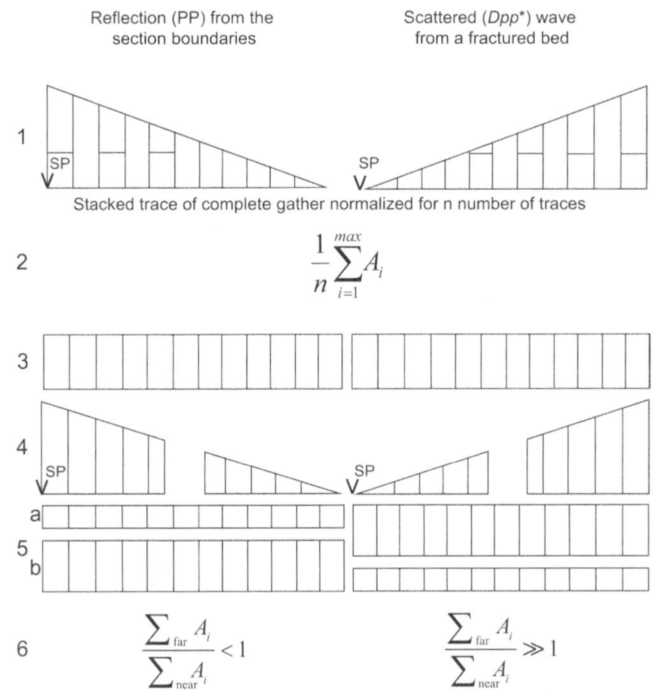

Figure 5.26. Schematic explanation of a method for recognizing wave responses from long fractured beds using partial CMP stacks: (1) amplitude profiles of the reflected (left) and scattered (right) waves at all offsets; (2) full stacking; (3) horizontal profiles of amplitudes after full stacking of all offsets; (4) amplitude profiles for separate sets of near and far offsets; (5) horizontal profiles of amplitudes after partial stacking of far (a) and near (b) offsets; and (6) ratio of the amplitude of the far-offset stack to that of the near-offset stack, provided that the number of traces is equal in both cases.

A suitable method for recognizing scattered compressional waves generated by fractured beds would be to form partial CMP stacks (Figure 5.26(5)) using separately far and near offsets (Figure 5.26(4)). The amplitude relation of these partial stacks may indicate the wave type. If the ratio of the amplitude of the far-offset stack to that of the near-offset stack is much less than unity, this indicates a specular PP-reflection. Otherwise, if the ratio is substantially higher than unity, this indicates a scattered compressional *Dpp** wavefront from a fractured bed. This ratio can be used as a recognition parameter. The reliability of the method and its stability to the impact of noise are rather high when large numbers of near and far offsets are used. The number of offsets for both partial stacks should be identical. It is important to note that the relative amplitude of a reflected and a scattered wave produces no impact on the analysis. The choice of a suitable range of offset depends on local conditions, including the presence of noise at midrange and far offsets. Optimal parameters may be found by testing several offset ranges (Table 5.6).

Table 5.6. Dependence of amplitudes (in relative units) of CMP stacked *Dpp** and PP waves on offset range for model A typical of East Siberia.

Offset range (km)	*Dpp** amplitude (r.u.)	PP amplitude (r.u.)
–4.0 to 4.0	15	9
–3.5 to 3.5	14	9.5
–3.0 to 3.0	13	10
–2.5 to 2.5	10	10.5
–2.0 to 2.0	9	11

Figure 5.27 presents the main elements of seismic CMP data analysis to detect fractured reservoirs using model A of a low-velocity-contrast medium with identical parameters for the fractured bed and host rocks, which is typical of East Siberia.

Figure 5.27a shows that, in the full CMP stack, the amplitudes of the PP-reflection and the scattered *Dpp** diffraction are almost equal, being 19 and 13 r.u., respectively. The predominance of the former over the latter decreases with decreasing the maximum offset to zero for the offset of 2.5 km. This phenomenon reduces the overall capability of standard CMP stacking of all offsets for identifying waves scattered by fractured beds.

When stacking only far offsets, the amplitude of the scattered *Dpp** wave is almost twice as large as that of the PP-reflection (Figure 5.27b) in our model. Such stacks may be used for preliminary identification of fractured reservoirs.

Figure 5.27c displays a substantial difference between the amplitudes of near- and far-offset stacks, both in the reflected wave and the scattered wave. With the PP-reflection, the difference is double, while it is close to twentyfold with the *Dpp** diffraction. Even greater is the amplitude difference between near- and far-offset stacks for reflected and scattered waves. In Figure 5.27d, these ratios are, respectively, 16 and 0.4, which indicates a fortyfold difference.

In some geologic settings, the target *Dpp** and PP waves can be contaminated by interfering noise at far offsets equal to or greater than 3 km, which reduces the confidence in estimates obtained using this method. Nevertheless, the data in Figure 5.15 and Table 5.7 suggest that the method might be effective with shorter (2.0 to 2.5 km) offsets in conditions similar to those of model A. In this case, the difference in the recognition parameter between the reflected PP wave and the scattered *Dpp** wave is greater than 5 to 10 times.

In addition to the extensive fractured beds discussed above, another valuable application is the detection of bed-bounded local fractured zones in low-contrast host rocks, which are represented by model B (Figure 5.15). For this model, on the full stack, a high-amplitude coherent event is identifiable at the time corresponding to a fractured zone (Figure 5.28a). On seismic images, however, reflections from lithologic lenses and other local features also can be present. An additional indicator that can help distinguish scattered waves generated by fractured zones from reflections related to local geologic features may be a higher amplitude ratio of these waves on a far-offset stack than that on a full stack.

Figure 5.28b shows that the amplitude ratio of the *Dpp** wave to the PP wave rises almost triple from 14.20 to 25.13. Accordingly, the ratio of amplitudes on the far-offset stack to those on the full stack is equal to 15 for the scattered wave and less than unity for the reflected wave.

The wave response from a fractured reservoir bed with a substantial velocity contrast, when the bed top and the bed base are acoustic boundaries (model C in Figure 5.15), is shown in Figure 5.17a and 5.17b. Such a geologic medium is typical of the Volga-Ural and Timan-Pechora petroliferous provinces. In this case, the basic feature of the response from the bed top is that it is formed as a synchronous sum of the scattered *Dpp** wave and the PPtop wave reflected from the top. As a result, the ratio of the response amplitude to the reflection amplitude (Table 5.8) is greater than for model A (Table 5.6).

The amplitude profile of the PPtop reflection from the bed top does not differ from that of the PP-reflection from the underlying boundary, i.e., it has maximum values in the vicinity of the source location and decreases away from the source. This appreciably smooths the amplitude profile of the total wave response from a fractured bed and mitigates the difference between amplitudes at near and far offsets. For model C, as compared with model A, the difference in the ratio of far-offset amplitudes to near-offset amplitudes between reflected and scattered waves reduces four to fivefold (see and compare Table 5.9 and Table 5.7).

These trends are illustrated in Figure 5.29, which provides the primary elements of CMP data analysis used to identify fractured reservoir beds. A tenfold dominance of the recognition parameter for the sum of the *Dpp** and PPtop waves over that for the PP wave decreases with decreasing maximum offset (Table 5.9).

The primary features of the response of a fractured zone in a bed with reflectors at its top and base (model D in Figure 5.30) are as follows.

- The amplitudes of the response from the bed top of the fractured zone have relatively high values on the full CMP stack (Figure 5.30a). On the far-offset

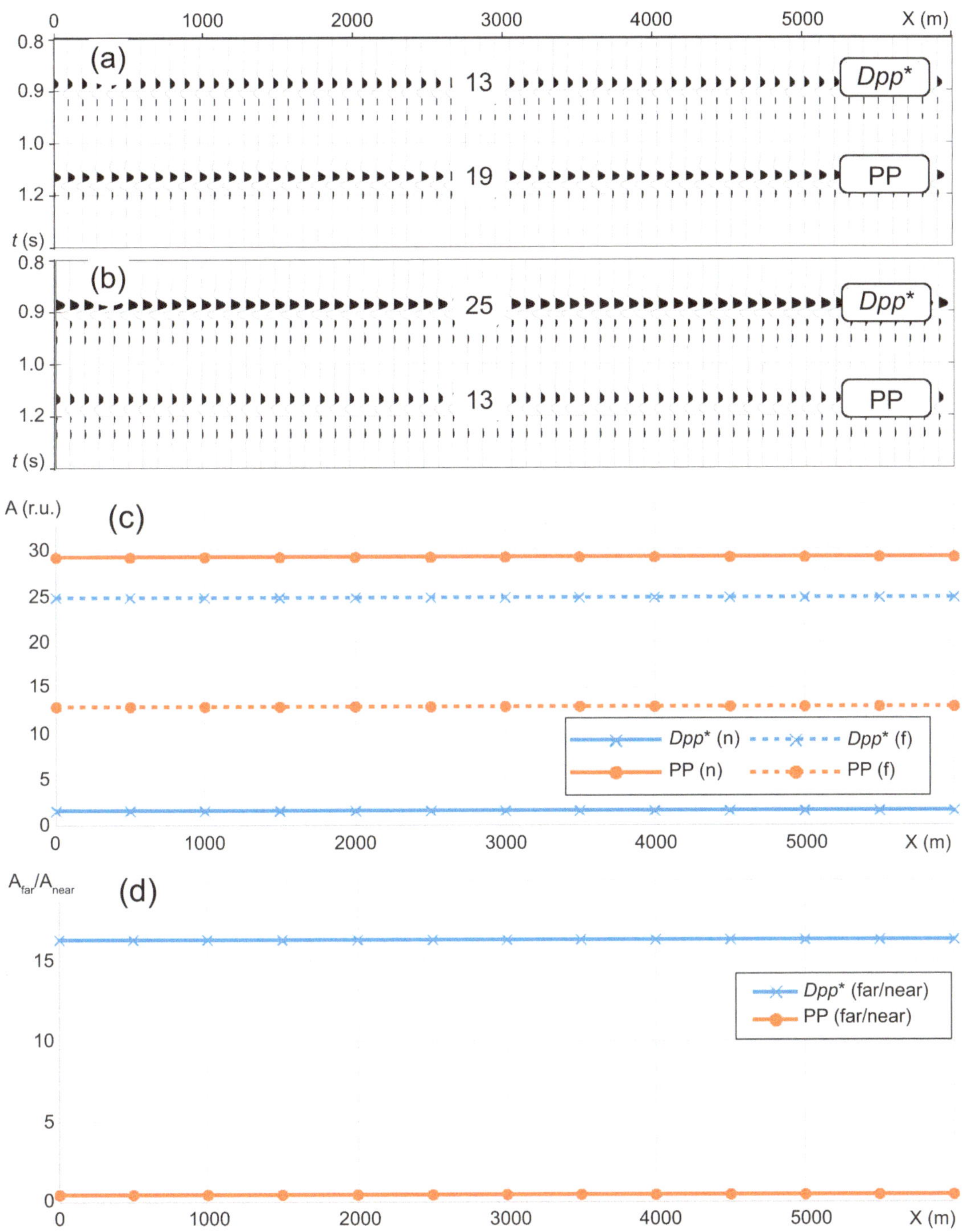

Figure 5.27. Primary elements of seismic CMP data analysis to identify fractured reservoirs for model A typical of East Siberia: (a) full stack along a test line; (b) partial stack with far offsets of –3 through –2 km and 2 through 3 km; (c) amplitudes of reflected PP and scattered *Dpp** waves stacked in the two different offset ranges; and (d) amplitude ratio of the far-offset stack and the near-offset stack for reflected PP and scattered *Dpp** waves. The circled numbers denote amplitudes.

stack, these amplitudes are more than double as much as those of the PP-reflection (Figure 5.30b).

– On the far-offset stack, i.e., beyond the fractured zone, the PPtop reflection exhibits amplitudes 3 times lower than those of the full response from the fractured zone, while the amplitude of the PP-reflection from below the fractured zone decreases by 35% to 40% (Figure 5.30c).

– The maximum ratio of amplitude on far-offset stacks to that on the near-offset stack is 4 for the full wave response of the fractured zone and does not exceed 0.6 for the PP-reflection (Figure 5.30d).

Table 5.7. Amplitudes of the *Dpp** and PP waves on near-offset and far-offset CMP stacks for model A typical of East Siberia.

Near-offset range (km)	Far-offset range (km)	*Dpp** (n) r.u.	PP (n) r.u.	*Dpp** (f) r.u.	PP (f) r.u.	*Dpp** far/near	PP far/near
−1.0 to 1.0	−3.0 to −2.0 and 2.0 to 3.0	1.5	13	25	5	16	0.35
−1.0 to 1.0	−2.5 to −1.5 and 1.5 to 2.5	1.5	13	18	7.5	13	0.5
−1.0 to 1.0	−2.0 to −1.0 and 1.0 to 2.0	1.5	13	13	95	9	0.6

In most cases, seismic surveys are performed to investigate near-horizontal layered geologic media. In order to assess the possibility of detecting bed-bounded fractured zones via their wave responses, we use a five-layer model. This model imitates a carbonate formation, with the parameters shown in Table 5.10 and Figure 5.31.

A bed-bounded fractured zone is embedded in layer 1. A fractured layer located in bed 2 extends below the entire seismic spread.

Figure 5.32a and 5.32b present, respectively, a full CMP stack and partial stack for far offsets ranging from 2000 to 3000 m. It is clear that, in the far-offset partial stack, the responses from the fractured features become noticeably stronger, while the reflections from the acoustic boundaries become weaker.

Figure 5.32c provides quantitative estimates of the amplitudes of scattered waves from the fractured targets and reflections from the acoustic boundaries. The far-offset stack shows a remarkable predominance of amplitudes associated with the fractured features and approximately a 10% decrease of *Dpp** amplitudes over the fractured zone, i.e., in the interval of 2500 to 5500 m. This decrease is related to a shadow zone formed by the overlying fractured medium, which is discussed in Subsection 5.3.2. Figure 5.32d presents estimates of the scattered-wave recognition parameter for the fractured zone, fractured bed, and acoustic boundary with a maximum reflection coefficient of 11%. The manifestation of both fractured features in this recognition parameter is much more pronounced.

As has been mentioned, this method for recognizing scattered waves coming from fractured beds is effective for geologic media with positive changes in impedance and little variation in the *Vp*/*Vs* ratio. It exploits opposite signs of the AVO gradients inherent to scattered and reflected waves.

We consider tight rocks that lie at considerable depths and have small changes of either sign in the *Vp*/*Vs* ratio (Figure 5.18a) and impedance at reflecting boundaries. For these geologic settings, the AVO curves are gentle and AVO gradients are small. In such a case, therefore, the recognition parameter may reach several tens for scattered waves, while deviating insignificantly from unity for specular reflections.

The factors that may reduce the effectiveness of the second recognition method for fractured beds and bed-bounded fractured zones include: (1) noise interference at large offsets and (2) the presence of boundaries with significant negative changes in impedance. As in the case of *Dpp** waves, they produce reflections with positive AVO gradients, which mask the scattered waves.

In order to overcome the adverse effect of the first factor, it is advisable to employ the first method to recognize the wave type, using an AVO gradient estimated at midrange offsets, where amplitudes of both PP and *Dpp** waves vary linearly. For the models used, a 1000 to 2500 m offset range is optimal. For scattered waves, the AVO gradient is always positive, and for reflections it is negative and insignificant.

To recognize scattered waves in the context of the second factor, additional independent information is required, such as the horizontal component of these waves. This will be discussed in the subsection that follows.

From the preceding analysis, we can conclude the following.

- The results presented demonstrate an important property of the AVO gradient of the seismic response from fluid-filled zones of fracturing, i.e., its positive sign. This phenomenon automatically filters out all waves with a negative gradient.
- From among the waves of a positive AVO gradient, the selection for analysis should be only those that have (1) an appreciable AVO gradient of both compressional and converted waves, (2) substantial asymmetry of traveltimes and amplitudes of converted waves, (3) a horizontal shift of the zero value of their AVO curves (for sub-horizontal boundaries), and (4) azimuthal anisotropy of velocities.

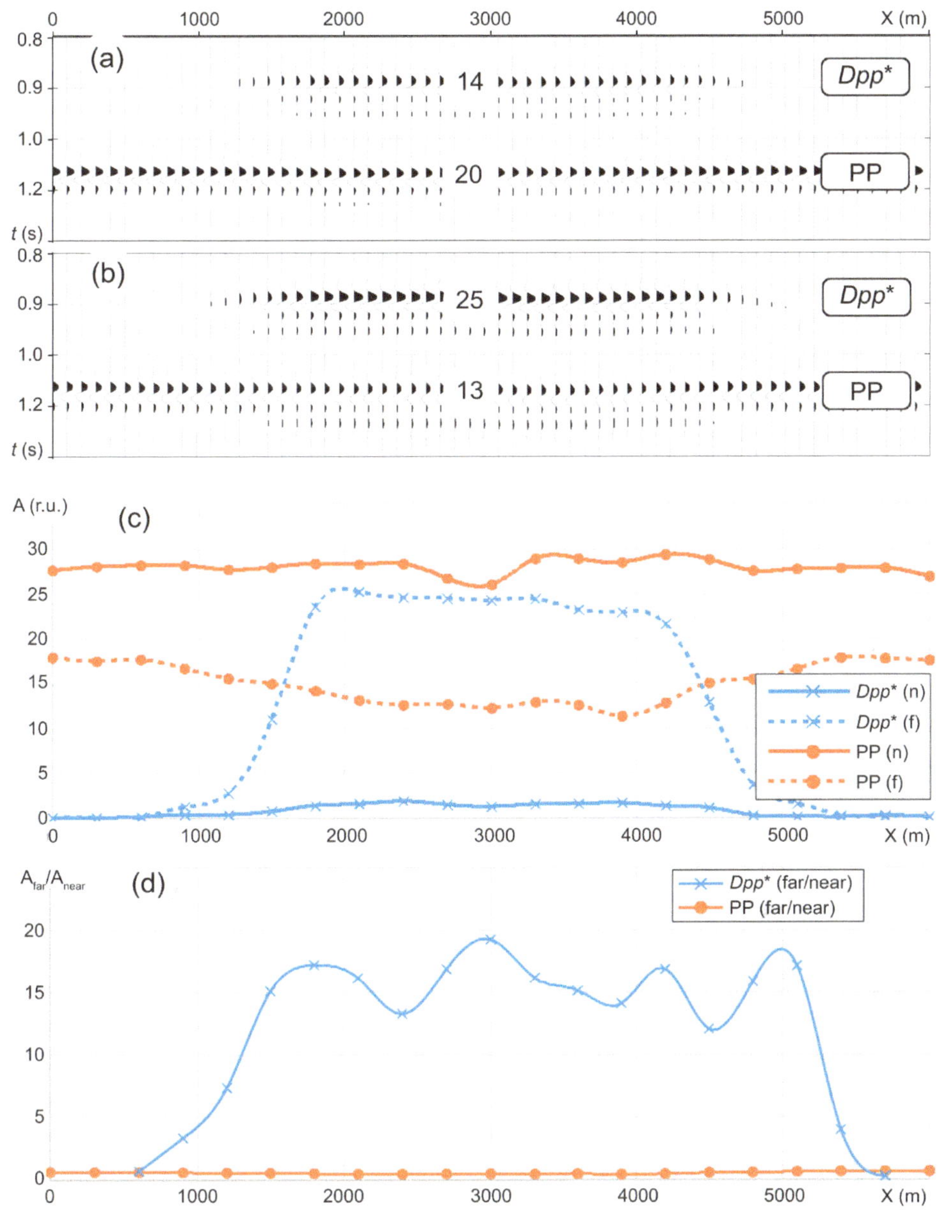

Figure 5.28. Primary elements of seismic CMP data analysis to identify fractured reservoirs for model B: (a) full stack along a test line; (b) partial stack with far offsets of –3 through –2 km and 2 through 3 km; (c) amplitudes of reflected PP and scattered *Dpp** waves stacked in the two different offset ranges; and (d) amplitude ratio of the far-offset stack and the near-offset stack for reflected PP and scattered *Dpp** waves. The circled numbers denote amplitudes.

Table 5.8. Amplitudes of waves on CMP stacks with different maximum offsets for model C typical of the Volga-Ural and Timan-Pechora basins.

Offset range (km)	*Dpp** + PPtop (r.u.)	PP (r.u.)
–4.0 to 4.0	22	15
–3.5 to 3.5	20	17
–3.0 to 3.0	19	18
–2.5 to 2.5	17	21
–2.0 to 2.0	15	23

5.3.1.1. Conclusions from Subsection 5.3.1

1) A substantial AVO gradient of scattered compressional waves from fluid-filled fractured beds, which is caused by maximum amplitudes at large offsets, allows us to propose two methods for recognizing this kind of wave.
2) The first method exploits the difference between two AVO gradients with identical signs. The method is suitable in the search for fractured zones occurring generally in tight rocks at depths of interest to the oil

Table 5.9. Amplitudes of the *Dpp** and PP waves on near-offset and far-offset CMP stacks for model B typical of the Volga-Ural and Timan-Pechora basins.

Near-offset range (km)	Far-offset range (km)	*Dpp** (n) (r.u.)	PP (n) (r.u.)	*Dpp** (f) (r.u.)	PP (f) (r.u.)	*Dpp** far/near	PP far/near
–1.0 to 1.0	–3.0 to –2.0 and 2.0 to 3.0	7	28	27	12	4.0	0.4
–1.0 to 1.0	–2.5 to –1.5 and 1.5 to 2.5	7	28	23	16	3.3	0.6
–1.0 to 1.0	–2.0 to –1.0 and 1.0 to 2.0	7	28	18	20	2.7	0.7

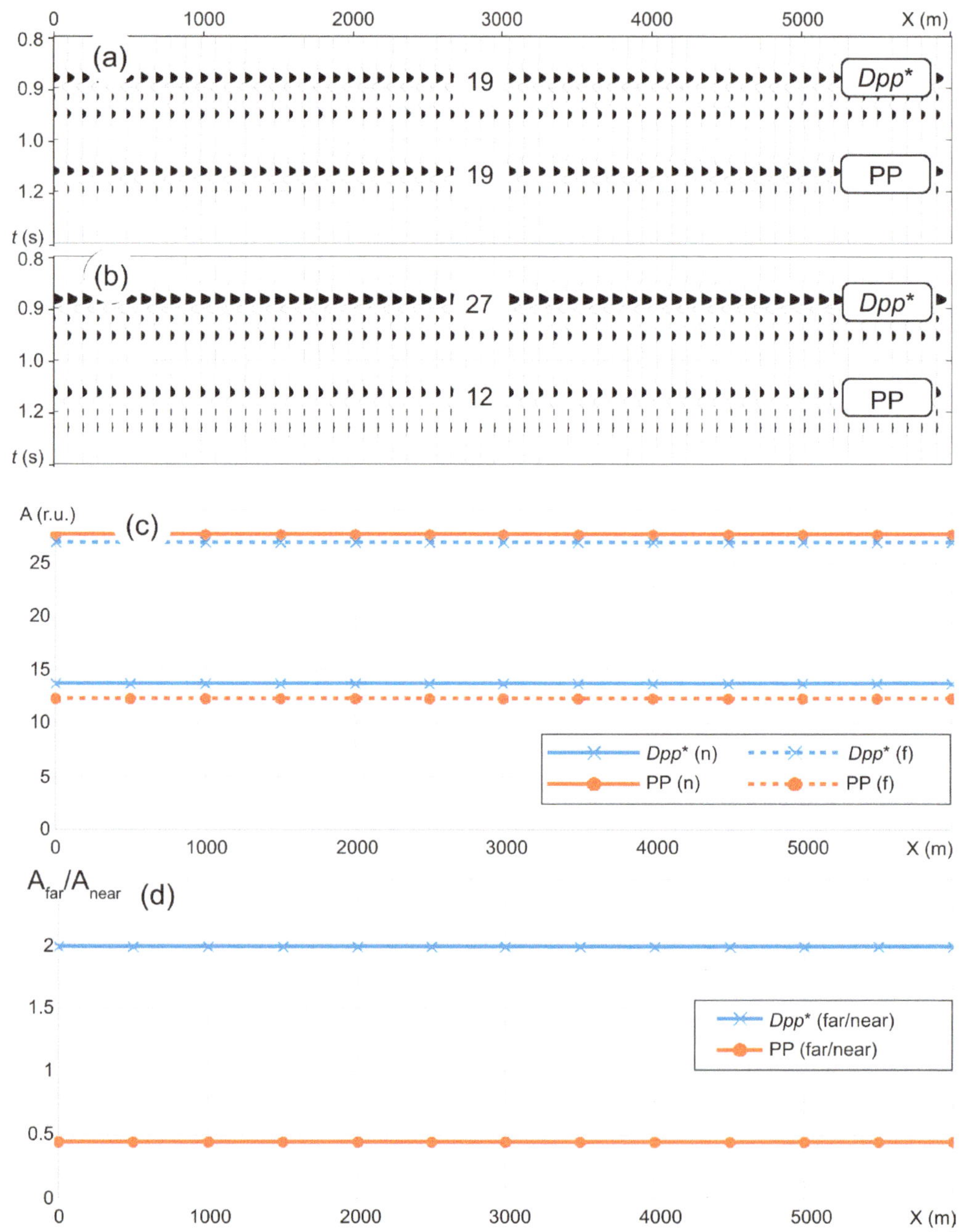

Figure 5.29. Primary elements of seismic CMP data analysis to identify fractured reservoirs for model C typical of the Volga-Ural and Timan-Pechora basins: (a) full stack along a test line; (b) partial stack with far offsets of –3 through –2 km and 2 through 3 km; (c) amplitudes of reflected PP and scattered *Dpp** waves stacked in the two different offset ranges; and (d) amplitude ratio of the far-offset stack and the near-offset stack for reflected PP and scattered *Dpp** waves. The circled numbers denote amplitudes.

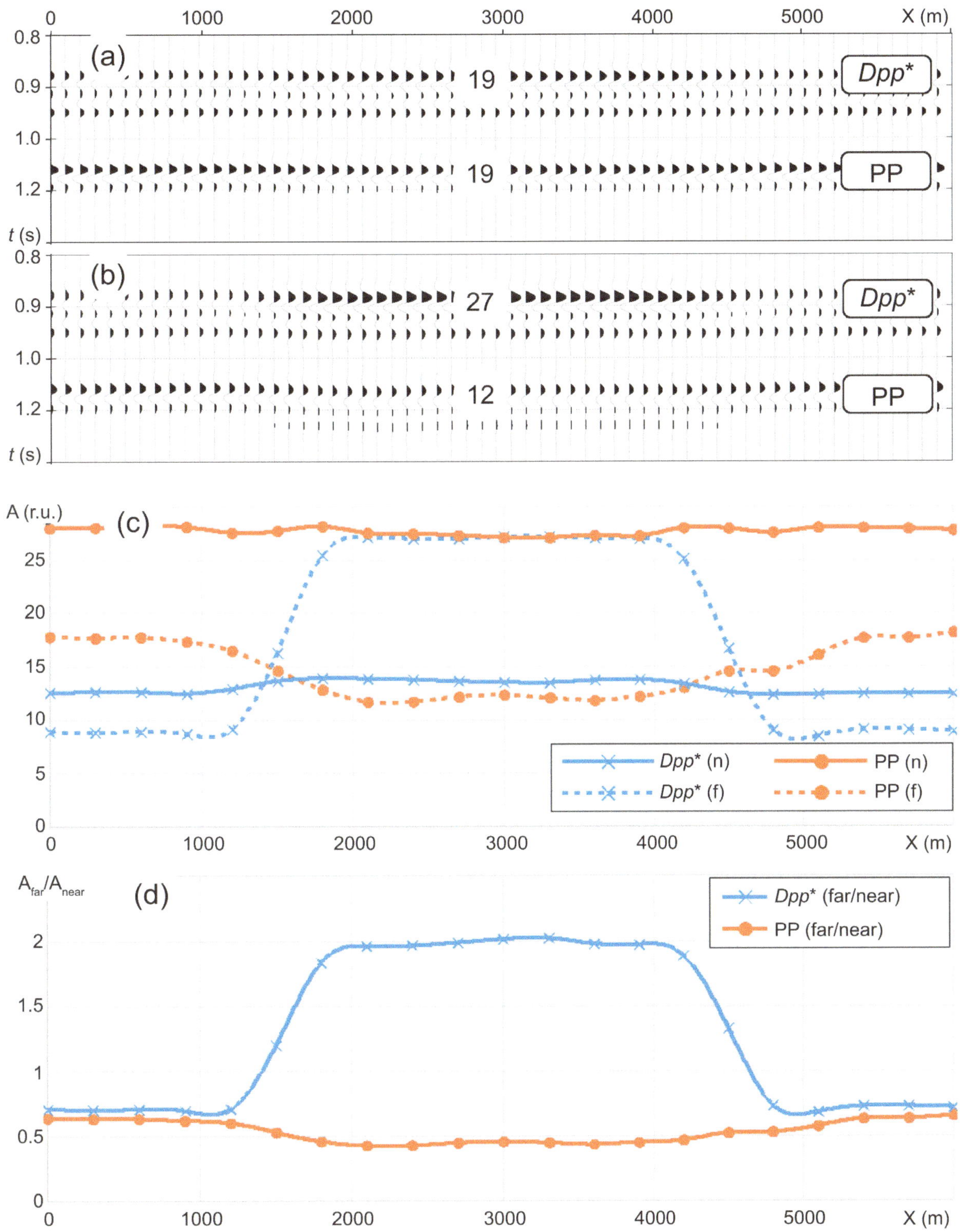

Figure 5.30. Primary elements of seismic CMP data analysis to identify fractured reservoirs for model D typical of the Volga-Ural and Timan-Pechora basins: (a) full stack along a test line; (b) partial stack with far offsets of –3 through –2 km and 2 through 3 km; (c) amplitudes of reflected PP and scattered *Dpp** waves stacked in the two different offset ranges; and (d) amplitude ratio of the far-offset stack and the near-offset stack for reflected PP and scattered *Dpp** waves. The circled numbers denote amplitudes.

Table 5.10. Parameters of a carbonate formation model.

Layer	*Vp* (m/s)	*Vs* (m/s)	ρ (g/cm^3)	*H* (m)	*R*	Fractures
1	4130	2150	2.50	1800	–	Yes
2	4540	2420	2.55	400	0.109	Yes
3	4800	2540	2.60	600	0.072	No
4	5150	2650	2.65	800	0.085	No
5	5660	2800	2.70	400	0.108	No

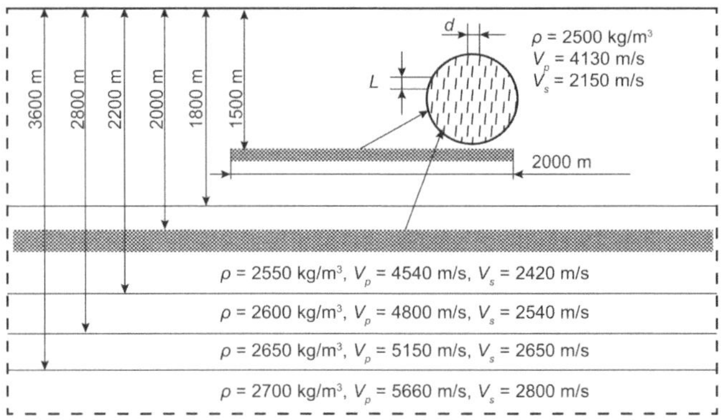

Figure 5.31. Five-layer model with a fractured zone and fractured bed (model E).

industry. The boundaries in this case have very low changes in both *Vp/Vs* ratio and impedance of either sign. This causes gentle AVO curves and low values of AVO gradient of any sign.

3) The AVO gradient has the following characteristics.
   - It is positive and considerable for scattered waves from fluid-filled fractured beds.
   - It is negative for reflections from boundaries with a positive change in impedance and a small change in Poisson's ratio (or *Vp/Vs* ratio).
   - It is insignificant and of either sign for reflections in the conditions described in item 2.
4) The second method uses opposite signs of gradients inherent to reflected and scattered waves and can be applied in conditions similar to those modeled here. The method compares the amplitudes in a far-offset partial stack with those in a near-offset stack. The amplitude ratio is supposed to be more than unity for scattered waves and less than unity for reflected waves. The reliability of the method and its stability to the impact of noise are rather high when large numbers of near and far offsets are used. The ratio may be regarded as a recognition parameter for fractured reservoirs.
5) For low-velocity-contrast media, when acoustic parameters of a fractured bed are almost the same as those of the host rock (which is typical of East Siberia), the proposed recognition parameter can be quite effective. This is because it may differ by up to a few tens of times between a scattered wave and a reflection, being much more than unity for the former and slightly less than unity for the latter.
6) For high-contrast media, when acoustic boundaries are present at the top and base of a fractured bed and, therefore, there are reflections from them, the difference between minimum and maximum amplitudes of the entire response decreases. As a result, the recognition parameter will differ by only a few times between a scattered wave and a reflection, being more than unity for the former and less than unity for the latter.
7) Identification of local fractured zones in productive intervals may proceed in the same manner, i.e., by comparing far-offset and near-offset amplitudes.

5.3.2. Examples of using scattered converted waves to identify bed-bounded mesofracture zones on *X*-component data

The above estimates of amplitudes of a converted wave from a bed-bounded zone of mesofractures are obtained with a plane-wavefront energy source (Figure 5.3) and in the absence of any type of noise. In real conditions, however, noise is inevitably present, and the recorded amplitude of a scattered converted wave may be insufficient to permit confident detection. For confident detection, CMP stacking is worthwhile.

5.3.2.1. Bed-bounded zones of mesofractures in a carbonate formation

Common-source gathers for CMP stacking are computed for the two versions of the base model shown in Figure 5.15b and 5.15d. They imitate, respectively, a zone of mesofractures embedded in tight (homogeneous) low-velocity-contrast rocks, typical of East Siberia, and a bed of mesofractures with reflectors at its top and base embedded in high-contrast rocks, typical of the Volga-Ural and Timan-Pechora basins. The common-source gathers are computed with a 100 m receiver spacing along a 6 km line. After NMO correction and muting (Figure 5.33), common-source stacking of the *X*-component is performed.

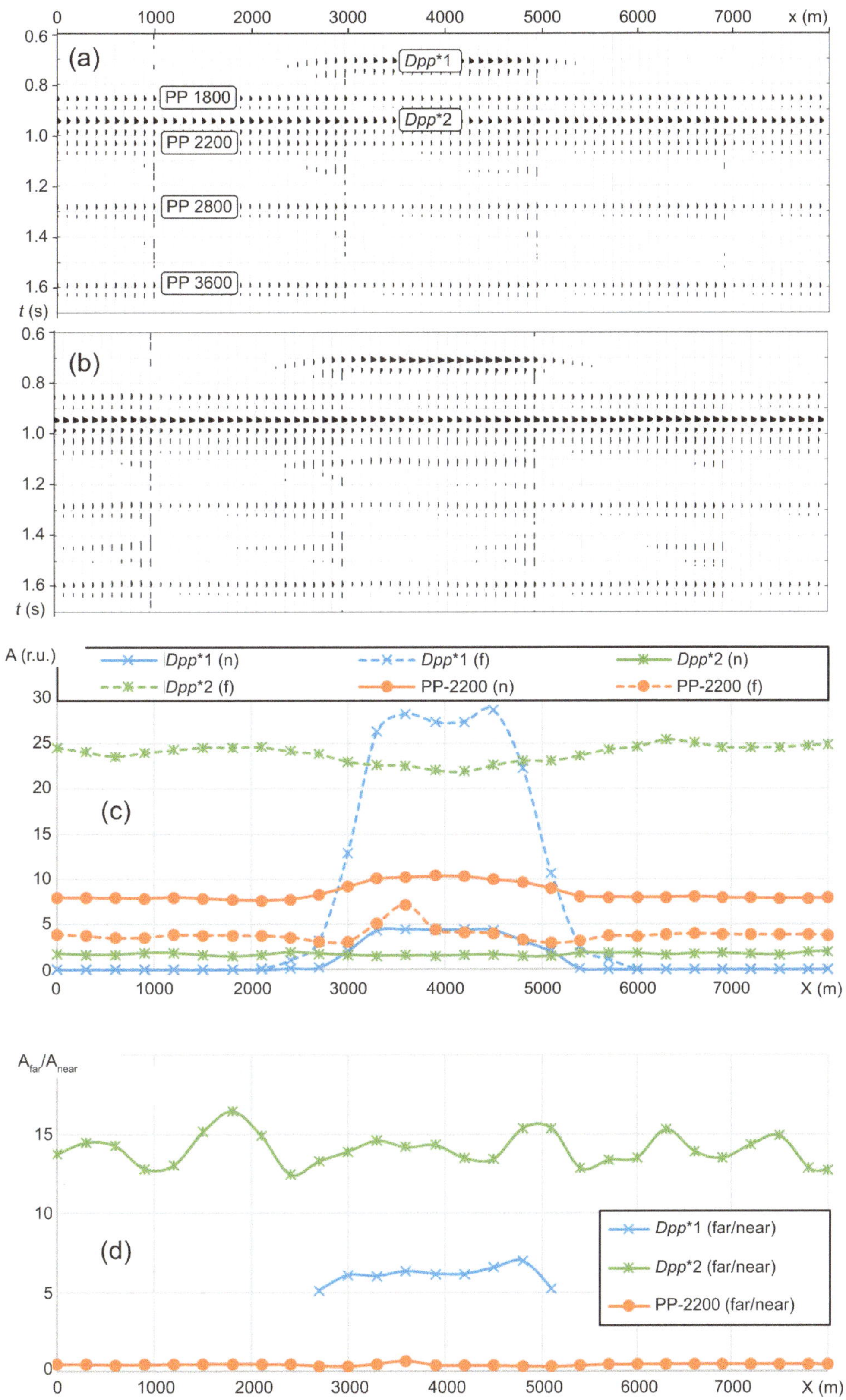

Figure 5.32. (a) Full CMP stack with offsets of up to 4 km; (b) partial stack for far offsets ranging from 2000 to 3000 m; (c) amplitudes of scattered waves from the fractured targets and of reflections from the acoustic boundaries for near-offset and far-offset partial stacks; and (d) estimates of a scattered-wave recognition parameter for the fractured zone, fractured bed, and acoustic boundary with a maximum reflection coefficient of 11%.

For the first model—fractured zone in a massive (homogeneous) medium—the wave response on the stacked image (Figures 5.34 and 5.35) includes:

- a scattered compressional wave (*Dpp**) at 0.90 to 0.95 s;
- a compressional wave (PP) at 1.08 to 1.12 s, reflected from a horizontal boundary;
- a scattered converted wave (*Dps**) at 1.24 to 1.30 s;
- a secondary scattered converted wave (PP*Ds*) at 1.4 s, formed by the PP-reflection coming from below; and
- a converted wave (PS) at 1.50 s, reflected from the horizontal boundary and recorded outside the zone of mesofractures.

Considering a polarity reversal at the source location, stacks are produced separately for each side of the spread. The amplitude anomaly of the scattered converted wave shows up clearly in the stacks from both sides of the spread (Figure 5.34a and 5.34b). Figure 5.34c depicts an almost double difference in the scattered converted *Dps** wave amplitude between the two sides of the spread.

On the *X*-component, the compressional PP-wave and converted PS-wave, unlike the scattered converted *Dps** wave, exhibit almost equal amplitudes (14 and 14 r.u. for PP, 28 and 29 r.u. for PS) on different flanks. The amplitude of the compressional PP-reflection varies only slightly along the entire line (Figure 5.34c). The amplitude anomalies of the *Dps** wave in Figures 5.34 through 5.37 experience minor displacements relative to the fractured zone, whose boundaries are denoted by the vertical lines. This is caused by a deviation of the mesofractures from the vertical.

Complete stacks of both the left and right split-spread sides shown in Figure 5.35 are obtained in different ways. The stack in Figure 5.35a is obtained with a polarity reversal of one of the split-spread sides, which is an approach commonly used when processing converted reflections. The amplitudes of all waves on the image are increased considerably because of stacking of both split-spread sides.

In this case, the amplitude of the scattered converted *Dps** wave is 6 times greater than that of the converted PS-wave (Table 5.11), which is a significant potential noise. In the image in Figure 5.35, this difference, however, does not seem so large, due to a limited range of visualization. This may create a risk of overlooking such anomalies originating from fractured zones.

To enhance significantly the relative amplitude of the target scattered converted wave, the *X*-components of PP, PS, and PP*Ds* waves, reflected from plane boundaries, should be maximally suppressed. For this purpose, following advice given to us by N.Y. Marmalevskyi (personal communication), full stack is obtained by summing both the left and right spread sides without a polarity reversal of one side. This approach is expected to reduce

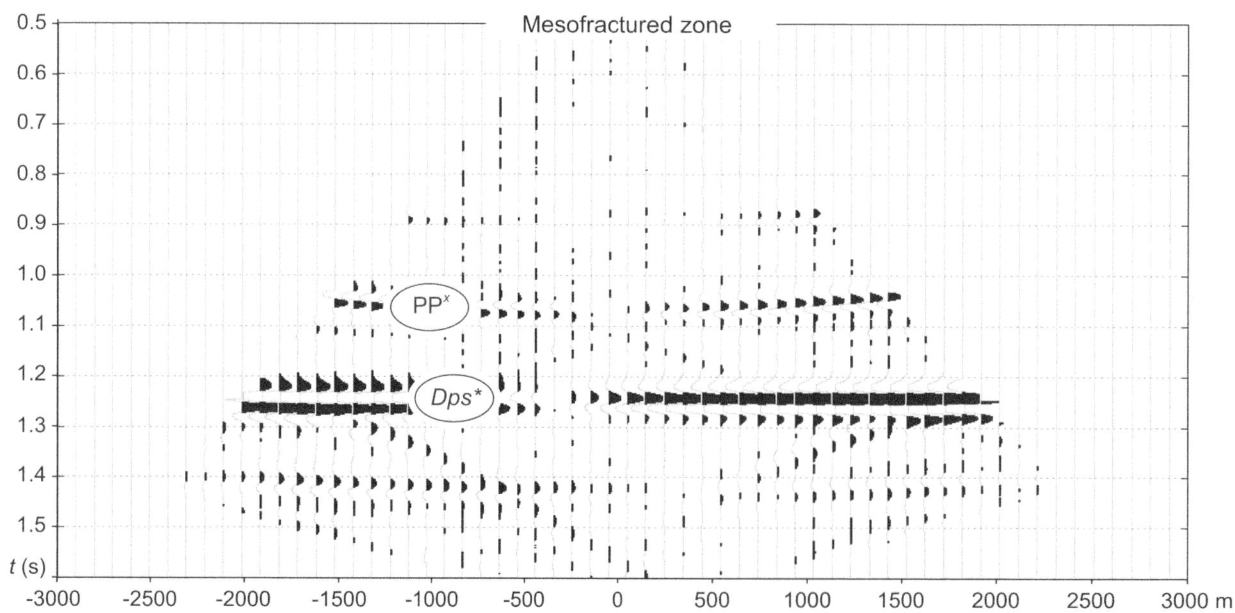

Figure 5.33. *X*-component of a common-source gather after NMO correction and muting. The latter takes into consideration the region of maximum amplitudes of the *Dps** wave. PP^x = *X*-component of PP-wave.

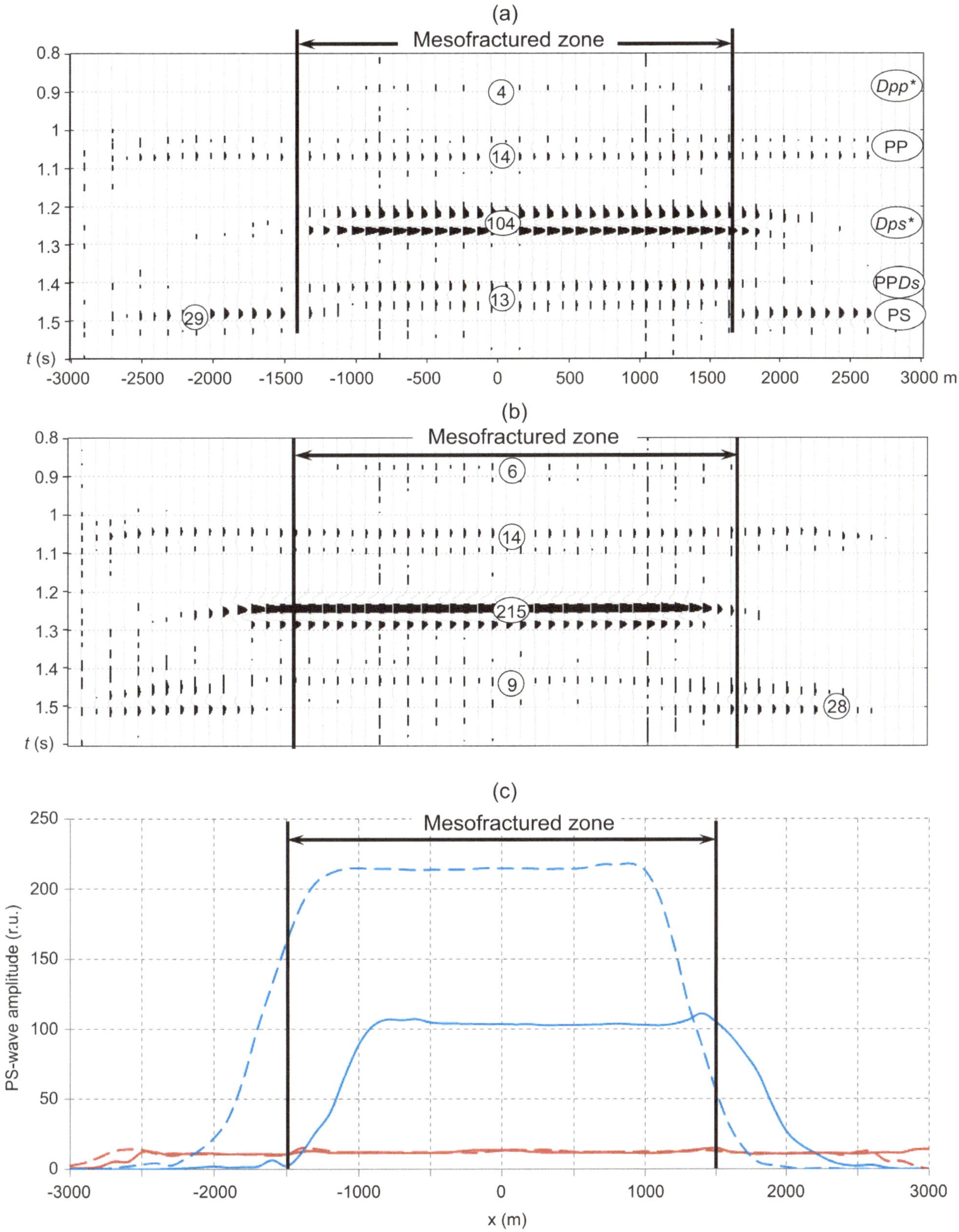

Figure 5.34. Common-source stacks of the *X*-components of data from the (a) left and (b) right sides of the split-spread for the first model that imitates a fractured zone in tight rocks. (c) Amplitudes of the scattered converted *Dps** wave at 1.24 s (blue) and PP-reflection at 1.05 s (red); solid lines correspond to (a) left side waves and dashed lines to (b) right side waves; vertical lines denote the mesofracture zone boundaries. The circled numbers are amplitude values of respective waves.

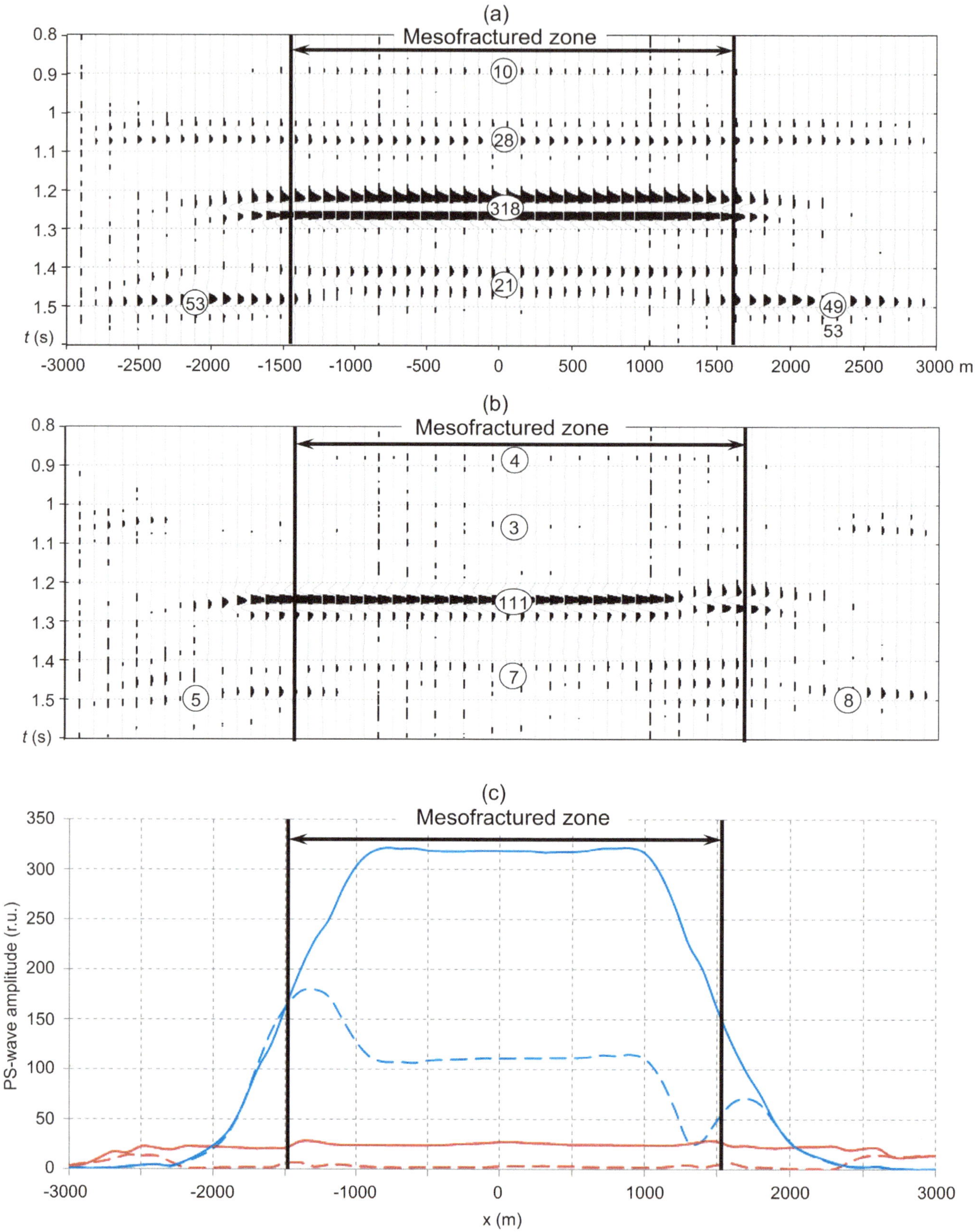

Figure 5.35. Complete common-source stacks of the *X*-components of data from both the left and right sides of the split-spread (a) with and (b) without a polarity reversal for the first model that imitates a fractured zone in tight rocks. (c) Amplitudes of the scattered converted *Dps** wave at 1.24 s (blue) and PP-reflection at 1.05 s (red). For designations, see Figure 5.34.

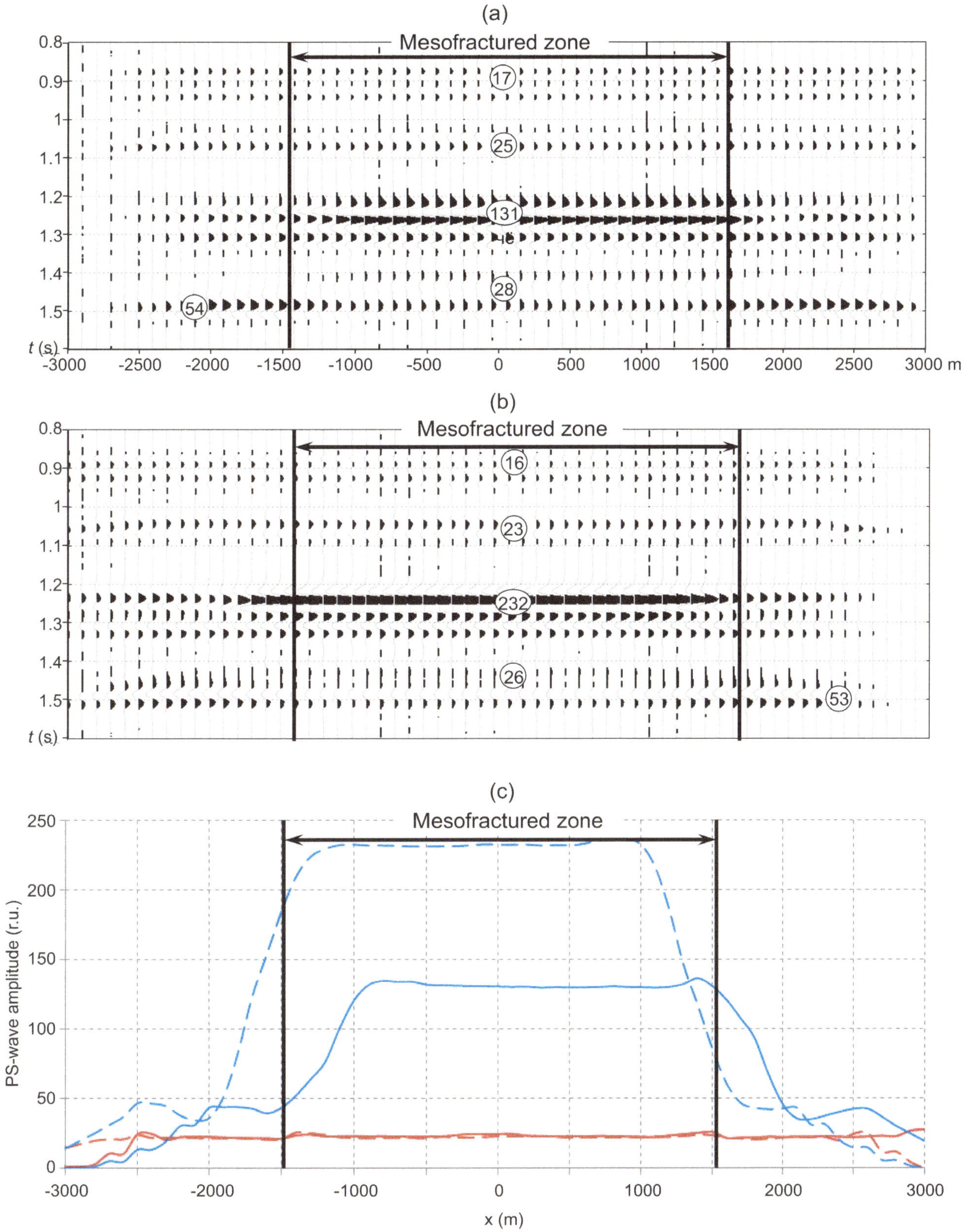

Figure 5.36. Common-source stacks of the *X*-components of data from the (a) left and (b) right sides of the split-spread for the second model that imitates a fractured bed with reflecting boundaries. (c) Amplitudes of the scattered converted *Dps** wave at 1.24 s (blue) and PP-reflection at 1.05 s (red); solid lines correspond to (a) left side waves and dashed lines to (b) right side waves; vertical lines denote the mesofracture zone boundaries.

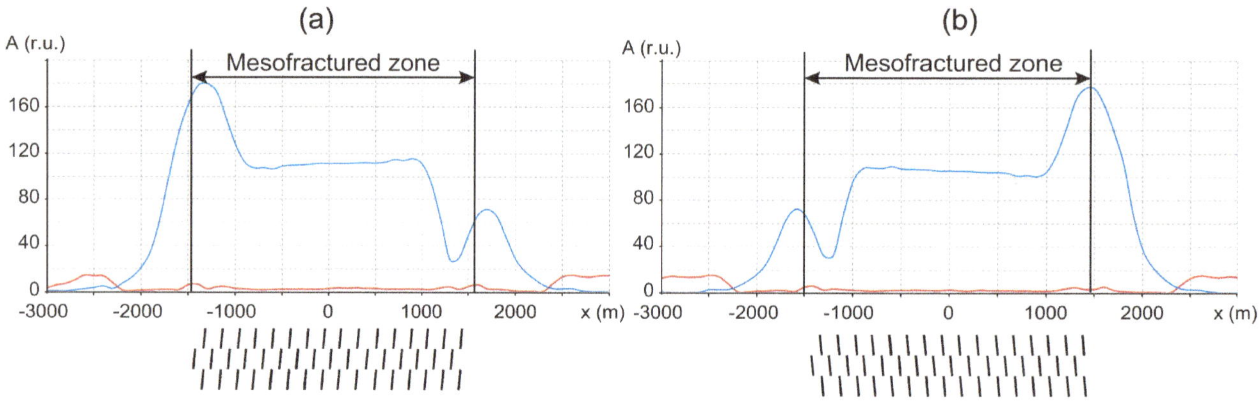

Figure 5.38. Wave amplitudes on the complete common-source stack of the *X*-components of data from both the left and right sides of the split-spread without a polarity reversal for 5° deviations to the (a) right and (b) left from the vertical.

sumed relationship of these effects to a deviation of mesofractures from the vertical is confirmed by recomputing with a deviation of the opposite sign (Figure 5.38). A local amplitude maximum of the scattered converted *Dps** wave from a mesofracture zone is found over the zone edge opposite to the deviation of fractures from the vertical.

To summarize, this approach to detecting and mapping mesofracture zones in tight, predominantly carbonate formations, using the *X*-component of scattered converted *Dps** waves (Section 5.2.2), requires

- employment of multifold three-component seismic surveys with midrange offsets that provide maximum *Dps** wave amplitudes;
- optimal application of NMO correction and muting to scattered converted waves;
- separately stacking the left- and right-side offsets in order to account for a polarity reversal at the source location; and
- completely stacking both the left- and right-side offsets performed in two ways:
  - with a polarity reversal applied to one of the spread sides, which allows the relative amplitude of these waves to be increased double, and
  - without such a reversal. This procedure appreciably reduces (practically to zero at times) the amplitudes of corruptive waves such as converted reflections and the *X*-components of compressional reflections from plane boundaries, while the energy difference between the scattered converted waves on both split-spread sides may remain sufficiently significant. This is most likely to occur where the response from a fractured bed is asymmetric due to a minor deviation of fractures from the vertical.

The above numerical 2D-simulation of wave responses from zones of near-vertical fluid-filled mesofractures in typical fractured reservoirs for the two types of carbonate formations has helped us establish the possibility of directly detecting such zones using scattered converted waves on three-component (3C) seismic data.

5.3.2.2. Investigating the possibility of mapping thin zones of mesofractures in the Bazhenov Formation using scattered converted waves

Relevance of the challenging issue.—The importance of the oil potential of the Bazhenov Formation, following commercial oil discoveries in many areas of the West Siberian Basin (1 million km^2), cannot be overestimated. However, no effective prospecting and exploration methodology other than those based on structural factors has as yet been developed for hydrocarbons in this formation. In the last decade, seismic explorationists have focused their efforts on developing methods for hydrocarbon prediction from dynamic characteristics of compressional reflections by seeking statistical relationships between these characteristics and productivity of existing wells. But, a statistical relationship that works successfully in one area is not generally applicable in others.

In the last five to seven years, from prolific oil wells in several areas, detailed well data characterizing the structural, lithologic, and petrophysical properties of the Bazhenov Formation have been obtained. Almost all producing intervals include either a layer or set of thin carbonate or siliciclastic layers (each 1 to 2 m thick), totaling 3 to 8 m and penetrated by near-vertical fractures.

In recent 3D surveys (e.g., Kunin et al., 2013; and Revenko et al., 2013), these detailed well data acquired using FMI and XMAC are used to simulate numerically signals from fractured beds in the vicinity of wells. Then this information, in combination with other data, is used to predict fracturing within the Bazhenov Formation. These predictions are generally satisfactory in the surveyed areas but application of the findings to other areas is usually unsuccessful.

The preceding discussion has shown that the *X*-components of direct responses from zones of near-vertical mesofractures in beds a few meters thick contain fronts of scattered (diffracted) converted waves. These results and the internal structure of the Bazhenov Formation suggest the potential for mapping fractured zones in the formation without using well data for support.

Internal structure models of the Bazhenov Formation.—Well log data and core investigations in commercially productive wells of various areas (Salym, Nazym, Ust-Kut, etc.) indicate a sufficient lateral consistency of the inner structure of the Bazhenov Formation. Bazhenites, a source rock, comprise a fractured carbonate layer, a stack of rigid carbonated interlayers, or a stack of siliciclastic radiolarites. Although a carbonate layer or a unit of rigid interlayers may typically be 3 to 10 m thick, the thickness of an interlayer may vary from 0.5 to 2 m. The acoustic characteristics of bazhenites are: $Vp = 2400$ to 2600 m/s, $Vs = 1100$ to 1200 m/s, and $\rho = 2200$ to 2300 kg/m^3. The host rocks typically have the parameters: $Vp = 3400$ to 3700 m/s, $Vs = 1600$ to 1700 m/s, and $\rho = 2600$ to 2700 kg/m^3. No reliable measurements of such parameters for rigid carbonates and radiolarites are available. Most researchers take the values for these parameters as being equal to or 10% greater than those of the host rocks when characterizing the internal structure of the Bazhenov Formation.

Based on this information, the following three models are conceived for the Bazhenov internal structure, depending on the structure of the rigid fractured layers and interlayers:

1) One layer of carbonate or carbonated fractured rock, 5 m thick, in the middle of the Bazhenov Formation, by analogy with sections shown in Usmanov et al. (2006);
2) A stack of four radiolarite or carbonated interlayers 1.5 m thick with near-vertical fracturing. These interlayers are separated by 1 m thick bazhenite layers. The unit is located in the middle of the Bazhenov Formation, according to sections shown in Revenko et al. (2013); and
3) Rigid fractured 2 m thick carbonate interlayers uniformly embedded in the entire 32 m thick Bazhenov Formation. This is based on the Bazhenov composite lithology section for the Salym field shown in Gogonenkov and Nekrasova (2012).

Tables 5.13 through 5.15 contain the velocity and density parameters for each layer in each model. For these models, common-source gathers are computed for split spreads 5000 and 7000 m long with a 50 m receiver spacing. The energy source at the surface is either a point source of the "expansion center" type or a plane compressional wave of the plane-wavefront type with a 30 Hz Berlage wavelet.

The fractures are supposed to be filled with fluid. This is achieved due to setting a free sliding condition at all fracture boundaries, because any ideal fluid creates only an orthogonal force at a boundary.

Computational results are presented as snapshots of wavefields and common-source gathers recorded on the earth's surface.

Wave responses from fractured layers and interlayers of the Bazhenov Formation for plane-wavefront energy sources.—Schematic computational result of a wave response for model 1 (Figure 5.39) is presented as the *X*- and *Z*-components of a common-source gather (recorded on the earth's surface) in Figure 5.40b and 5.40c, respectively. Figure 5.40b shows a single-phase event recorded

Table 5.13. Model 1 parameters.

Bed number	Bed thickness (m)	ρ (kg/m^3)	*Vp* (m/s)	*Vs* (m/s)	Presence of fractures
1	1500	2500	3000	1400	no
2	1500	2600	3500	1600	no
3	15	2300	2400	1100	no
4	5	2700	3800	1700	yes
5	10	2300	2400	1100	no
6	470	2600	3500	1600	no

Table 5.14. Model 2 parameters.

Bed number	Bed thickness (m)	ρ (kg/m³)	Vp (m/s)	Vs (m/s)	Presence of fractures
1	1500	2500	3000	1400	no
2	1500	2600	3500	1600	no
3	10	2300	2400	1100	no
4*	1.5/1.0	2700/2300	3800/2400	1700/1100	yes/no
4**	1.5/1.0	2700/2300	3800/2400	1700/1100	yes/no
4***	2.0/1.0	2700/2300	3800/2400	1700/1100	yes/no
4****	2.0/1.0	2700/2300	3800/2400	1700/1100	yes/no
5	10	2300	2400	1100	no
6	470	2600	3500	1600	no

Table 5.15. Model 3 parameters.

Bed number	Bed thickness (m)	ρ (kg/m³)	Vp (m/s)	Vs (m/s)	Presence of fractures
1	1500	2500	3000	1400	no
2	1500	2600	3500	1600	no
3	2.0/2.0	2700/2300	3800/2400	1700 /1100	yes/no
4	2.0/2.0	2700/2300	3800/2400	1700 /1100	yes/no
5	2.0/2.0	2700/2300	3800/2400	1700/1100	yes/no
6	2.0/2.0	2700/2300	3800/2400	1700 /1100	yes/no
7	2.0/2.0	2700/2300	3800/2400	1700/1100	yes/no
8	2.0/2.0	2700/2300	3800/2400	1700/1100	yes/no
9	2.0/2.0	2700/2300	3800/2400	1700/1100	yes/no
10	2.0/2.0	2700/2300	3800/2400	1700/1100	yes/no
11	468	2600	3500	1600	no

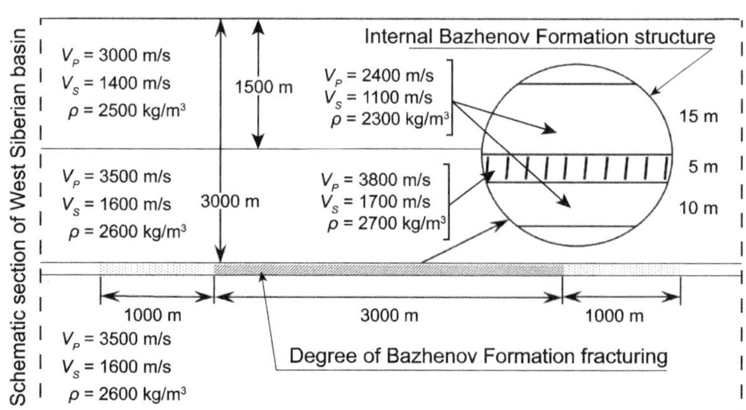

Figure 5.39. Model 1 used to compute seismic responses from a zone of mesofractures within the Bazhenov Formation. Refer to Table 5.13 for the model parameters.

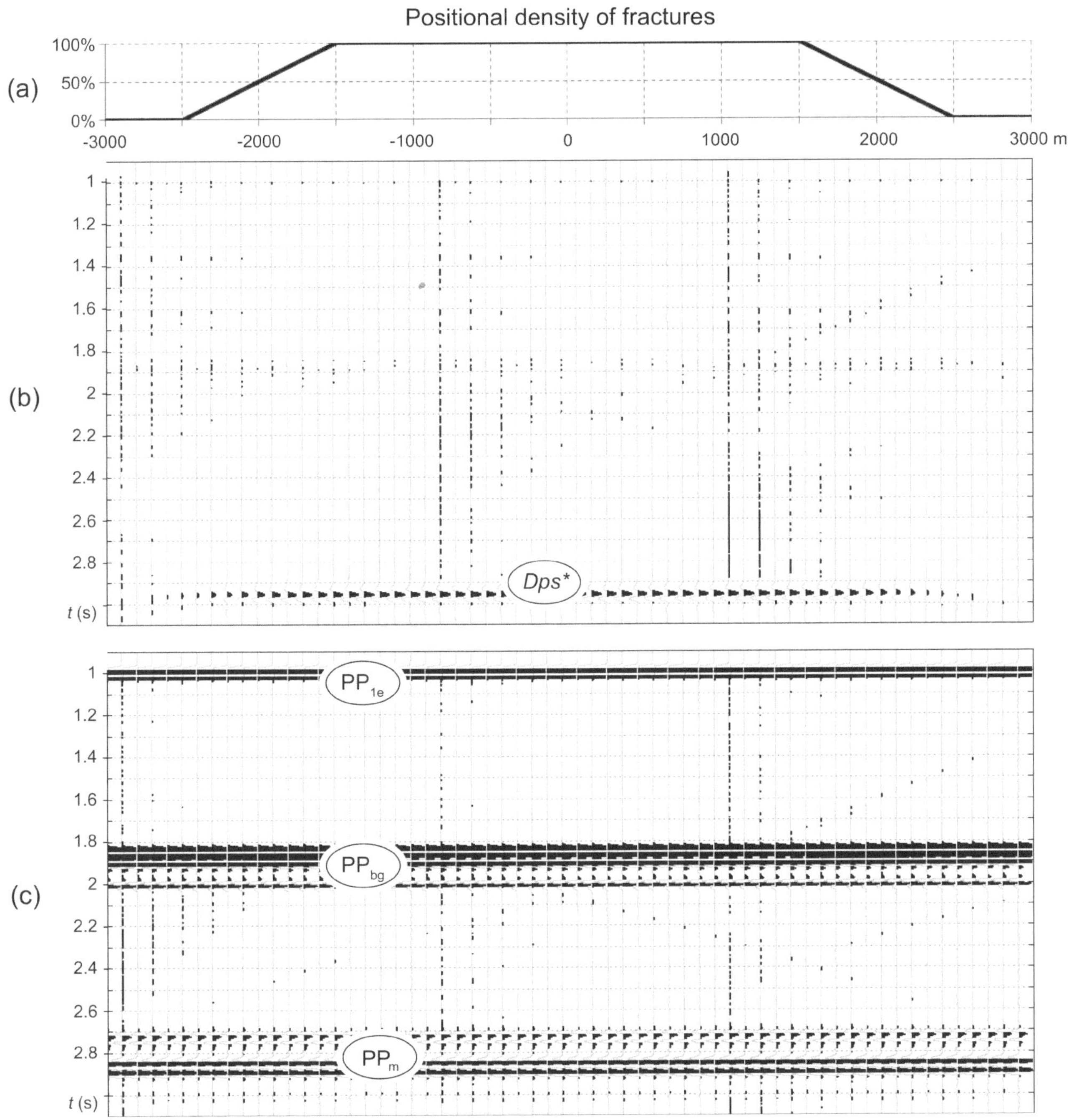

Figure 5.40. (a) Fracture density for model 1 of the Bazhenov Formation; (b) X-component and (c) Z-component of a corresponding common-source gather. PP_1 is upper reflection horizon, PP_{bg} is Bazhenov Formation top, and PP_m is a multiple wave.

at 2.93 s with quite strong amplitudes coincident with the position of a fractured zone (see Figure 5.40a). This event is a scattered converted wavefront from the zone of near-vertical fractures in the carbonated layer in the middle of the Bazhenov Formation. Figure 5.40c shows three horizontal events recorded at 0.98, 1.86 to 1.92, and 2.70 to 2.88 s. They are a reflection from the intermediate boundary, a complex reflection from the Bazhenov Formation, and a double multiple, respectively.

Figure 5.41 shows the target interval on the X- and Z-components and amplitudes within different time intervals of the complex Bazhenov reflection wavetrain. The amplitude anomaly of scattered converted waves emanated from the fractured zone in model 1 shown in Figure 5.41d is replicated in Figure 5.42, which shows the analogous curves for models 2 and 3. Note that the noise-free scattered converted wavefront of interest in Figures 5.40 and 5.42 is clearly identifiable on the

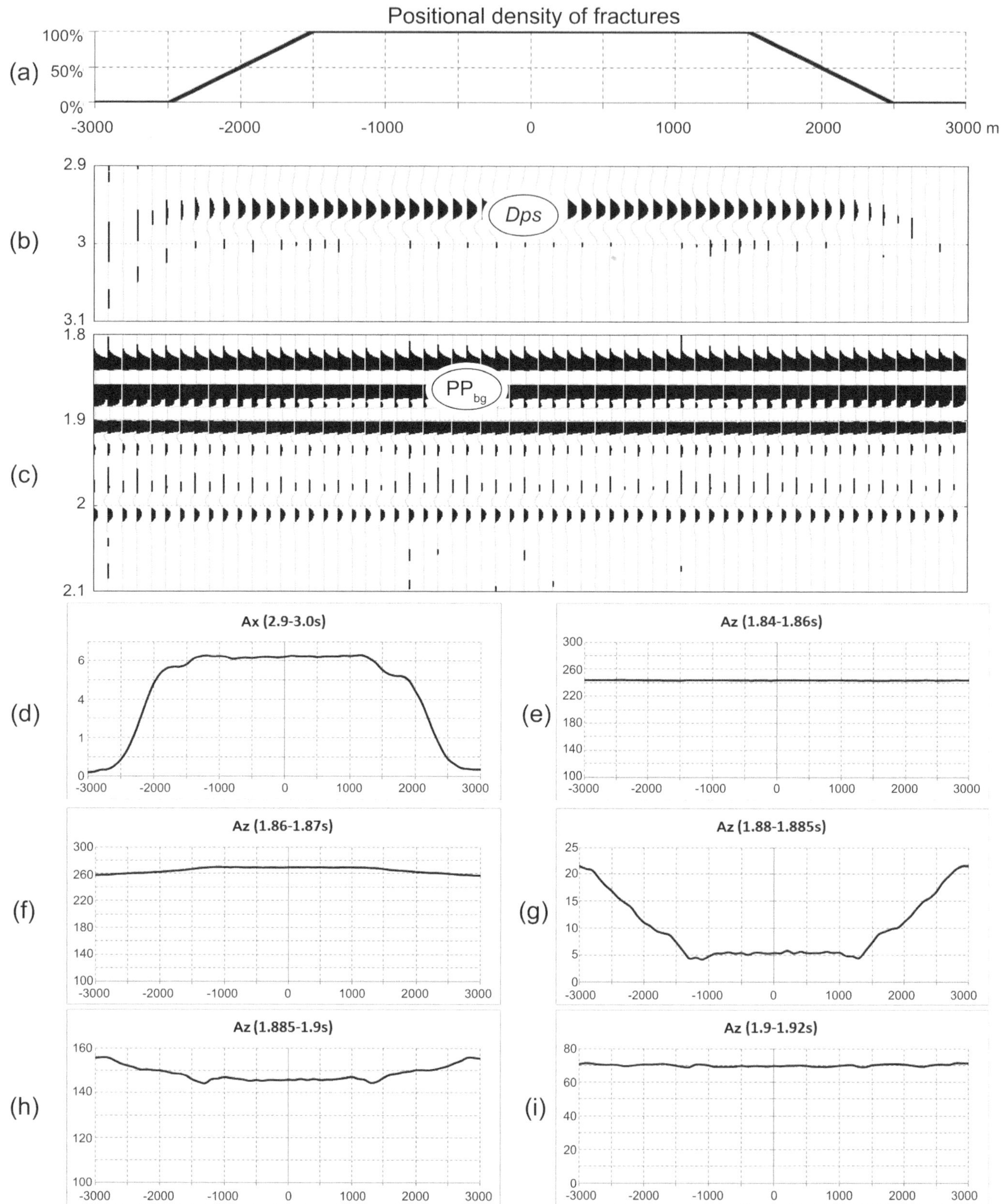

Figure 5.41. (a) Fracture density for model 1 of the Bazhenov Formation; (b) *X*-component and (c) *Z*-component of the Bazhenov wavetrain. Amplitudes within the time intervals of (d) 2.9 to 3.0 s, (e) 1.84 to 1.86 s, (f) 1.86 to 1.87 s, (g) 1.88 to 1.885 s, (h) 1.885 to 1.9 s, and (i) 1.9 to 1.92 s.

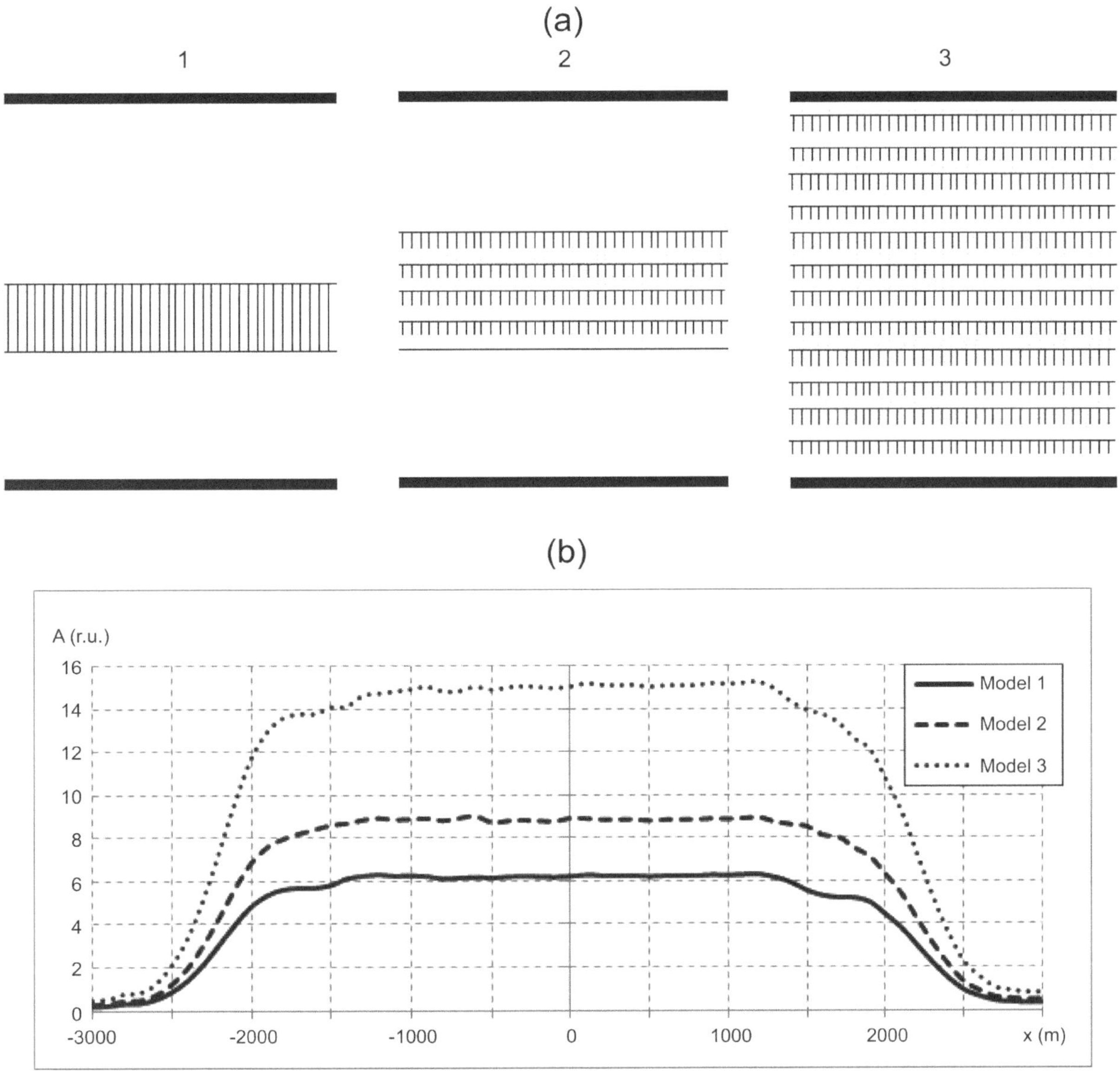

Figure 5.42. (a) Three models of fracturing for the Bazhenov Formation; and (b) amplitudes of scattered converted wavefronts for the three models.

X-component and corresponds to the spatial position of the mesofracture zone. At the same time, this wavefront amplitude comprises only 2.5% to 6% of the Bazhenov maximum amplitude on the *Z*-component. Examination of the amplitudes within different time intervals of the complex Bazhenov wavetrain reveals the influence of the mesofracture zone on the amplitudes of one of the compressional reflection cycles. This influence manifests itself as a fourfold amplitude decrease at 1.881 to 1.865 s (Figure 5.41g), corresponding to the fractured zone position. The decrease apparently can be attributed to scattering a portion of the reflection from the Bazhenov base when this wave travels upward through the fractured zone.

Thus, when using a plane-wavefront energy source, a fractured zone in the 5 m thick bed of the Bazhenov Formation (model 1) is identifiable on the *X*-component of scattered converted waves (Figures 5.40b and 5.42b).

Computations conducted for models 2 and 3 of fractured zones (Figure 5.42a) show that their respective response amplitudes exceed by 1.5 and 2.5 times those of model 1 (Figure 5.42b).

Wave responses from fractured layers and interlayers of the Bazhenov Formation for point energy sources.—When using point energy sources for the fracture models of the Bazhenov Formation, the same waves (Figure 5.43) as those identified when using plane-wavefront sources (Figure 5.40c) are present on the *Z*-component.

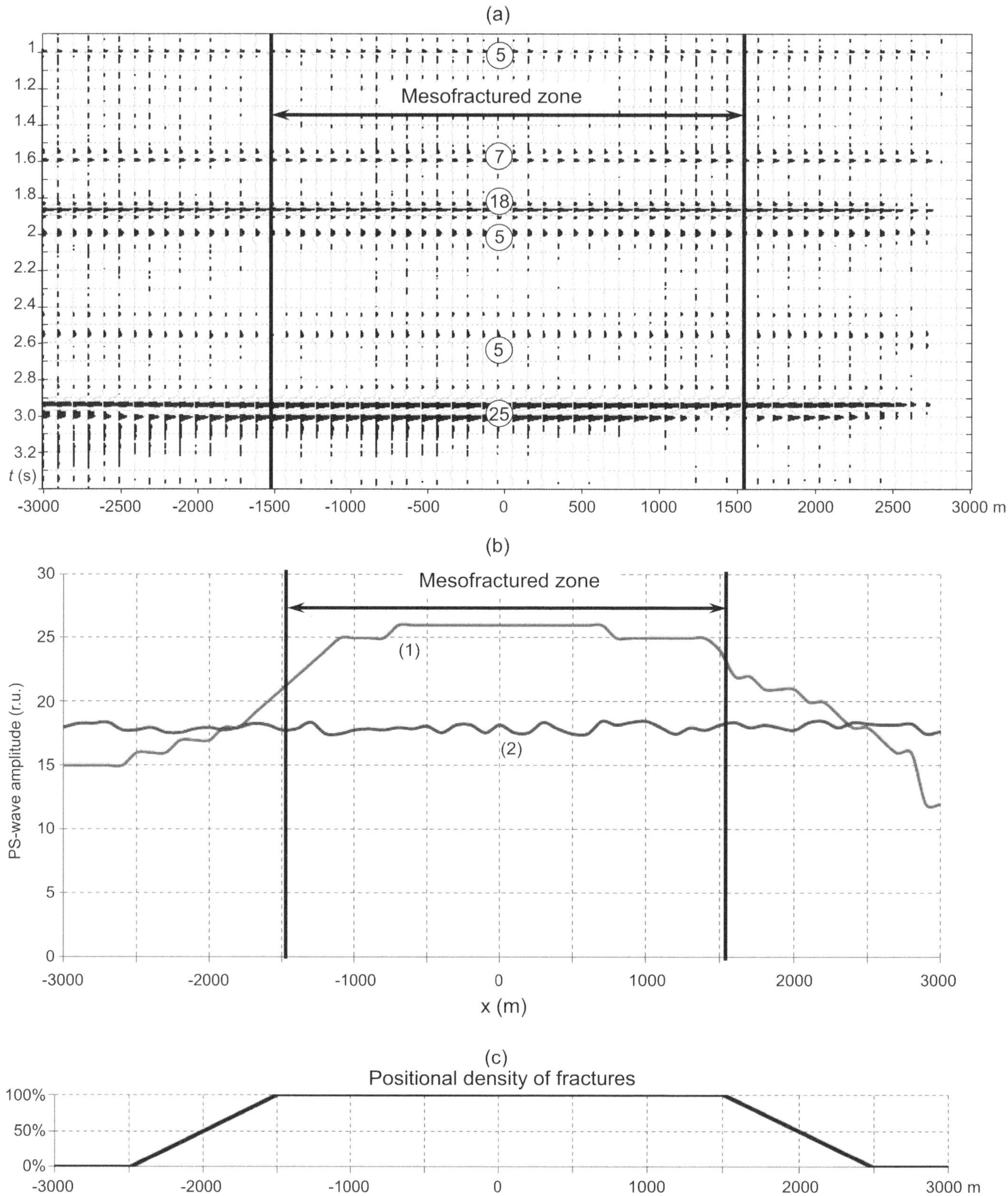

Figure 5.46. (a) *X*-component of the stack of the right split-spread sides for a point source; (b) amplitudes of waves scattered at a fractured zone in (1) and reflected from (2) the Bazhenov Formation; and (c) mesofracture density along the line.

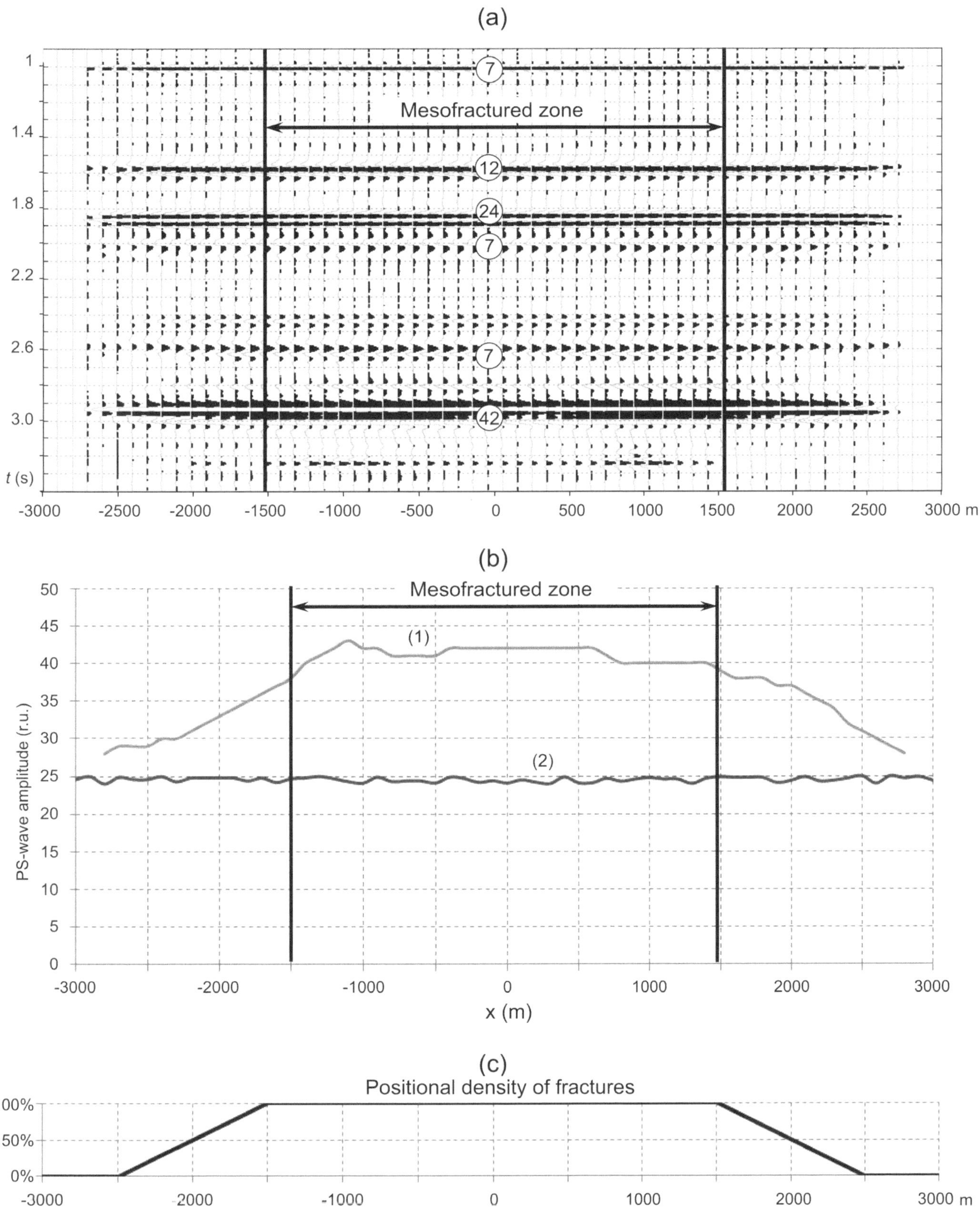

Figure 5.47. (a) *X*-component of the stack of both split-spread sides with a polarity reversal of one of them; (b) amplitudes of waves scattered at a fractured zone in (1) and reflected from (2) the Bazhenov Formation; and (c) mesofracture density along the line.

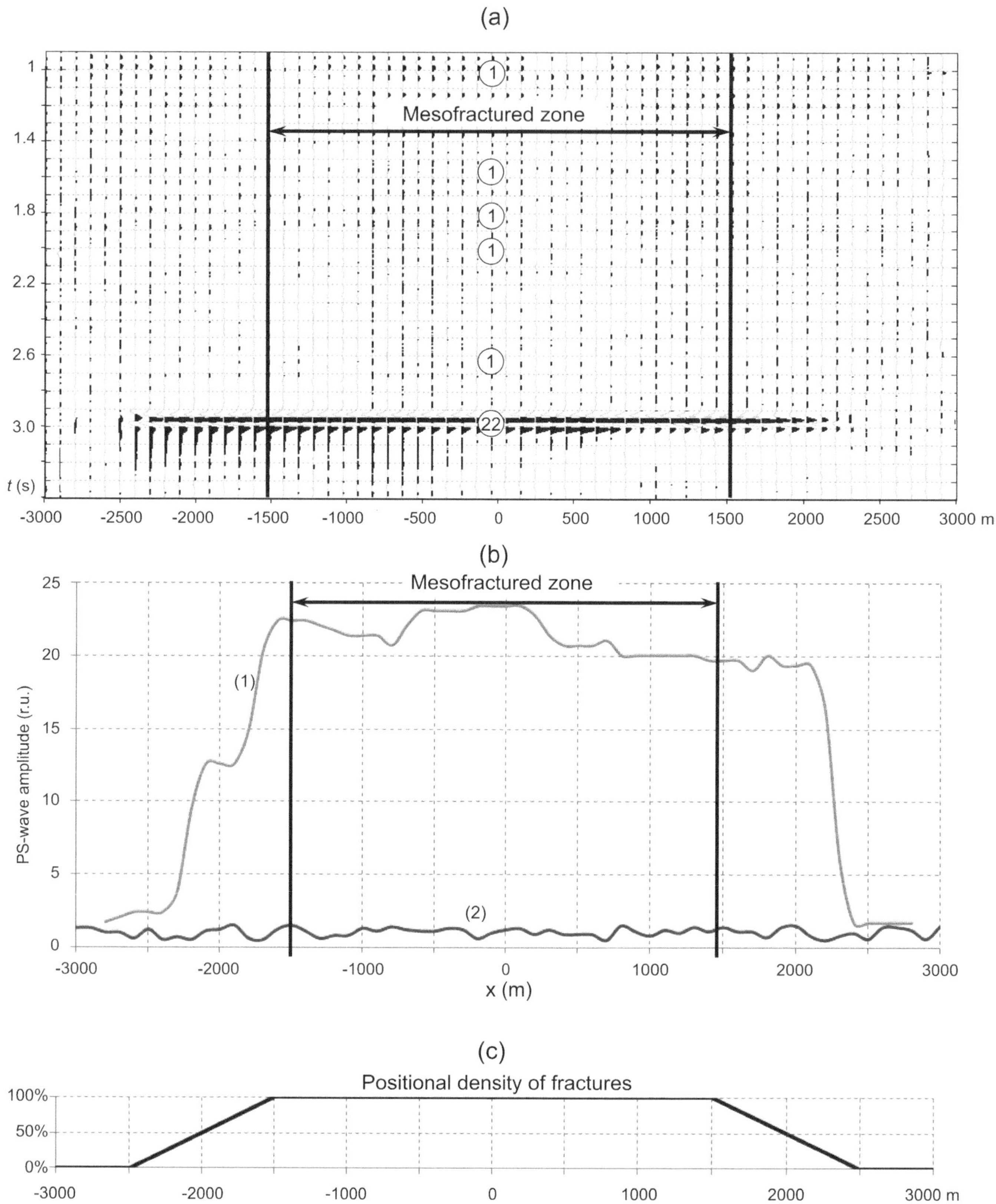

Figure 5.48. (a) *X*-component of the stack of both split-spread sides without a polarity reversal of one of them; (b) amplitudes of waves scattered at a fractured zone in (1) and reflected from (2) the Bazhenov Formation; and (c) mesofracture density along the line.

Table 5.16. Amplitudes (in relative units) of waves recorded on the *X*-component.

Amplitude estimation conditions	PP_1	PS_1	PP_{bg}	*Dps**	PS	*Dps**/ PP_{bg}
For a single point source	12	6	9	5	5	0.55
For stack of left spread side only	4	7	12	26	4	2.0
For stack of right spread side only	5	7	18	25	5	1.4
For stack of both spread sides with a polarity reversal	7	13	24	42	7	1.75
For stack of both spread sides without a polarity reversal	1	1	1	22	1	22.8

wave provides a triple increase (up to 150%) in the level of this wave relative to that of the PP_{bg} reflection from the Bazhenov Formation. This, in turn, enables a satisfactory identification of an amplitude anomaly in the converted wave scattered by a fractured zone.

3) Differences in the character of the scattered converted wavefield between the two split-spread sides, due to a slight deviation of fractures from the vertical, are further increased by the complexity of the reflections from the fine-layered Bazhenov Formation. This entails variations in arrival time, signal shape, and amplitude of scattered converted waves on both split-spread sides.
4) Confidence in identifying the scattered converted wave from a fractured zone can be enhanced significantly by stacking the *X*-component data from both split-spread sides without applying a polarity correction. This process almost eliminates the converted reflection noise and the *X*-component of compressional reflections from the plane boundaries.
5) The energy of the scattered converted wave stacked without a polarity reversal may be rather high. This is very possible because of the asymmetry of the response from a fractured bed, which is attributed to a minor deviation of fractures from the vertical. The difference between the scattered converted waves on both split-spread sides is further increased by instability in the reflectivity of the fine-layered Bazhenov Formation.

A difference between the scattered waves on the two sides of a spread is determined not only by a difference in total amplitude but also by differences in arrival time, extreme amplitude, and wavefront shape.

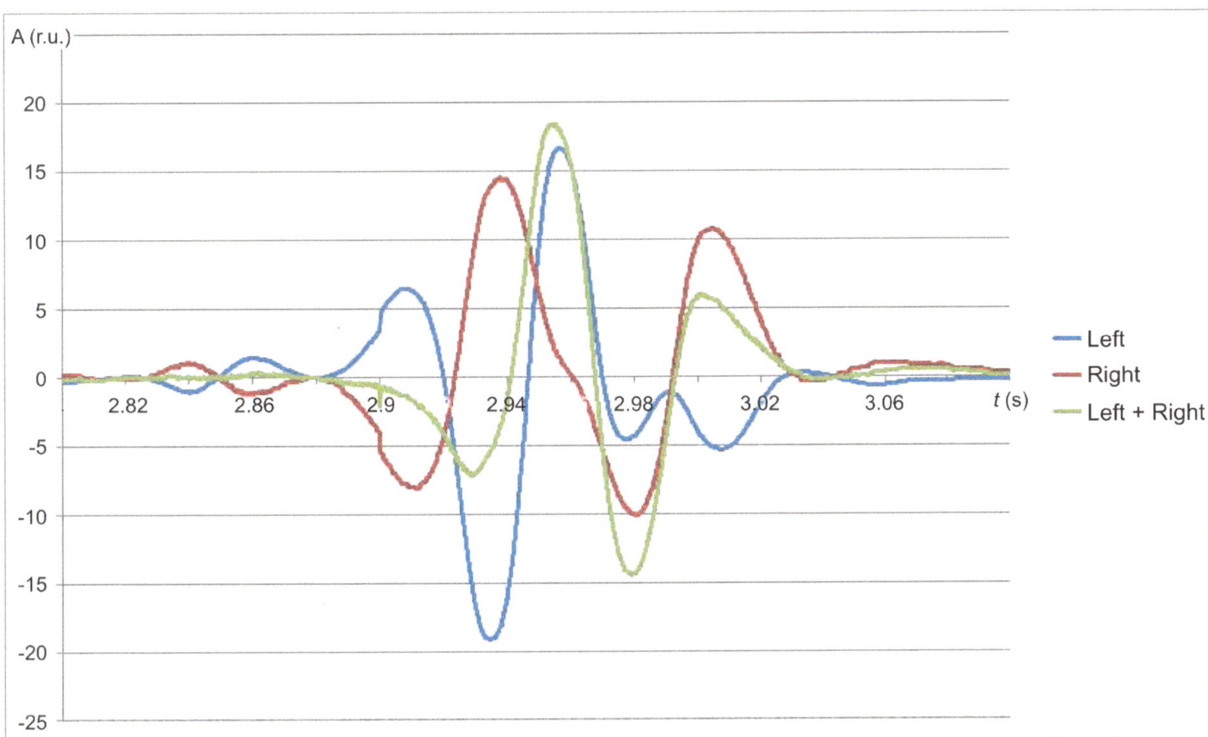

Figure 5.49. Comparison between scattered converted *Dps** waves on the left (blue) and right (red) split-spread sides, and their stack (green) obtained without a polarity reversal.

Thus, the numerical 2D-simulation of the wave responses from fluid-filled mesofracture zones in thin rigid beds and stacks of thin interbeds has helped us establish that direct detection of the mesofracture zones is possible using the horizontal component of the converted waves scattered at those zones. The numerical simulation has demonstrated that direct detection of fractured zones in the Bazhenov Formation is possible in principle, but further investigations are needed to assess the effects of real operational conditions and take into consideration all of the factors in applying the method. It also should be noted that near-horizontal reflecting boundaries, a thin low-velocity layer (weathering), and extensive permafrost areas in West Siberia create favorable conditions for both converted waves to be sufficiently strong and noise elimination.

Assessment of shadow effects produced by a mesofractured bed.—Abnormally high scattering of shear waves, when they propagate through a mesofractured bed, is characterized and analyzed in Section 5.2. In this section, we investigate the process of propagating compressional and shear waves through the mesofractured bed by generating waves with a plane-wavefront energy source, which simplifies the wavefield. However, the results, though significant, are rather hypothetical, as this method of energy generation is not used in seismic exploration practice.

The preceding analysis examines the possibility of identifying mesofracture zones under more realistic exploration conditions, with point sources and multifold coverage acquisition, and defines how converted scattered waves can be used for this purpose. The character and relative amplitude of upgoing scattered converted Dps^* and compressional Dpp^* waves generated by fractured beds have received considerable attention in our discussion. Therefore, it is worthwhile to assess the effects of a mesofractured medium multi-n converted and compressional waves passing through a fractured bed with parameters and in a setting typically encountered during seismic exploration.

The degree of shadow produced by mesofractured beds to various types of waves propagating from depth to receivers on the earth's surface is an important aspect of those effects.

To determine the significance of the shadow effect, we use a numerical model based on the base model from Figure 5.1, complemented with a second 100 m thick fractured bed positioned 600 m deeper than the first bed. The horizontal reflector, as in Figure 5.1, is 300 m deeper than the second bed. In this model, the thickness of the first bed of mesofractures is variable. Wavefields are computed for thicknesses of 0, 5, 10, 15, 20, 25, and 50 m. Results are presented as the *X*- and *Z*-components of common-source gathers on the earth's surface, with the energy source positioned in the middle of a 6 km seismic spread.

Figure 5.50 presents both components for the case when the first bed has no fractures. The scattered converted Dps_2^* wave from the second bed is shown to dominate the *X*-component data. The wave has asymmetry and a polarity reversal. The variation in the amplitudes of scattered converted Dps_2^* waves with thickness of mesofractured beds is shown in Figure 5.51. The *X*-components for 5, 15, and 50 m thicknesses of the first bed show an appreciable and gradual decrease in the amplitude of the scattered converted Dps_2^* wave from the second (lower) bed as the thickness of the first bed increases. This demonstrates how the shadow effect for the Dps_2^* wave grows as the thickness of the fractured bed increases. Note that the amplitude of the scattered converted Dps_1^* wave from the first bed barely changes and remains maximally high.

A corresponding set of the *Z*-components for the same thickness values (5, 15, and 50 m) of the first bed is shown in Figure 5.52. In this case, the character of the compressional waves Dpp_1^*, Dpp_2^*, and *PP* does not change significantly, which indicates that the mesofractured beds are transparent to compressional waves.

The data of all computations (except for $H = 50$ m in Table 5.17) are used in Figure 5.53 to demonstrate how the amplitudes of the compressional and converted waves scattered by the lower bed and also of the PP-reflection from the horizontal boundary depend on the thickness of the upper bed. For commonality, the amplitude estimates are given as the percentage of the amplitudes of respective waves, rather than in relative units, when the top fractured bed is absent and, therefore, no shadow effect is present. The decrease in relative amplitude is 35% to 50% for $H = 5$ m, 45% to 62% for $H = 10$ m, and 56% to 77% for $H = 20$ m. The dependence of the compressional waves on the thickness of the top bed is rather weak, and the decrease in their amplitudes is less than 15% to18%.

Figure 5.54 shows common-source stacks of the *X*-components of left sides, right sides, and both sides of split spreads with and without a polarity reversal. The stacks are obtained similarly to those shown in Figures 5.51 and 5.52, but with the use of two fractured beds, the first of which is 15 m thick and the second is zero m thick. The scattered converted wave Dps_2^* from the second mesofractured bed, although weaker than the Dps_1^* wave from the first bed, is fairly well identifiable on the complete stack with a polarity reversal (Figure 5.54c).

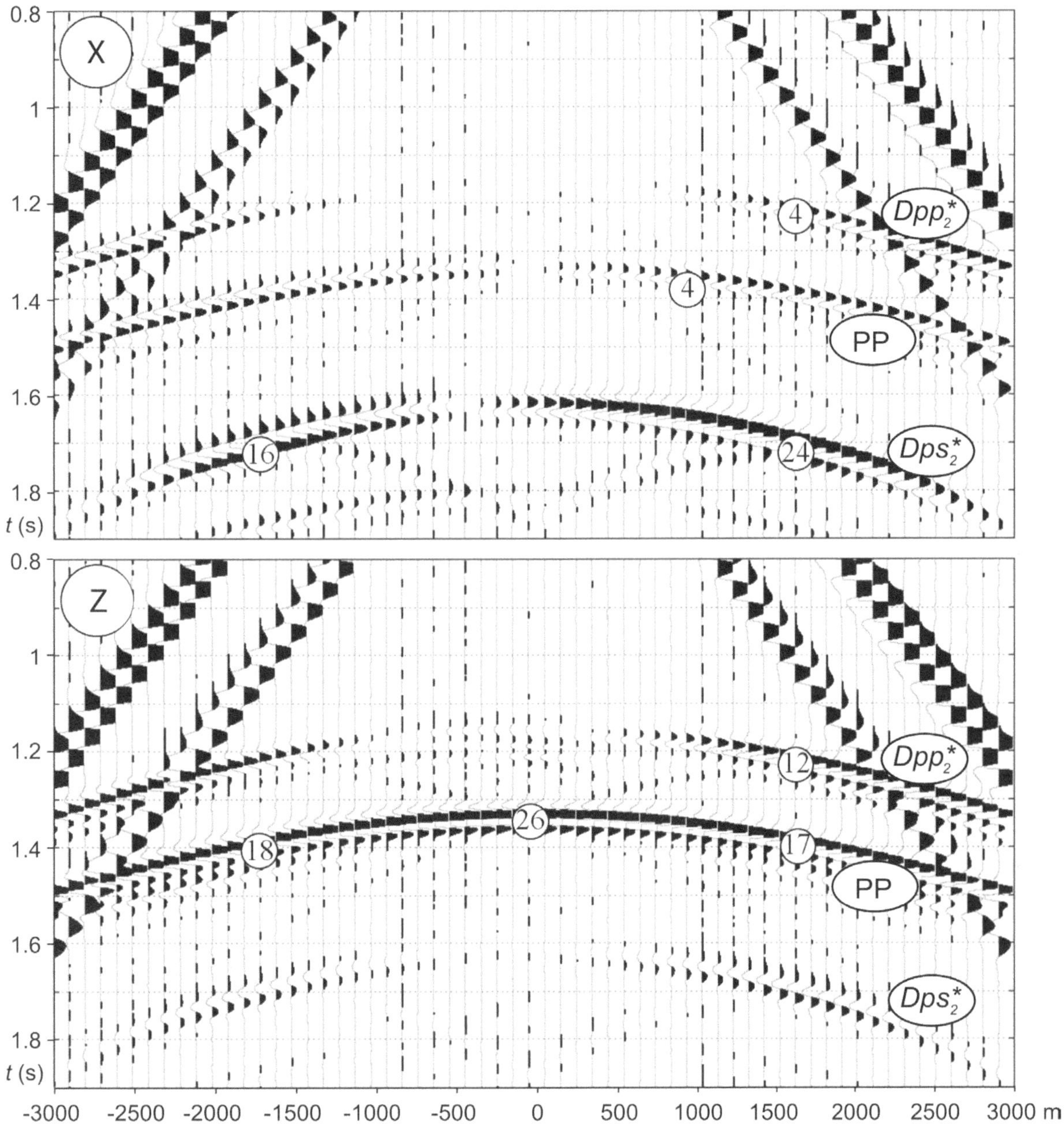

Figure 5.50. *X*-component (top) and *Z*-component (bottom) of a common-source gather for zero thickness of the top mesofracture bed.

From these findings we can conclude the following.

- With principal elements of modern seismic exploration such as point sources and multi-fold coverage acquisition, the great decrease in the energy of shear waves propagating in a medium with aligned mesofractures, discussed in Section 5.1, is confirmed. This phenomenon may be attributed to scattering of inter-fracture reflections.
- This effect is especially significant for reflected and scattered waves propagating from beneath through a mesofractured bed.
- The amplitudes of scattered and reflected compressional waves traveling through a mesofractured bed decrease approximately 5 times less than those of converted waves.
- The degree of shadow effect is determined by two factors: (1) the thickness of a fractured shadow bed and (2) the sign of a deviation of fractures from the vertical in the shadow bed. The effect of the first factor may vary between 0 and 100%, while the effect of the second factor may vary on different split-spread sides between 0 and 30%.

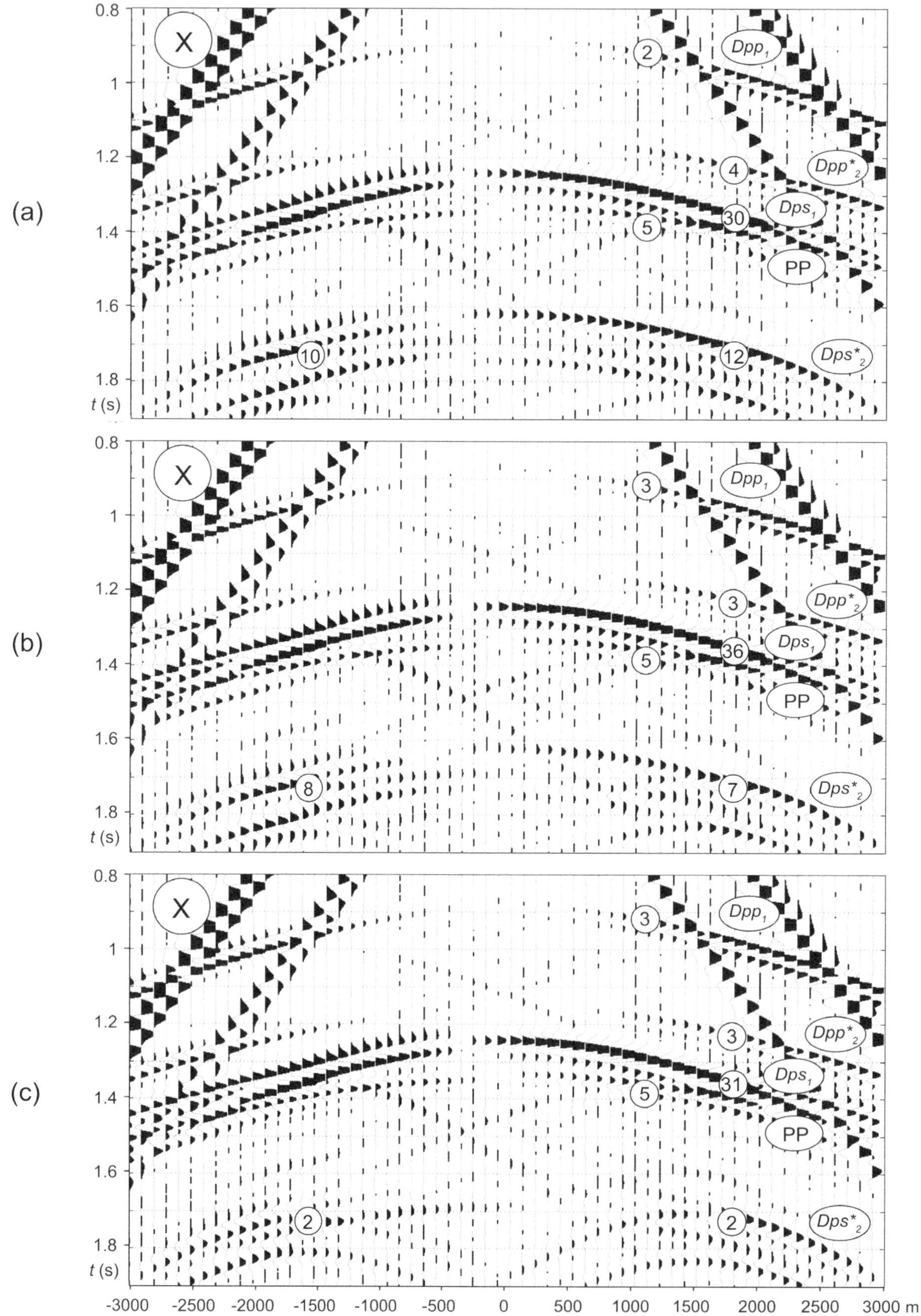

Figure 5.51. *X*-components of common-source gathers for (a) 5, (b) 15, and (c) 50 m thicknesses of the top mesofracture bed.

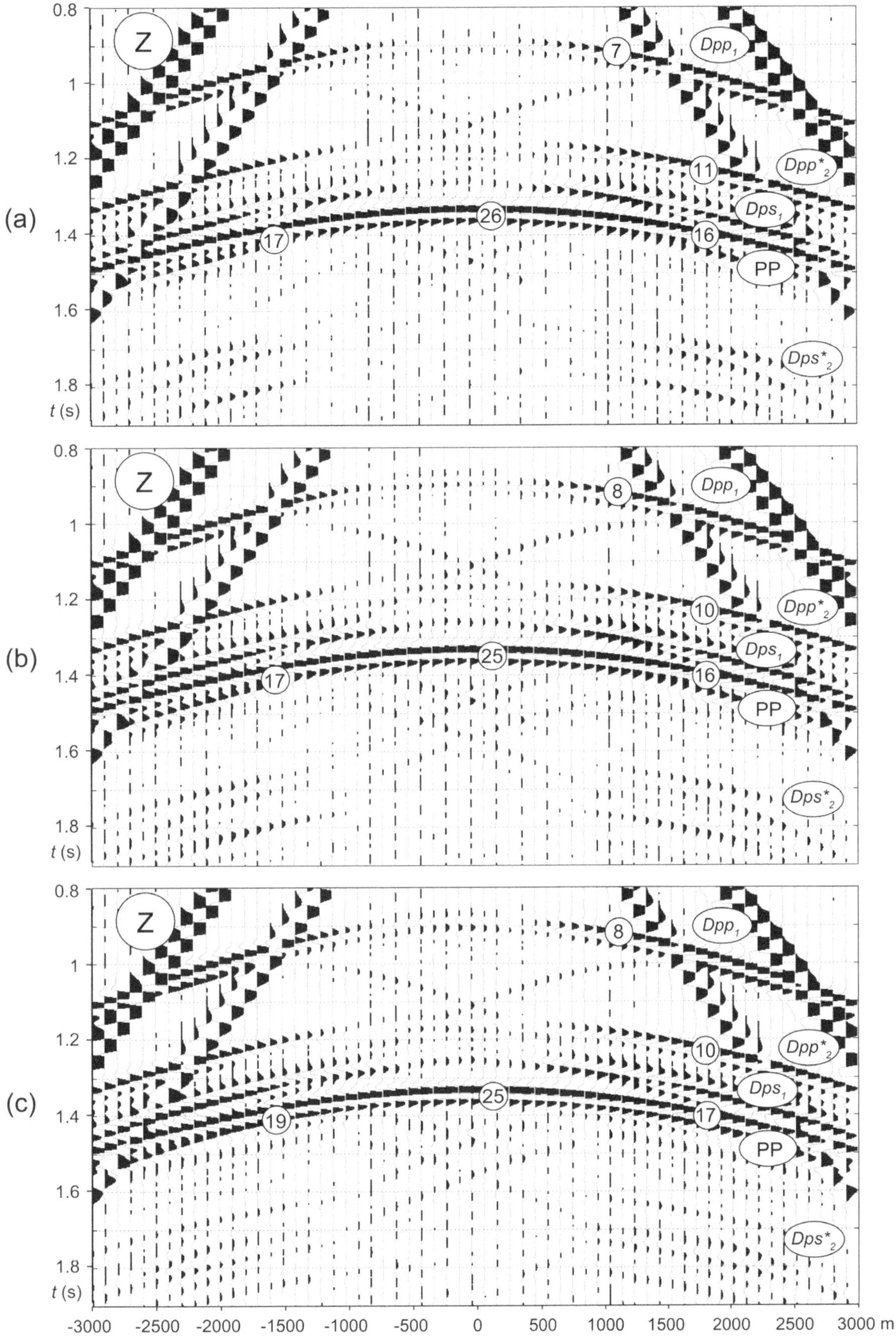

Figure 5.52. *Z*-components of common-source gathers for (a) 5, (b) 15, and (c) 50 m thicknesses of the top mesofracture bed.

Table 5.17. Maximum amplitudes (in relative units) of all wave types for all computations.

H, m	*X*-component					*Z*-component			
	Dpp^*_2 (1.16 s, 1700 m)	Dps^*_1 (1.24 s, 1700 m)	PP (1.33 s, 1000 m)	Dps^*_2 (1.62 s, 1700 m)	Dps^*_2 (1.62 s, –1700 m)	Dpp_2 (1.16 s, 1700 m)	PP (1.33 s, 1700 m)	PP (1.33 s, 0 m)	PP (1.33 s, –1700 m)
0	4.3	5.6	4.1	24.1	16.1	11.6	16.9	26	18
5	3.8	29.9	5.3	12.3	10.3	10.6	16.1	25.8	16.8
10	3.6	34.2	5.1	9.1	9.1	9.8	15.7	25.5	16.7
15	3.4	35.7	4.8	7.4	8.1	9.5	15..7	25.1	16.8
20	3.5	35.2	4.8	5.9	7.1	9.6	16	25.3	1.9
25	3.4	34.5	4.8	4.7	5.9	9.4	16.3	25.2	17.1
50	3.4	31.2	4.8	1.7	2.3	9.4	16.9	25	18.4

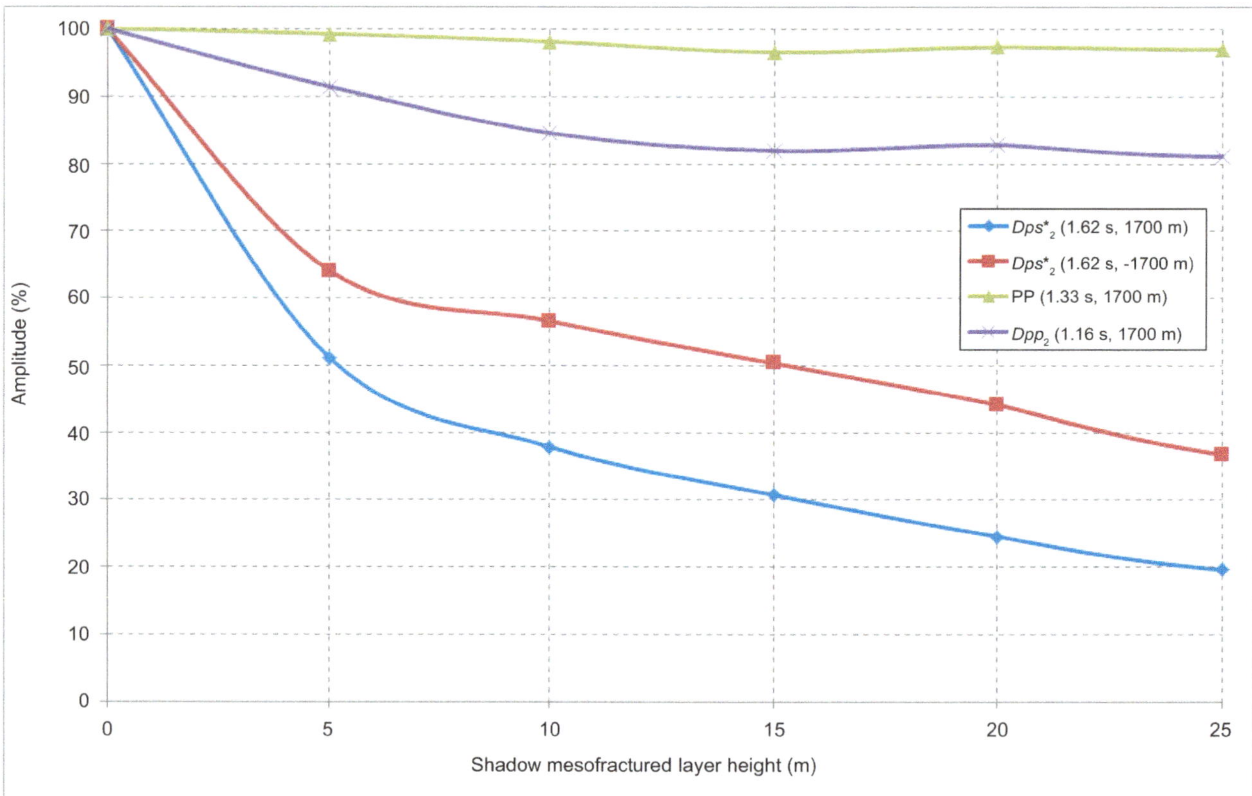

Figure 5.53. Amplitudes of different waves as a function of the thickness of the top mesofractured bed.

5.3.2.3. Conclusions from Subsection 5.3.2

Features of the *X*-component of scattered converted waves from fractured reservoir zones, established in Subsection 5.2.2, such as pronounced amplitudes and asymmetry relative to the source location, are proposed as recognition parameters for these waves.

Seismic data acquisition and processing must meet the following requirements.

1) Split-spread and horizontal component data acquisition should be used.
2) Offsets should not exceed the depth of the target prospect.

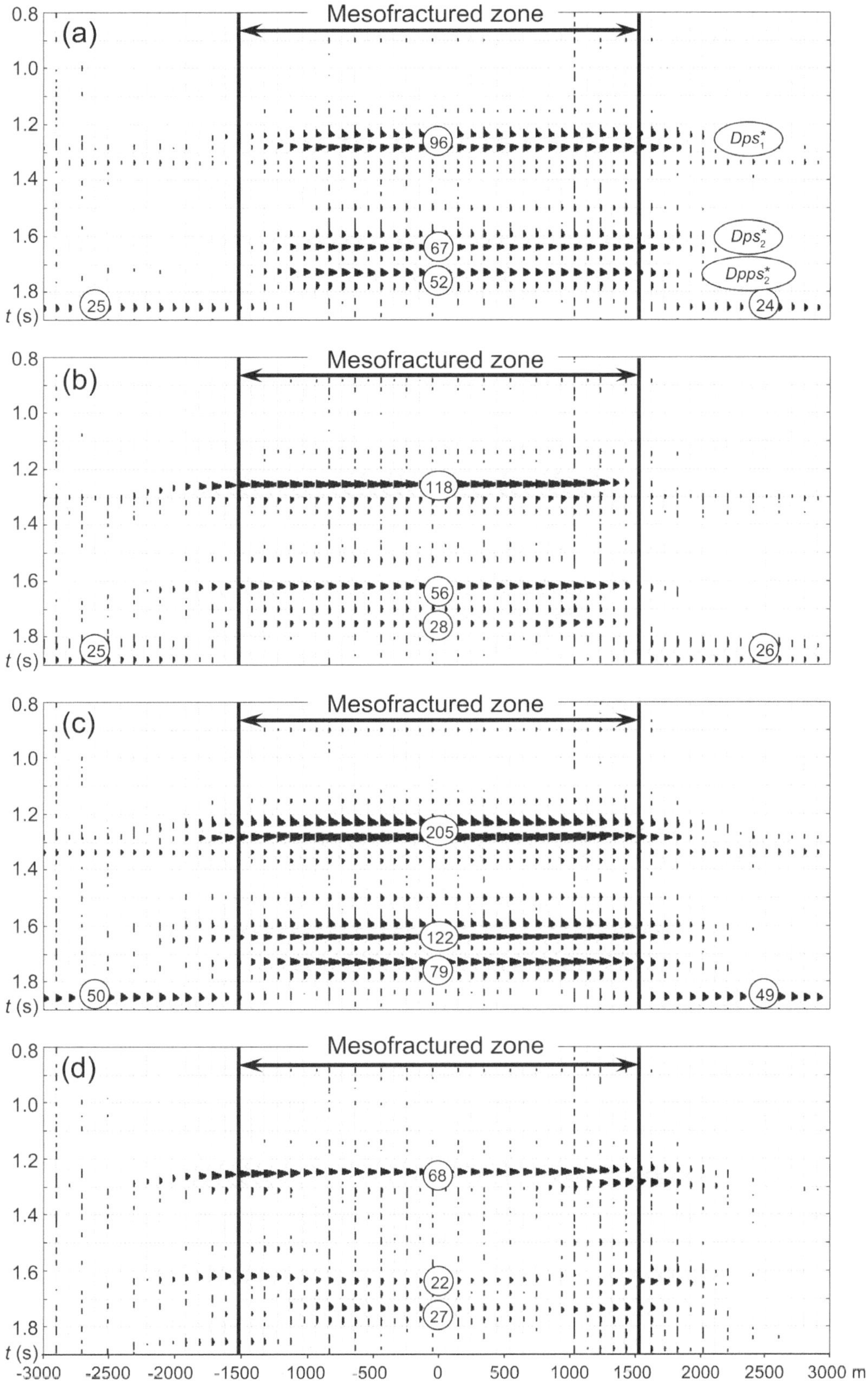

Figure 5.54. Common-source stacks of the *X*-components of (a) left sides, (b) right sides, and both sides of split spreads (c) with and (d) without a polarity reversal. Common-source gathers are computed for a 15 m thick fracture zone embedded in a tight medium. The vertical lines denote the zone boundaries.

3) NMO correction should be performed optimally for converted waves along with an effective mute, while the stacking process should be performed on different split-spread sides separately, to allow for a polarity reversal.
4) Complete stacks of split-spread data should be obtained using two methods, i.e.,
    - by stacking the data from both split-spread sides with a polarity reversal, which is conventional when processing converted waves (on average, this procedure doubles the amplitude of a target scattered converted wave relative to those of other waves), and
    - by stacking the data from both split-spread sides without a polarity reversal. (This considerably reduces, sometimes to zero, types of noise such as converted waves and the X-components of compressional waves reflected from plane boundaries. At the same time, the energy of the difference between the scattered converted waves from different split-spread sides may be rather high. This effect is attributed to the asymmetry of the scattered wave caused by a deviation of fractures from the vertical, which is usually small).

Test examples for two types of carbonate formations and the Bazhenov Formation in West Siberia confirm the feasibility in principle of using the X-components of scattered converted waves for detecting and mapping bed-bounded zones of mesofractured reservoirs.

Application of this technology to synthetic data to determine the possibility of mapping thin bed-bounded mesofracture zones in the fine-layered Bazhenov Formation shows:

- the originally low level of the scattered converted wave from fractures in the rigid interlayers within the Bazhenov Formation can be raised to be comparable with that of the X-component of the PP-reflection from the formation top;
- a significant difference between wavefields on different split-spread sides is created both by a deviation of fractures from the vertical and an unstable nature of the reflectivity of the fine-layered Bazhenov Formation; and
- a scattered converted wave from a fractured zone can be identified reliably via stacking without a polarity correction applied, due to a considerable difference between the scattered waves on different spread sides and the almost complete suppression of other waves.

In mapping fractured reservoirs with the use of scattered converted waves, the shadow effect of fractured beds or zones above the target interval, if any, should be taken into consideration.

5.4. Conclusions from Chapter 5

Using 2D numerical simulation of wave responses from fractured reservoir beds and bed-bounded zones of fluid-filled mesofractures, we have established that the direct use of scattered compressional and converted waves is overall a promising method for detecting and mapping these geologic features.

In particular, we have established the following.

1) When a compressional plane wave encounters a bed or bed-bounded zone of near-vertical fluid-filled mesofractures, this geologic feature produces a response containing both scattered compressional and scattered converted wavefronts. They are formed by interfering individual diffractions from mesofractures in top rows and are kinematically similar to respective hypothetical waves reflected and converted by the top boundary of the bed.
2) When initiated by a plane wavefront, the scattered converted wave on the X-component is strongest relative to the others. Its one-pulse signal is almost independent of the bed thickness because it is formed only by the top rows of mesofractures. This is due to anomalously high attenuation of shear motions in scattered waves propagating through the fractured bed.
3) When a spherical wavefront, generated by a point source, impinges on a bed-bounded zone of near-vertical fluid-saturated mesofractures that zone produces a response of fronts of scattered compressional and converted waves. The wavefield of these wavefronts corresponds, according to kinematic indicators, to quasi-reflected compressional and converted waves emanating from the boundaries coincident with the top of the mesofractured bed.
4) When reflecting boundaries are present at the top and base of a fractured bed or bed-bounded zone, the compressional and converted waves reflected from the bed top overlap respective scattered waves, with due regard for phase relationships.
5) In zones of their maximum values, scattered compressional waves show relatively high amplitudes, which are comparable to those of reflections from smooth boundaries with a rather high reflectivity (e.g., 22%) and are several times higher than those caused by a weak reflectivity (e.g., 5% to 10%).

6) The *Z*-component of a scattered compressional wave, when generated with a point source and recorded at rather long offsets, shows a considerable AVO gradient for fluid-filled fractures. This property manifests itself in an extensive zone of severely reduced amplitudes in the vicinity of the source location, the amplitudes rising sharply away from the source. This character of the AVO curve for the scattered compressional wave differs sharply from that of the amplitude profiles of most reflections.
7) Reflections from boundaries with a positive change in impedance and a minor change in *Vp*/*Vs* ratio have maximum amplitudes in the vicinity of the source location and, therefore, a negative AVO gradient. Generally, according to typical relationships between the *Vp*/*Vs* ratio and impedance, the boundaries in tight rocks at depths relevant to oil exploration have very low changes in *Vp*/*Vs* ratio and impedance, of any signs. This entails slight variations in the AVO curve and, therefore, small values of the AVO gradient of either sign.
8) These effects are the basis for detecting the wave response from a fluid-filled fractured reservoir, using the following techniques.
   - By comparing the values and signs of the AVO gradient for waves being analyzed. For the scattered wave from a fluid-filled fractured bed, the gradient is positive and large in value, whereas it is negative for the reflection from a boundary with a positive change in the impedance and a minor change in Poisson's ratio or *Vp*/*Vs* ratio. In a general case, as mentioned, the gradient is negligible and of any sign.
   - By comparing the amplitude ratios of far-offset and near-offset partial stacks. For reflections without fracturing, the ratio is about unity, whereas it is much higher than unity for scattered waves.
9) On the *X*-component, the scattered converted wave from fractured features is strongest at midrange offsets that are less than the depth of the target features. The wave has these additional characteristics:
   - different structure and amplitude on opposite sides of the receiver spread;
   - a horizontal shift of the zero value of the AVO curve, which is related to a deviation of fractures from the vertical, even if it is minor (e.g., 5°);
   - a polarity reversal at the source location; and
   - appreciable attenuation when it propagates through a fractured medium.
10) The scattered compressional wave and its key characteristics (amplitude and asymmetry) are rather insensitive to variations of the fractured bed parameters in their realistic ranges.
11) The high amplitude of the *X*-component of the scattered converted wave at midrange offsets and the difference in the wavefield between the two sides of a spread are diagnostic indicators of a mesofractured bed-bounded zone. In combination with processing intended for converted reflections, specifically full stacking without a polarity reversal, they present a third independent method for detecting and mapping fractured reservoirs. Unlike the first two, described in (7), this method requires 3C data.
12) The numerical simulations have demonstrated the theoretical basis for the possible detection and mapping of fractured fluid-filled reservoirs and zones using waves scattered by them. For this purpose, multifold coverage acquisition, CMP processing, and methods for identifying scattered compressional and scattered converted waves, as described in (7) and (10), should be employed.

Chapter 6

Responses from Macrofractures and Macrofracture Systems: 2D modeling

According to the classification of MacBeth (1995) and Kozlov (2006), macrofractures are characterized by openings (apertures) of a few fractions of a millimeter to a few millimeters, heights of a few tens to hundreds of meters, and widely varying lengths. Their aspect ratio is prevailingly 1/100,000.

This class of fractures is widespread in tight carbonates and sandstones. As a rule, they do not extend beyond one bed.

Predominantly, macrofractures are encountered as systems of near-vertical unidirectional fractures caused by tectonic, frequently shear stresses (Timurziev, 2010). They are possible fluid conduits with a high vertical permeability, forming networks that may provide the basis for filtration models of hydrocarbon reservoirs. The considerable interest shown in the wave characteristics of macrofractures lies in their potential for detecting such fluid conduits.

6.1. Characteristics of seismic responses from a single near-vertical macrofracture

There were few studies of the seismic responses from macrofractures prior to the advent of methods for numerical simulation. For instance, Schoenberg (1980) proposes an analytical solution for reflection coefficients of infinitely long (high) fractures.

For macrofractures within actual oil reservoirs, where macrofractures have finite heights and lengths, analytical solutions for seismic responses are either approximate or specific (for SH-waves only), or are based on an unrealistic elliptical fracture form (Chen et al., 2010).

The properties of a wave response from a single near-vertical macrofracture of finite length comparable to a seismic wavelength are described more fully by Fang et al. (2010), who use numerical finite-difference simulation for a linear-slip fracture model (Hsu and Schoenberg, 1993). The limitations in this work are as follows.

a) Only *Z*-component data, i.e., only compressional waves, are analyzed.
b) Wave responses are considered as a cloud of scattered energy, rather than as a set of specific waves.
c) Using a finite-difference numerical simulation (Coates and Schoenberg, 1995) along with the application of an empirical coefficient such as "rock compliance" (Thomsen, 1986) and the linear-slip hypothesis (Hsu and Schoenberg, 1993) does not produce a strict solution in terms of continuum mechanics.

The grid-characteristic method is free from these drawbacks and allows us to investigate wave responses from macrofractures with a range of parameters and different properties of the host medium.

The modeling data in Chapter 4 for the seismic response from a fracture, depending on its inclination (see Figure 6.1, which contains a fragment from Figure 4.2), is evidence that only diffractions from the tips of near-vertical macrofractures carry information to the earth's surface. This indicates the need for a detailed

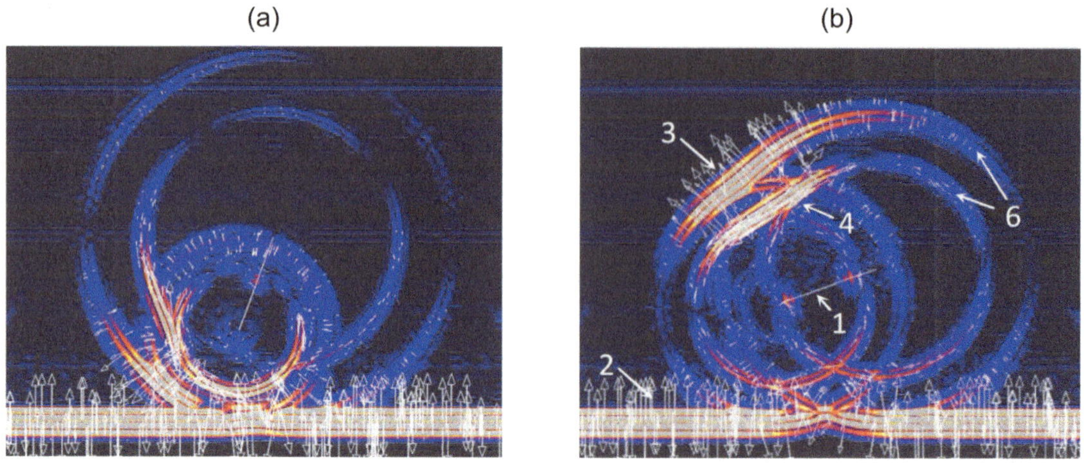

Figure 6.1. Snapshots of wave responses from macrofractures with different angles of deviation from the vertical: (a) 20° and (b) 70°. The macrofracture length (height) is 200 m. In (b): 1 = macrofracture; 2 = incident compressional wavefront; 3 = compressional reflection; 4 = converted PS-wave; and 6 = wavefronts of compressional diffracted and converted diffracted waves.

investigation of diffractions from near-vertical macrofractures in different aspects and conditions of use.

6.1.1. Composition and characteristics of the seismic response from near-vertical macrofractures in a half-space, using point and plane-wavefront sources

Numerical simulation of seismic wave propagation is performed on subsurface models containing a near-vertical fracture embedded in a half-space or a two-layer medium. The model parameters are as follows.

- The integration domain is a rectangle 4000 m wide by 3000 m deep.
- The center of the macrofracture is positioned at a depth of 2000 m in a host medium with the parameters $Vp = 4500$ m/s, $Vs = 2500$ m/s, and $\rho = 2.5$ g/cm^3.
- The wavelet frequency is 30 Hz.
- Minimal fracture thicknesses, in reality ranging from a few fractions of a millimeter to a few millimeters, are modeled using special contact conditions.
- The fracture is filled with gas or fluid.
- Point source and plane-wavefront sources are used.

The results are presented as a series of snapshots for fixed times and as common-source gathers recorded on the earth's surface.

Figure 6.2 shows the *X*- and *Z*-components of common-source gathers for a gas-filled macrofracture 40 m high. They are obtained with a point source located at 0, 500, 1000, and 2000 m from the projection point of the vertical macrofracture on the earth's surface. When the energy source coincides with the projection point, no waves associated with the fracture are observed. The absence of any response is explained by the directivity pattern of diffraction when the compressional wavefront propagates along the vertical fracture (Karaev et al., 2015). On the remaining source gathers, the position of hyperbolic traveltime minima does not change with the increase in source offset, which indicates that the recorded waves are diffractions.

Figure 6.3 shows the *X*-component and *Z*-component of a source gather and a snapshot at 0.2 s for a point source and a single fluid-filled macrofracture. The first wave to arrive is a compressional *Dpp* wave. This is evident from the particle motion along the direction of propagation of its wavefront (Figure 6.3b).

The subsequent events are converted diffracted *Dps* and *Dsp* waves with shear particle motions[1] (Figure 6.3b), which form an interference group. They are close in arrival times at the point of the traveltime minimum and different in the traveltime slope. The traveltime slope of *Dsp* is the same as that of *Dpp*, while its traveltime is dependent on the source offset from the fracture projection. The *Dps* traveltime slope is much steeper than that of the *Dpp* wave, but its time delay is constant relative to

[1] In the experiment, the surface source generates both compressional and shear wavefronts.

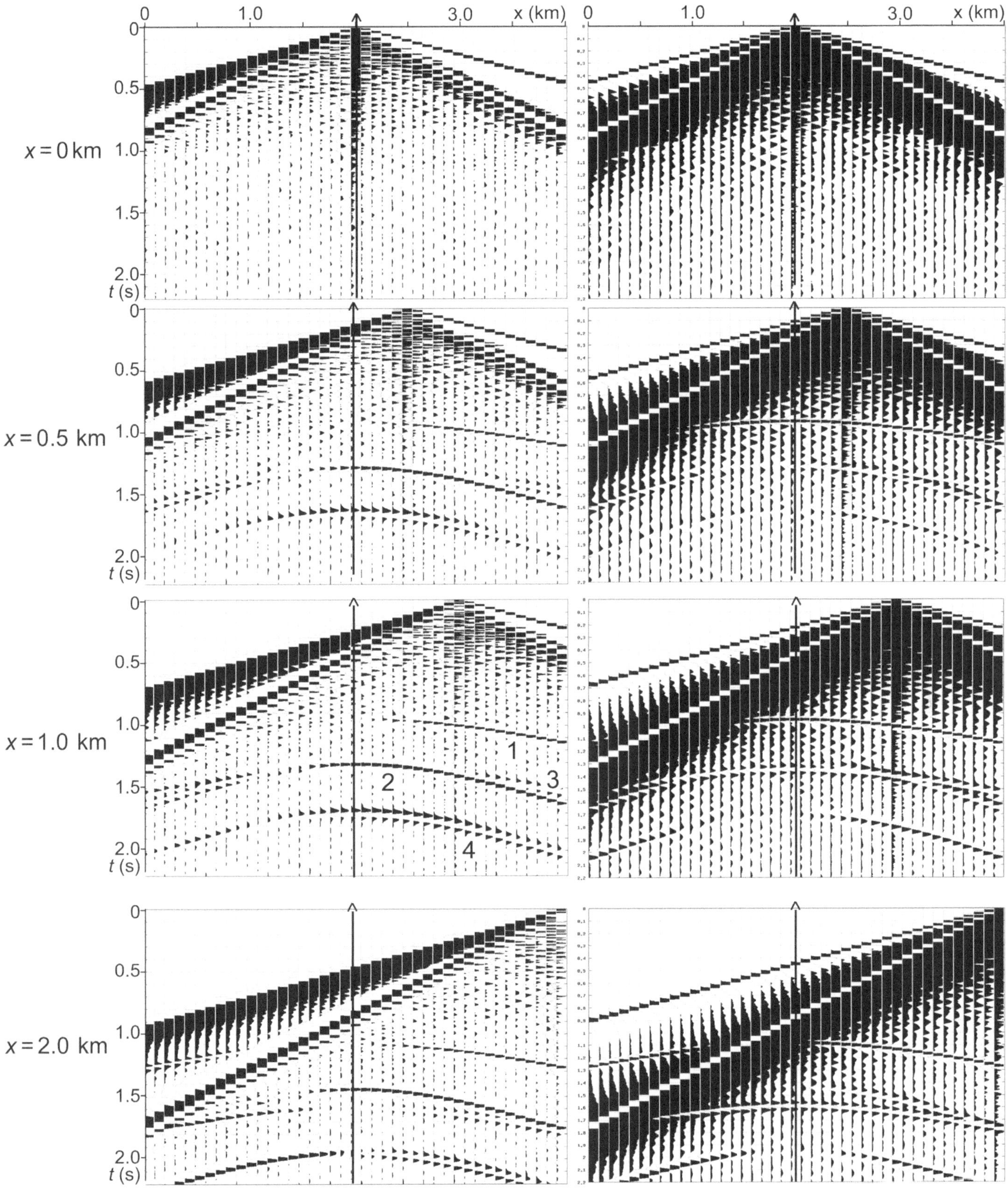

Figure 6.2. *X*-components (left column) and *Z*-components (right column) of common-source gathers for a gas-filled macrofracture with different source offsets *x* from the fracture projection on the earth's surface: ^ = macrofracture projection; 1 = compressional *Dpp* diffraction; 2 = converted *Dps* diffraction; 3 = converted *Dsp* diffraction; and 4 = shear *Dss* diffraction.

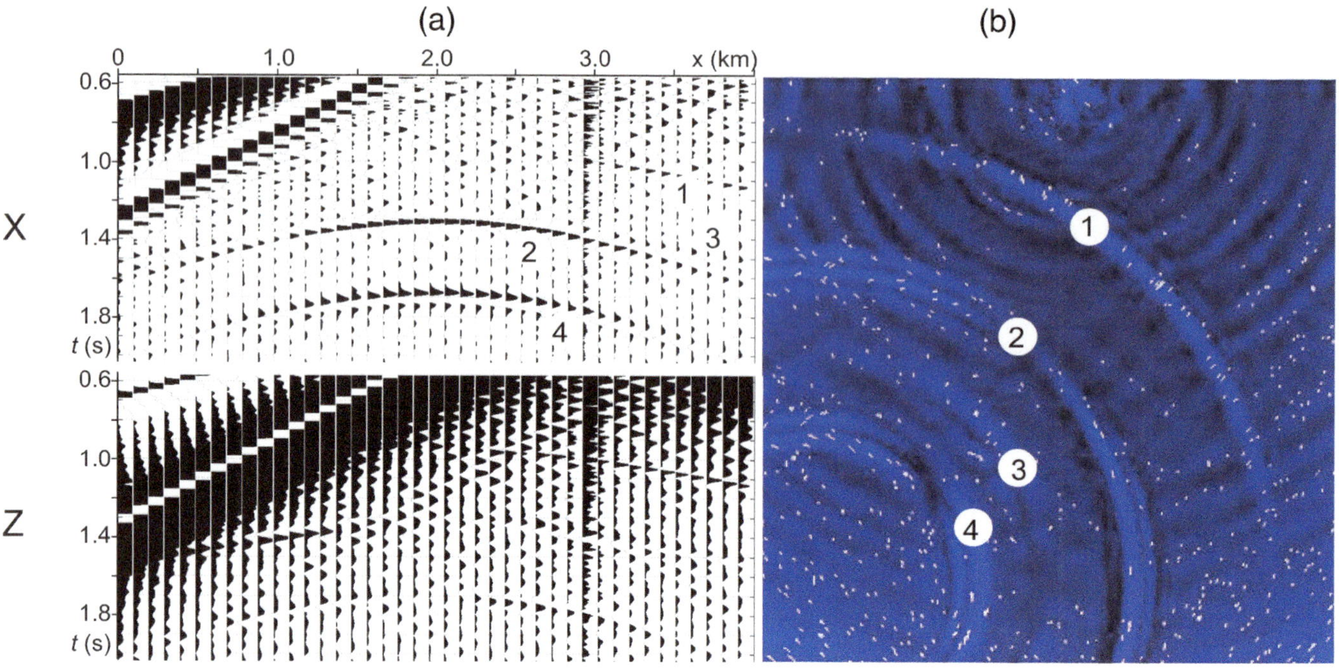

Figure 6.3. (a) X-component (top) and Z-component (bottom) of a common-source gather for a near-vertical macrofracture and a point source located 1 km away from the fracture projection. (b) Snapshot at 0.2 s: 1 = compressional *Dpp* diffraction; 2 = converted *Dps* diffraction; 3 = converted *Dsp* diffraction; and 4 = shear *Dss* diffraction.

Dpp. The last wave to arrive is a diffracted shear *Dss* wave, and its slope is the same as that of the *Dps* wave.

On the X-component (Figure 6.3a top), the converted *Dps* diffraction and shear *Dss* diffraction have no polarity reversal and have a considerably increased amplitude as compared with that of the compressional *Dpp* diffraction, especially over the fracture zone. The compressional *Dpp* wave amplitude falls in this zone almost to zero, whereas the *Dps* wave amplitude rises to its maximum.

To characterize the waves in the response from a near-vertical macrofracture comprehensively, we need better identification of the waves, which is achieved using a plane- wavefront source of compressional waves. Near-surface waves, normally observed when using a point source, are absent in this case, and we record the response from the macrofracture without any noise. Figure 6.4 shows both components of fragments of common-source gathers containing diffractions from a macrofracture in a half-space. The source gathers shown in Figure 6.4a are obtained for a 40 m high macrofracture, which is appreciably less than the signal length[2] of the compressional wave impinging on it. The source gathers shown in Figure 6.4b are obtained for a 200 m high macrofracture, which is greater than the signal length. The top line (I) corresponds to the gas-filled fractures; the bottom line (II) represents fluid-filled fractures.

From our analysis of the seismograms, we note the following.

1) The macrofracture that is shorter than the signal length (Figure 6.4a) appears as a single source of diffraction forming two types of diffraction, i.e., compressional *Dpp* and converted *Dps*.
2) The macrofracture whose height is comparable to or greater than the signal length (Figure 6.4b) generates four diffractions: two (compressional *Dpp*1 and converted *Dps*1) from the upper tip and two (*Dpp*2 and *Dps*2) from the lower tip of the macrofracture. Each of the wave types has a complex signal due to the overlapping two wavelets having opposite signs. The resolution of these wavelets depends on the delay of the arrival time from the lower tip of the macrofracture.

When the macrofractures are filled with fluid, the Z-components of both *Dpp* and *Dps* diffractions have a polarity reversal over the macrofractures. The X-component of the compressional *Dpp* diffraction has an extensive zone of low amplitudes over the fracture, whereas the converted *Dps* diffraction has relatively high amplitudes and no polarity reversal. This result is unexpected and

[2] The commonly accepted term "wavelength" is the product of dominant wave period and velocity, which corresponds to a single-pulse wave. For other cases (two- to three-pulse signals), we propose the term "seismic signal length" as the product of the signal duration in time and velocity.

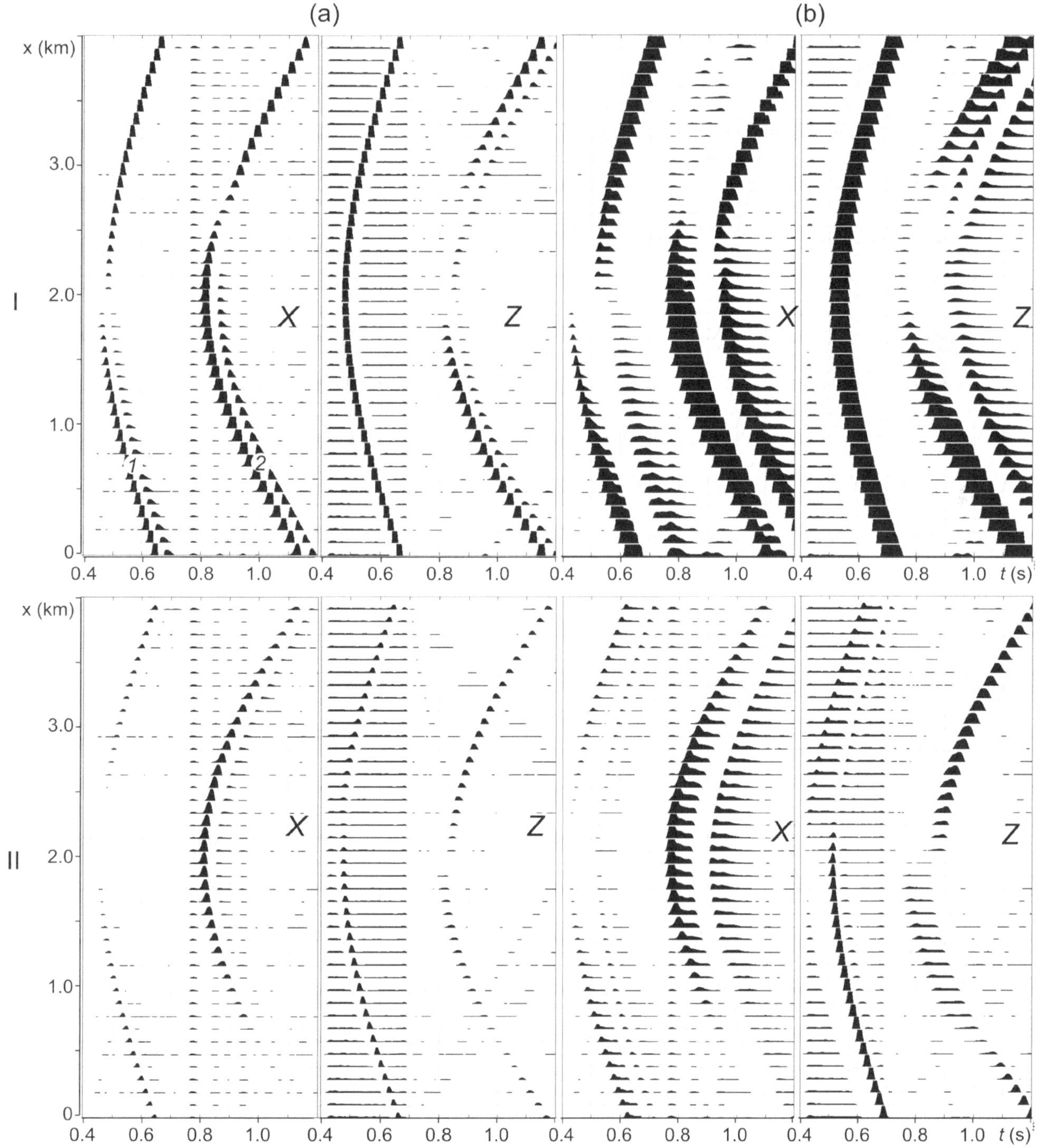

Figure 6.4. *X*- and *Z*-components of fragments of common-source gathers containing diffractions from single macrofractures of different heights: (a) 40 m and (b) 200 m. The top row (I) corresponds to the gas-filled fractures and the bottom row (II) to fluid-filled fractures. The wavefields are generated by a plane-wavefront source. The fractures deviate by 10° from the vertical. On each gather, the first wave is a compressional *Dpp* diffraction and the second wave is a converted *Dps* diffraction.

very different from commonly accepted results of finite-difference modeling.

When the macrofractures are filled with gas, the *Z*-components of both the compressional *Dpp* wave and converted *Dps* wave have no polarity reversal, although the latter has a zone of low amplitudes over the fractures. On the *X*-component, both diffractions are complicated by a polarity reversal over the macrofracture, which, unlike the preceding result, is quite consistent with the commonly accepted understanding.

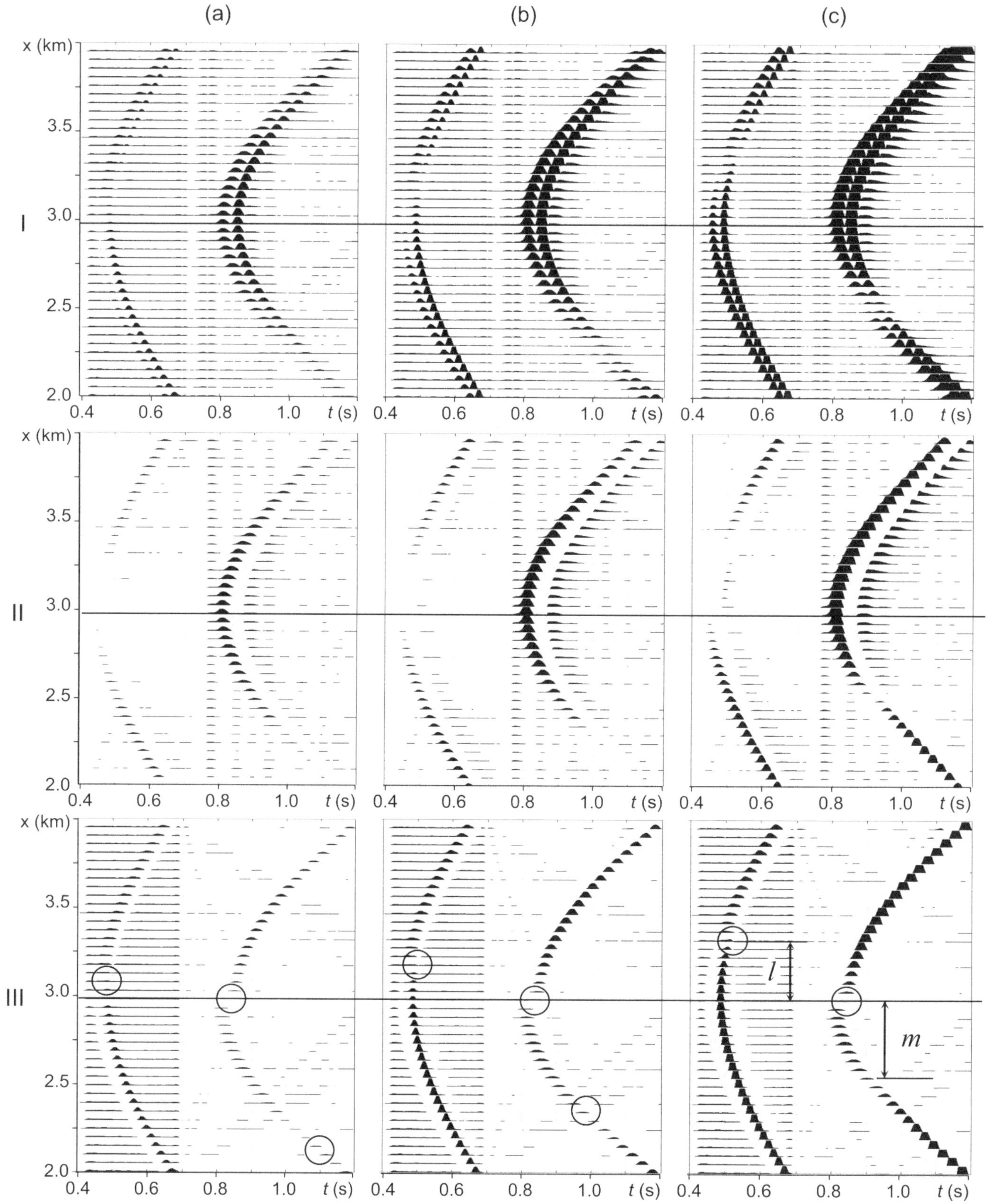

Figure 6.9. Common-source gathers showing the effects of different macrofracture deviations, (a) 5°, (b) 10°, and (c) 20°, on the character of diffractions: I = absolute values; II = X-component; and III = Z-component. Circles indicate polarity reversals.

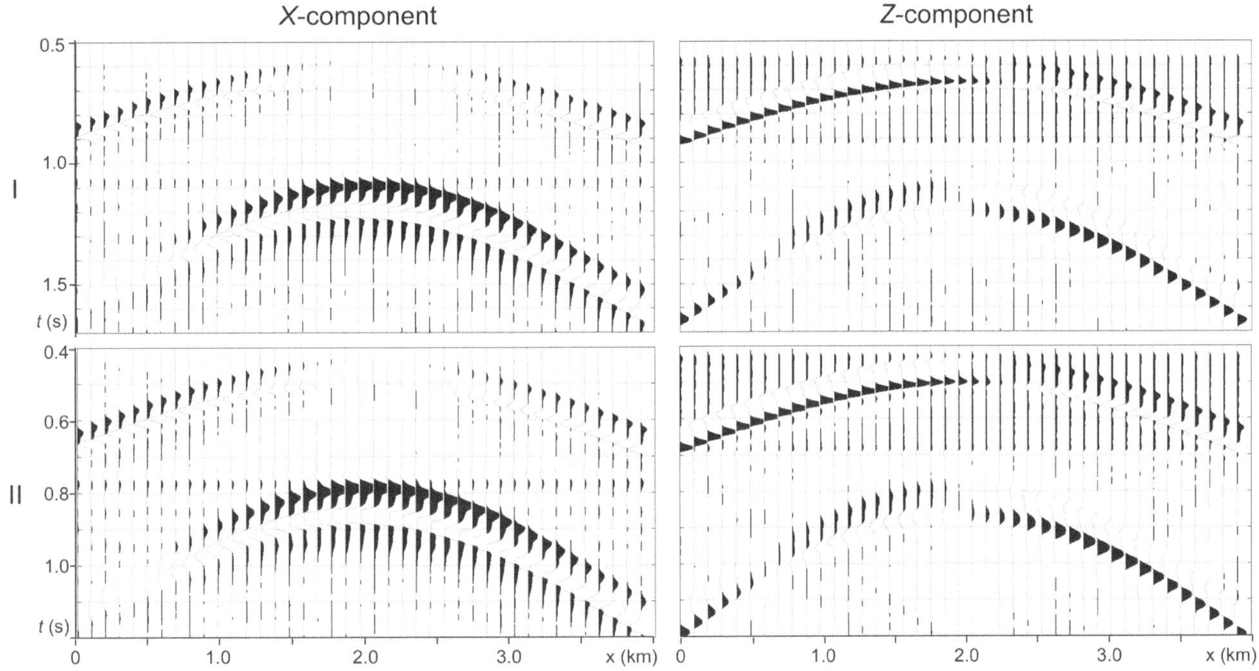

Figure 6.10. *X*- and *Z*-components of common-source gathers for a 140 m high macrofracture in different types of host rocks: I = siliciclastic rocks with $Vp = 3370$ m/s, $Vs = 1800$ m/s, and $\rho = 2.3$ g/cm^3; and II = carbonate rocks with $Vp = 4500$ m/s, $Vs = 2500$ m/s, and $\rho = 2.5$ g/cm^3. The energy source is a plane wavefront.

to zero in that zone, while the amplitude of converted *Dps* diffraction rises 2 to 3 times relative to the maximum amplitude of the compressional *Dpp* diffraction on the *Z*-component.

3) For a gas-filled macrofracture, both compressional and converted diffractions on the *Z*-component have no polarity reversal. The amplitudes of the converted diffraction, however, are greatly reduced over the macrofracture. On the *X*-component, both diffractions undergo a polarity reversal.
4) When the macrofracture deviation from the vertical is 5° or greater, the kinematic and dynamic characteristics of the diffractions show asymmetry on the split-spread sides.
5) When the height of a near-vertical macrofracture exceeds the signal length, two pairs of compressional and converted diffractions associated with the upper and lower tips of the fracture are observed.
6) If there is a reflector beneath the macrofracture, two pairs of secondary diffractions are generated: compressional and converted diffractions from both the upper and lower tips of the fracture. They are formed by a reflected wave passing through the fracture from below.
7) The type of host rocks has little effect on the compressional and converted diffractions from near-vertical macrofractures.

6.2. The seismic response from a system (cluster) of near-vertical uniformly oriented macrofractures

Macrofracture systems are, as a rule, associated with a certain formation or a section interval. In some cases, this interval is located in sufficiently homogeneous host rocks. In other cases, the macrofractures may be associated with a bed whose physical characteristics differ from those of the host rocks. The former type may be referred to as a macrofracture cluster in a homogeneous half-space and the latter as a cluster in a fractured bed with sufficiently contrasting boundaries coincident with the upper and lower tips of macrofractures.

6.2.1. Composition of the wave response from a macrofracture cluster using a plane-wavefront energy source

Wavefields are computed for the models shown in Figure 6.11. The half-space model is used to examine the response from a macrofracture cluster in its "pure" form, when no reflectors occur at the top and base of the fractured bed. The effects of reflections from the top and base boundaries of the fractured bed on identifying the response are discussed next.

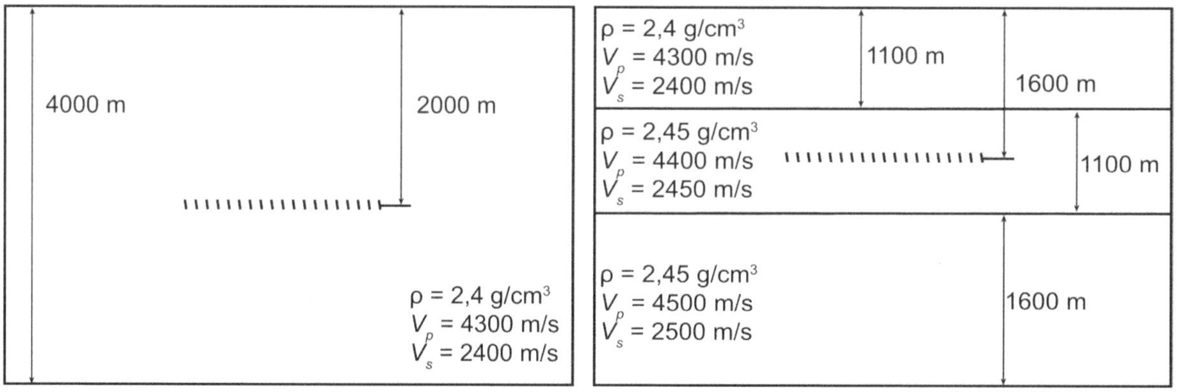

Figure 6.11. Half-space (left) and layered 2D (right) models used for computing wave responses from macrofracture systems.

Figure 6.12 shows common-source gathers for a single macrofracture and two macrofracture clusters that are 1000 and 2000 m wide and filled with fluid. The following features are evident.

1) The source gather for a single macrofracture illustrates the advantage of the *X*-component of the converted diffractions over the *Z*-component of the compressional diffractions.
2) On the *Z*-component, the scattered compressional wavefronts are barely discernible against the background noise.
3) On the *X*-component, the level of useful signals—responses from the clusters—is substantially higher. Here, at the time of compressional *Dpp* diffraction, there are two faintly discernible steep hyperbolic events from the cluster edges. At the arrival time of converted *Dps* diffractions, a wavefront consisting of several horizontal events is observed over the cluster, of which the first event is strongest. The horizontal portions of the first cycles approximately correspond to the position and width of the cluster. Outside the cluster boundaries, the signals attenuate abruptly and turn into hyperbolic slopes, which are diffracted waves from the cluster edges. The wavefronts are somewhat displaced relative to the clusters because of a 5° deviation of the fractures from the vertical. The amplitudes of the scattered converted diffraction wavefronts is 5 to 7 times higher than those of all other waves (Table 6.3).

Figure 6.13 shows common-source gathers for a single macrofracture and two macrofracture clusters that are 1000 and 2000 m wide and filled with gas. The following features are evident.

1) On the *Z*-component (Figure 6.13), the compressional *Dpp* diffraction from the single macrofracture has no polarity reversal and forms a multicycle wavetrain of five or six horizontal events. The amplitude of the events is at a maximum on the first cycle and gradually decreases toward the last cycle. The length of the horizontal portions of the events corresponds roughly to the width of the cluster. Beyond the cluster boundaries, the horizontal portions become hyperbola flanks and abruptly lose their amplitude strength. Practically no horizontal events of converted *Dps* diffraction are observed over the macrofracture cluster, although weak hyperbolic events—diffractions from the edge fractures of the cluster—are visible beyond that interval.
2) On the *X*-component (see Figure 6.13), over the single macrofracture, the compressional *Dpp* diffraction has a zone of significantly reduced amplitudes (down to zero), while the converted *Dps* diffraction undergoes a polarity reversal. At the arrival time of a compressional *Dpp* diffraction, no horizontal events are present over the cluster. Although two weak hyperbolic events—diffractions from the edge fractures of the cluster—are visible beyond that interval. On the converted *Dps* diffraction from the single fracture, a polarity reversal is observed. At the arrival time corresponding to converted *Dps* diffractions, the clusters produce a weak wavefront of a few near-horizontal events, whose width is close to the cluster width. The first cycle is strongest. The edge zones beyond the horizontal parts of the events are somewhat asymmetric because of a 5° deviation of macrofractures from the vertical, and turn into hyperbolic flanks.

To summarize, clusters of fluid-filled macrofractures are best recognized on the *X*-component by fronts of converted waves (that are stronger relative to *Dpp** on the *Z*-component) with a one-pulse signal. In turn, clusters of

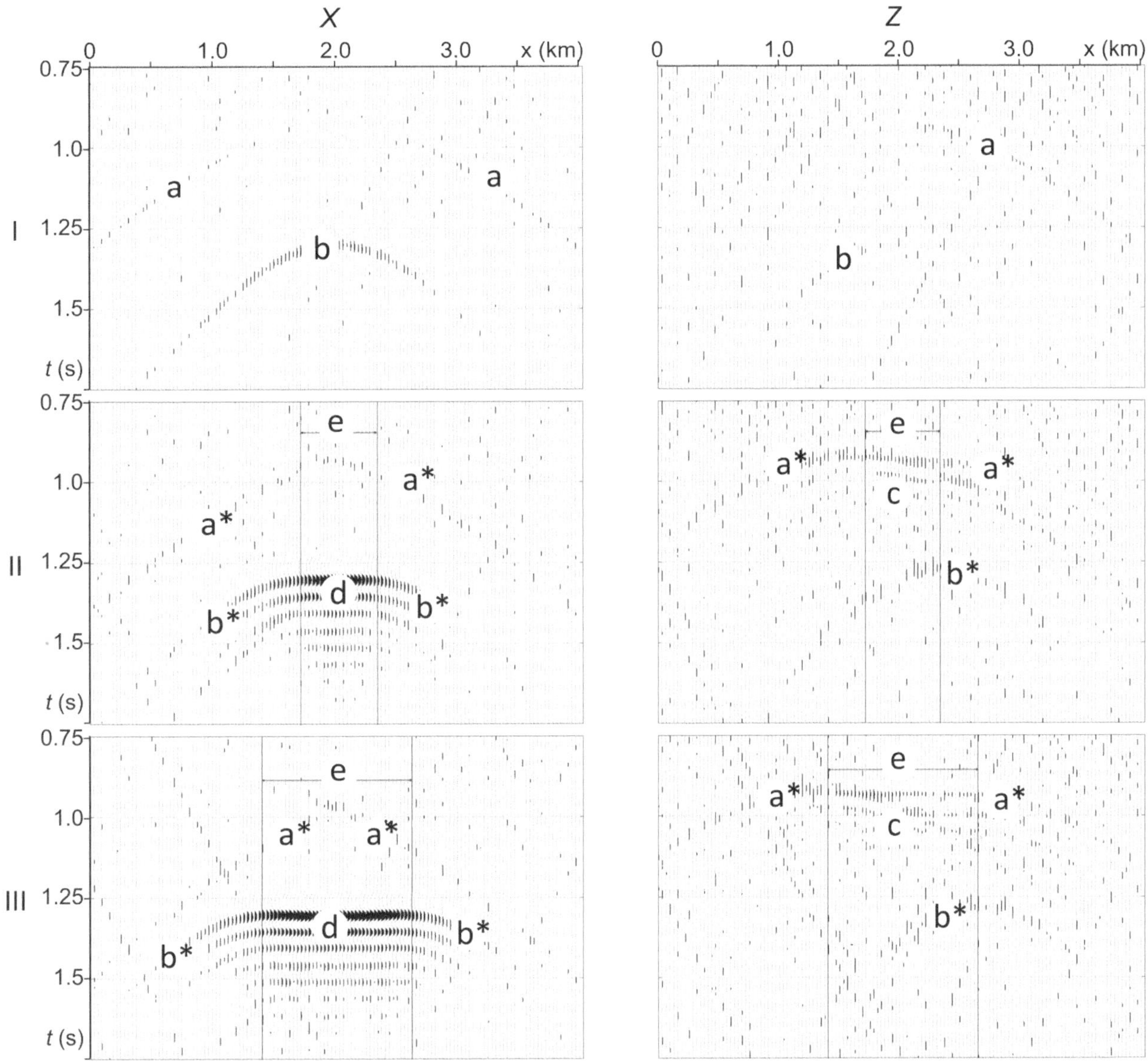

Figure 6.12. *X*- and *Z*-components of common-source gathers for clusters that consist of different numbers (*N*) of 100 m high water-filled macrofractures deviated by 5° from the vertical and embedded in a half-space with 100 m spacing: I = single macrofracture (*N* = 1); II = 1000 m wide cluster (*N* = 11); III = 2000 m wide cluster (*N* = 21); a = compressional diffraction; a* = compressional diffraction from edge macrofractures; b = converted diffraction; b* = converted diffraction from edge macrofractures; c = multicycle scattered compressional wavefront from cluster; d = multicycle scattered converted wavefront from cluster; and e denotes the cluster width.

Table 6.3. Maximum amplitudes (in relative units) of scattered wavefronts generated by macrofracture clusters.

Infill type	**Water**				**Gas**			
Component	***Z*-component**		***X*-component**		***Z*-component**		***X*-component**	
Wave	*Dpp*	*Dps*	*Dpp*	*Dps*	*Dpp*	*Dps*	*Dpp*	*Dps*
N = 1	0.8	0.8	0.8	2.0	5.0	3.0	5.0	7.0
Cluster *N* = 11	1.7	0.8	0.8	11.3	32.0	3.0	5.0	13.0
Cluster *N* = 21	1.9	0.8	0.8	10.8	34.0	3.0	5.0	13.0

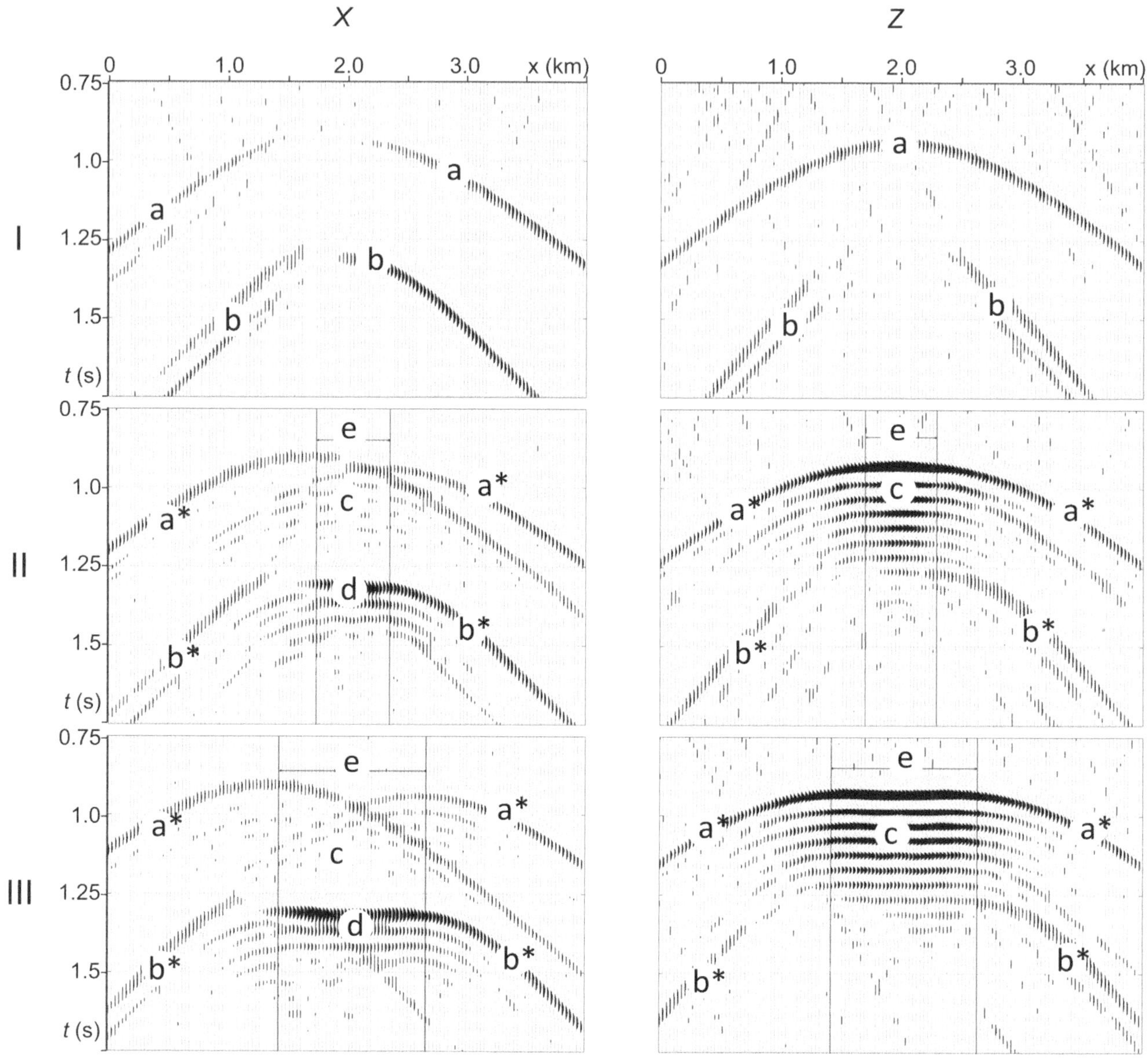

Figure 6.13. X- and Z-components of common-source gathers for clusters that consist of different numbers (N) of 100 m high gas-filled macrofractures deviated by 5° from the vertical and embedded in a half-space with 100 m spacing: I = single macrofracture ($N = 1$); II = 1000 m wide cluster ($N = 11$); III = 2000 m wide cluster ($N = 21$); a = compressional diffraction; a* = compressional diffraction from edge macrofractures; b = converted diffraction; b* = converted diffraction from edge macrofractures; c = multicycle scattered compressional wavefront from cluster; d = multicycle scattered converted wavefront from cluster; and e denotes the cluster width.

gas-filled fractures are most apparent on the Z-component as strong fronts of scattered diffracted compressional waves with a multi-cycle signal.

Amplitude estimates (in relative units) in Table 6.3 indicate the following.

1) The overall amplitude of wave response from the gas-filled macrofracture cluster is about 4 times greater than that from the water-filled macrofracture cluster.
2) The amplitude ratio of the X-component of the scattered converted *Dps* wave and the Z-component of the *Dpp* wave is 5 to 6 for the water-filled macrofracture cluster, whereas the amplitude ratio of the Z-component of the dominant *Dpp* wave to the X-component of the *Dps* wave is equal to 2.5 for the gas-filled cluster.

The fronts of scattered compressional waves have a more significant multicycle character than those of scattered converted waves (see Figures 6.12 and 6.13).

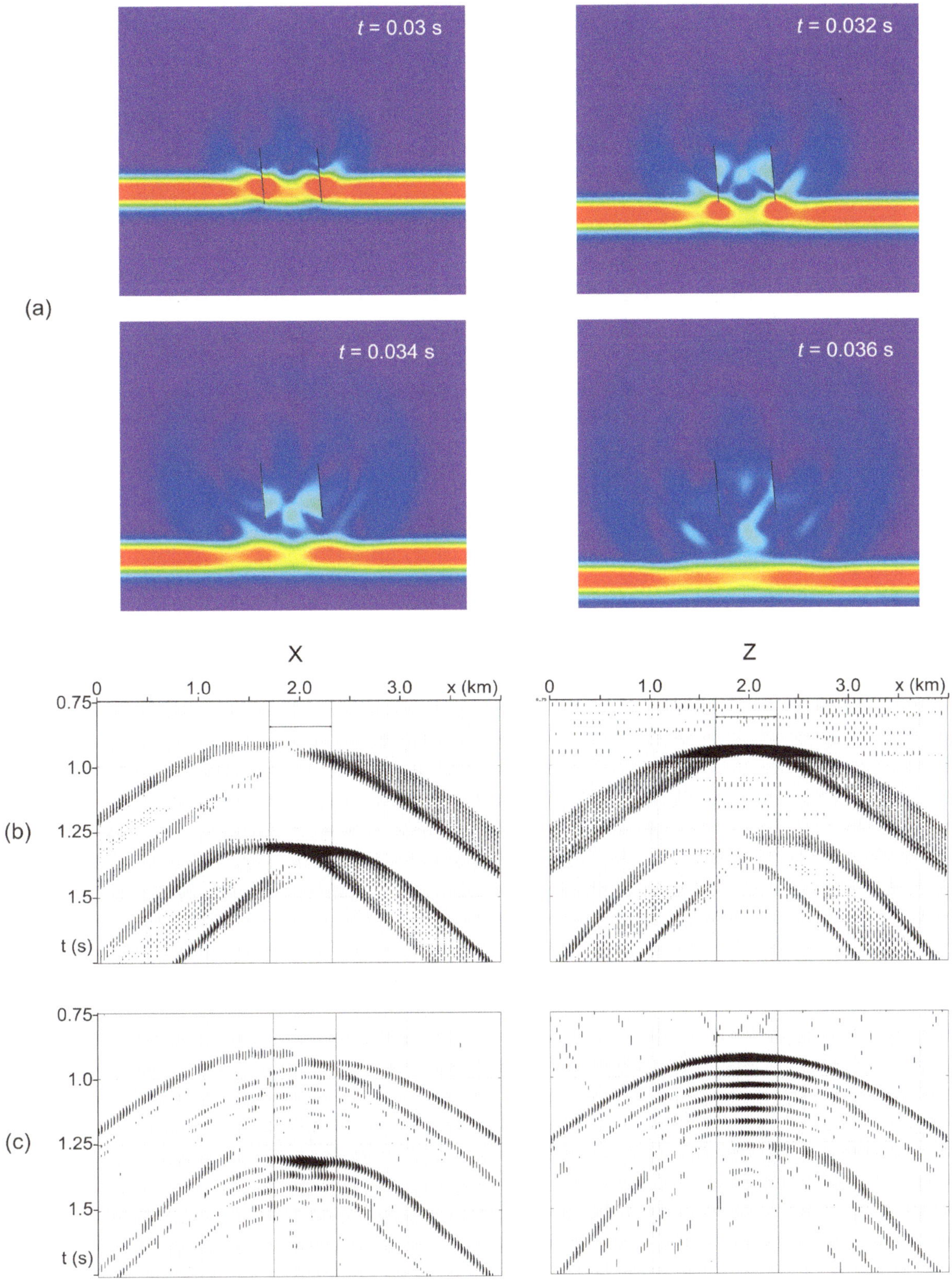

Figure 6.14. Explanation of the multicycle nature of wavefronts scattered at fracture clusters. (a) Snapshots illustrating the formation of interfractural reflections contributing to the wave response at different times. (b and c) Both components of the sum of separate responses from each of 11 macrofractures forming a cluster (b, top line) and of the response from the total cluster containing these 11 macrofractures (c, bottom line).

Establishing the cause of a multicycle profile of scattered waves and its variation has required special experimentation. In an attempt to determine how a response is formed when an incident plane wave passes through a cluster portion consisting of two macrofractures, a series of time-successive snapshots has been computed (Figure 6.14a). These snapshots, in addition to the well-known individual responses from each fracture, show the process of forming interfractural reflections.

To verify the impact of interfractural reflections on the multicycle response, we produce a sum of separate responses from each of the 11 macrofractures forming a cluster (Figure 6.14b, top row). In this case, as distinct from the response from the total cluster (Figure 6.14c, bottom row), the multicycle property is absent. Thus, we have shown that the multicycle property of wavefronts scattered by a cluster of near-parallel macrofractures is caused by interfractural reflections. The multicycle character of the scattered converted wavefront is 3 times less pronounced than that of the scattered compressional wavefront, which is apparently related to a faster scattering of the energy of shear particle motions normal to the fracture planes.

6.2.2. Character of seismic responses from a macrofracture cluster in a layered subsurface, using a plane-wavefront energy source

In most cases, seismic exploration for oil and gas is conducted in layered sedimentary basins. For this reason, here we consider reflectors that are near-parallel to a bed containing macrofractures (see Figure 6.11b).

It has been important for us, while using a plane-wavefront energy source, to assess the amplitude ratio of a response from a fracture cluster to reflections from usually near-horizontal boundaries, which might obscure the response from the fracture cluster.

Figures 6.15 and 6.16 show the results of numerical simulation performed to investigate this issue on three- and five-layer models of a carbonate setting, in which acoustic impedance increases with depth. Their thicknesses exceed five seismic wavelengths. The macrofracture clusters are located in the second and third layers, and the models have the parameters shown in Table 6.4.

When the macrofractures in the cluster are filled with fluid, the *Z*-component of scattered compressional waves is barely discernible against the background noise, as compared with reflections from the horizontal boundaries. On

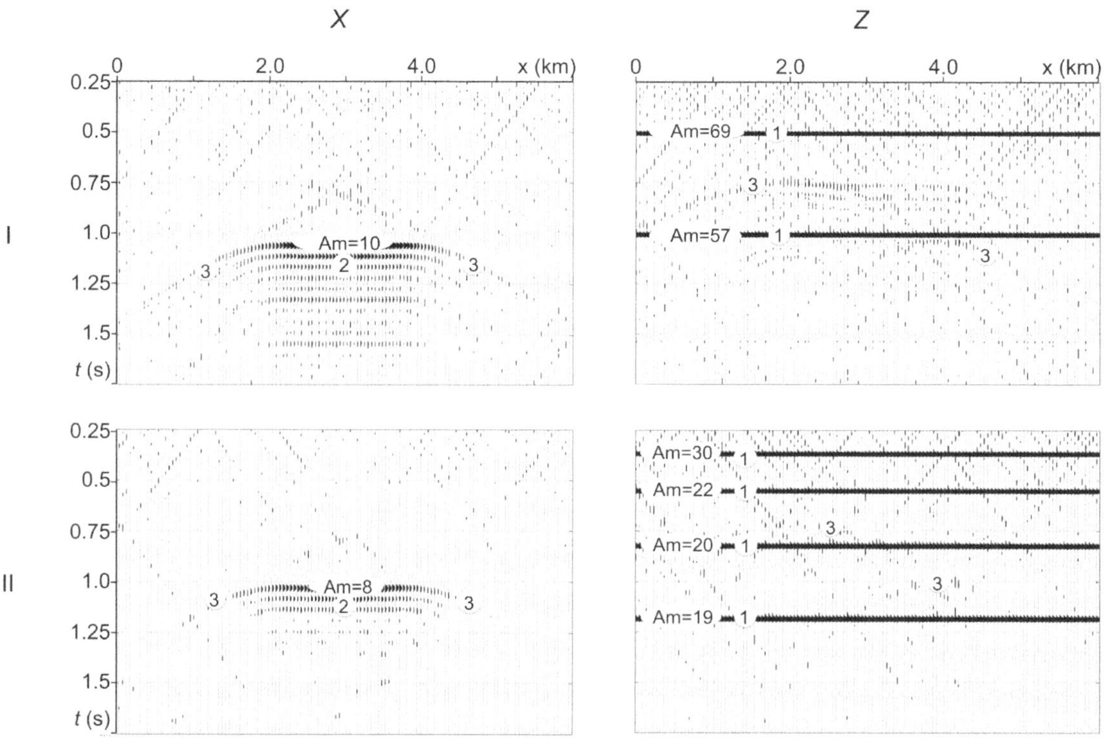

Figure 6.15. *X*- and *Z*-components of common-source gathers for a cluster of fluid-filled macrofractures in a three-layer model (row I) and a five-layer model (row II): 1 = reflections; 2 = multicycle wavefronts scattered from the clusters; 3 = *Dpp* and *Dps* diffractions from macrofractures at the cluster edges; and Am = maximum amplitudes.

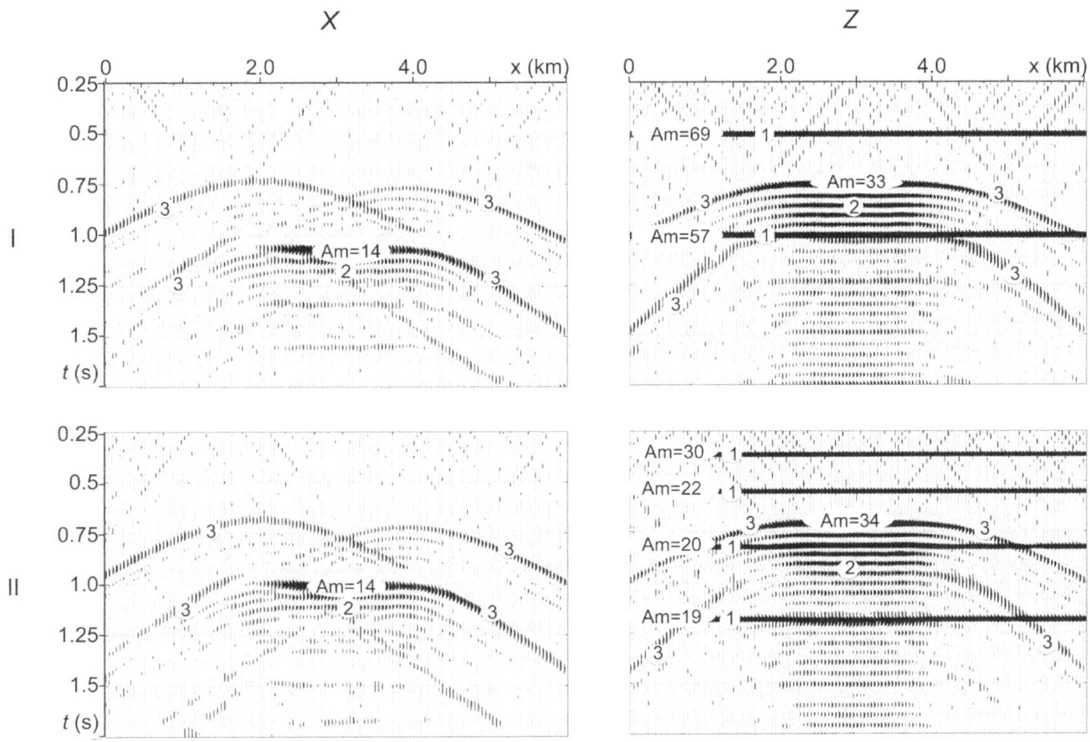

Figure 6.16. *X*- and *Z*-components of common-source gathers for a cluster of gas-filled macrofractures in a three-layer model (row I) and a five-layer model (row II): 1 = reflections; 2 = multicycle wavefronts scattered from the clusters; 3 = *Dpp* and *Dps* diffractions from macrofractures at the cluster edges; and Am = maximum amplitudes.

Table 6.4. Parameters for three- and five-layer models of a carbonate setting.

Three-layer model					Five-layer model				
Vp (m/s)	*Vs* (m/s)	ρ (g/cm^3)	*H* (m)	*R*	*Vp* (m/s)	*Vs* (m/s)	ρ (g/cm^3)	*H* (m)	*R*
4130	2150	2.05	1100		4130	2150	2.5	800	
4820	2450	2.6	1100	0.177	4540	2420	2.55	400	0.109
5660	2800	2.7	800	0.185	4800	2540	2.6	600	0.072
					5150	2650	2.65	800	0.085
					5660	2800	2.7	400	0.108

the *X*-component, a sufficiently strong front of scattered converted waves is observed that is 3 times or more stronger than the background noise. Here, there are practically no reflections from the horizontal boundaries or other types of coherent noise (Figure 6.15). In this situation, therefore, it is appropriate to use *X*-component data (in addition to conventional *Z*-component data) in order to detect macrofracture clusters using scattered converted wavefronts.

In the case of gas-filled fractures, the front of scattered compressional waves on the *Z*-component manifests itself as a multicycle (five to six cycles) sequence of horizontal events. The amplitude of this front of scattered waves is one-half that of reflections from horizontal boundaries with a reflection coefficient of about 18% (Figure 6.16, row I) and 1.5 times higher than that of reflections from boundaries with a reflection coefficient of about 10% (Figure 6.16, row II).

On the *X*-component, the multicycle nature of the scattered converted wavefront is substantially less pronounced (only two to three cycles). Furthermore, the wavefront is almost not distorted by reflections from the horizontal boundaries. Proper identification of the front is somewhat hindered by hyperbolic compressional and converted waves diffracted at edge fractures of the cluster. Their amplitude, however, is 5 times lower than that of the scattered converted wavefront.

6.2.3. Assessing the influence of geometry and infill type of a macrofracture cluster on its response

Cluster width D.—When fracture spacing d is constant, the cluster width manifests itself only in the length of the strong horizontal part of the front of both scattered compressional and scattered converted waves (Figure 6.17). Accordingly, cluster width can be estimated from the length of strong near-horizontal scattered wavefronts.

Macrofracture spacing d.—The impact of macrofracture spacing on the cluster response is substantially greater. The wave responses computed for 3000 m wide clusters with variable fracture spacing (75, 100, 150, 200, and 300 m) are shown in Figure 6.18 for fluid-filled fractures and in Figure 6.19 for gas-filled fractures.

When the macrofractures are filled with fluid, the dependence of the wave response on the fracture spacing is rather stable and simple. On the Z-component, the front of scattered compressional waves is barely discernible at

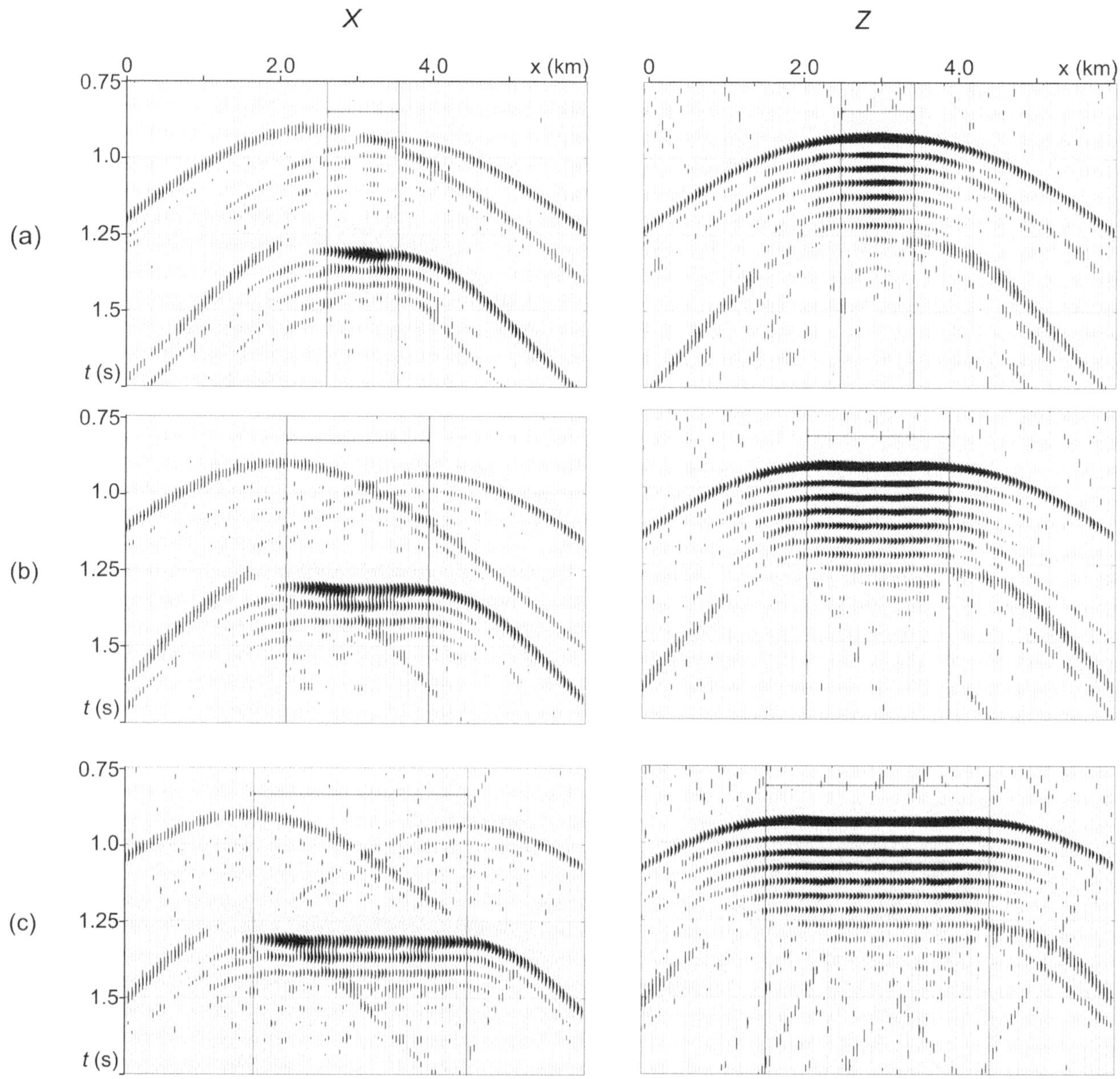

Figure 6.17. X- and Z-components of common-source gathers showing the dependence of the wave response on varying cluster width: (a) 1000 m, 11 macrofractures; (b) 2000 m, 21 macrofractures; and (c) 3000 m, 31 macrofractures. The macrofractures are 100 m high, filled with fluid, deviate by 5° from the vertical, and have 100 m spacing.

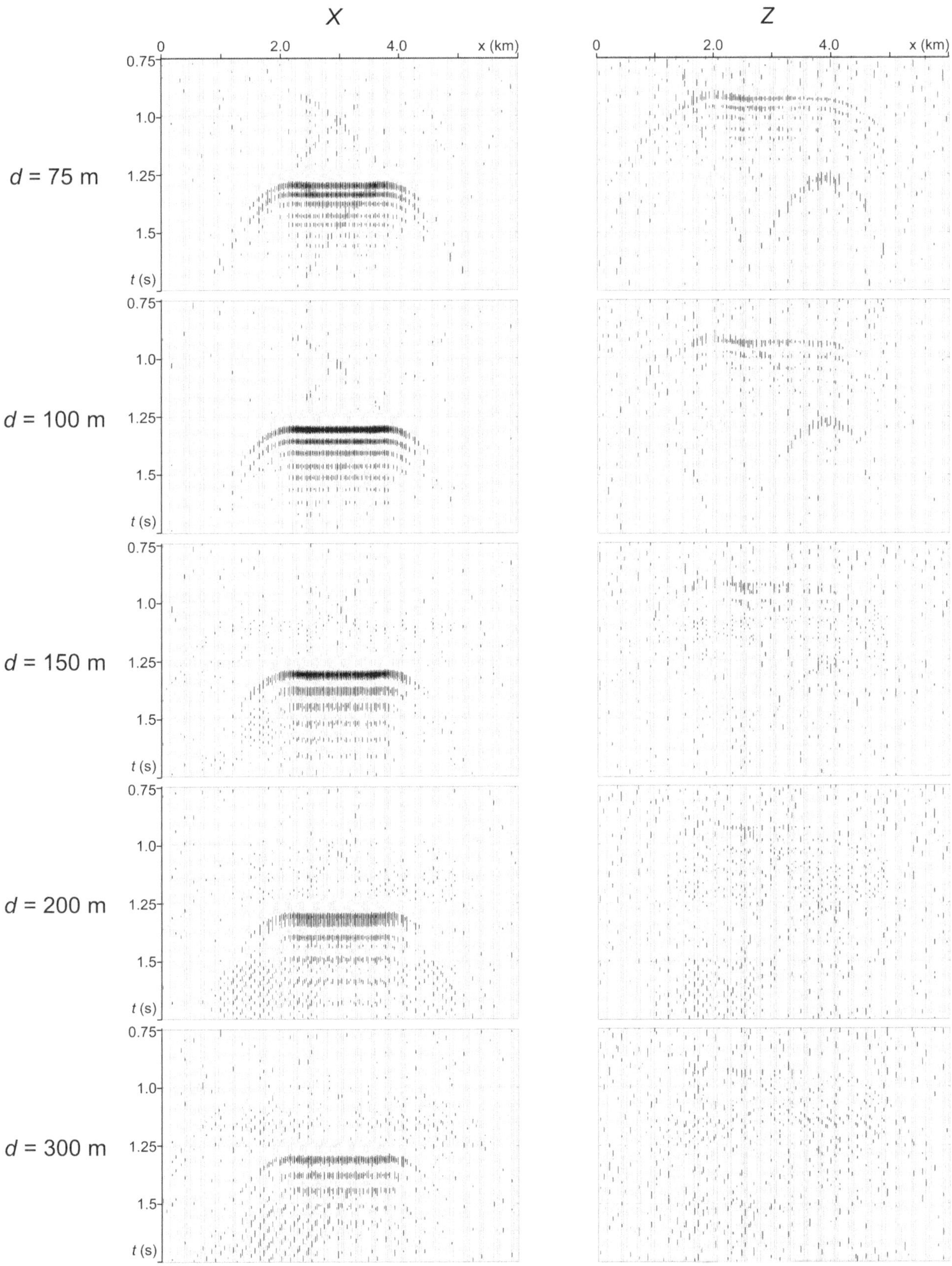

Figure 6.18. *X*- and *Z*-components of common-source gathers showing the dependence of the wave response from a 3000 m wide macrofracture cluster on varying fracture spacing *d*. The macrofractures are 100 m high, filled with fluid, and deviate by 5° from the vertical.

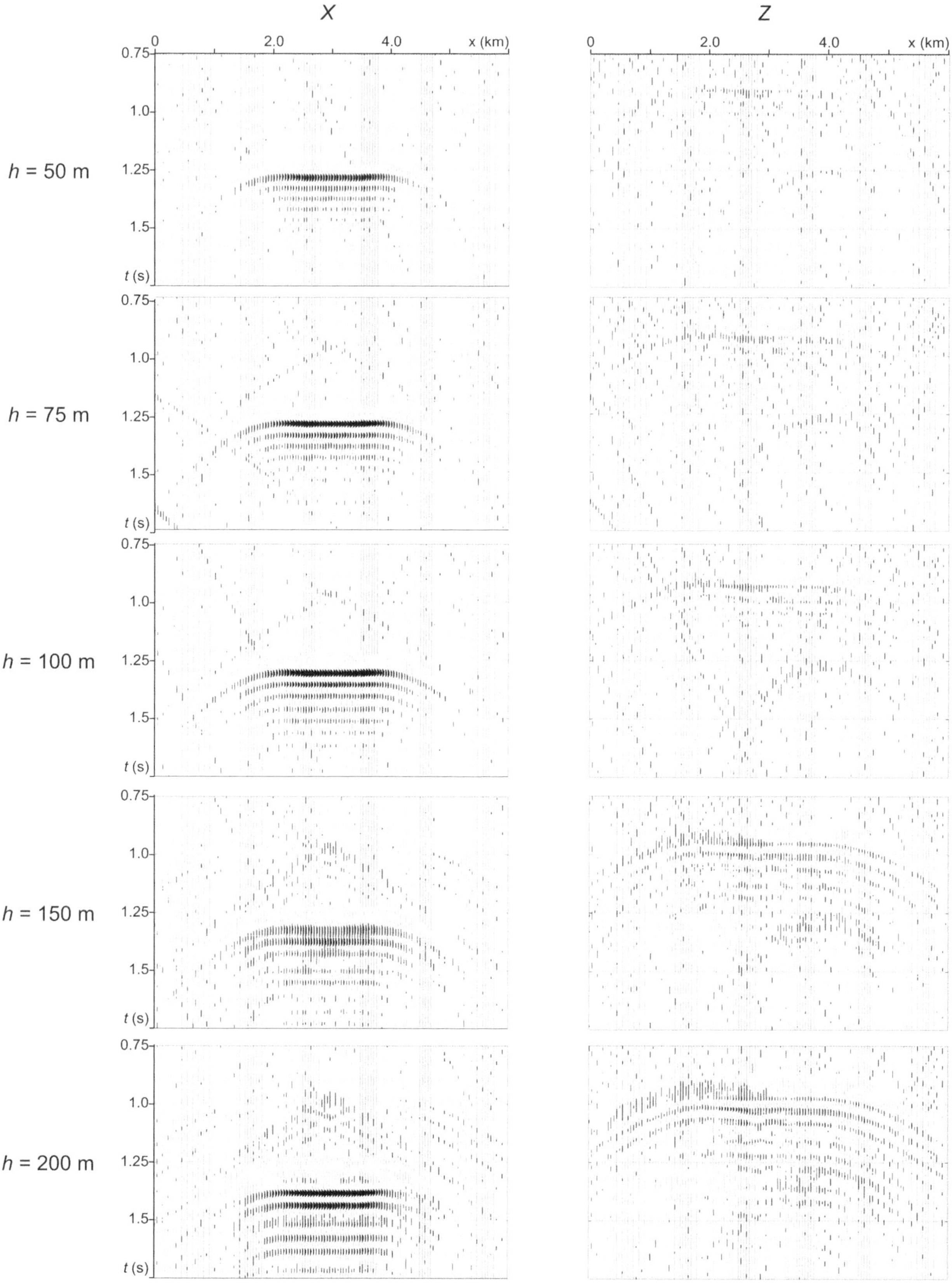

Figure 6.20. *X*- and *Z*-components of common-source gathers showing the dependence of the wave response from a 2000 m wide macrofracture cluster on varying fracture height *h*. The fractures are filled with fluid, deviate by 5° from the vertical, and have 100 m spacing.

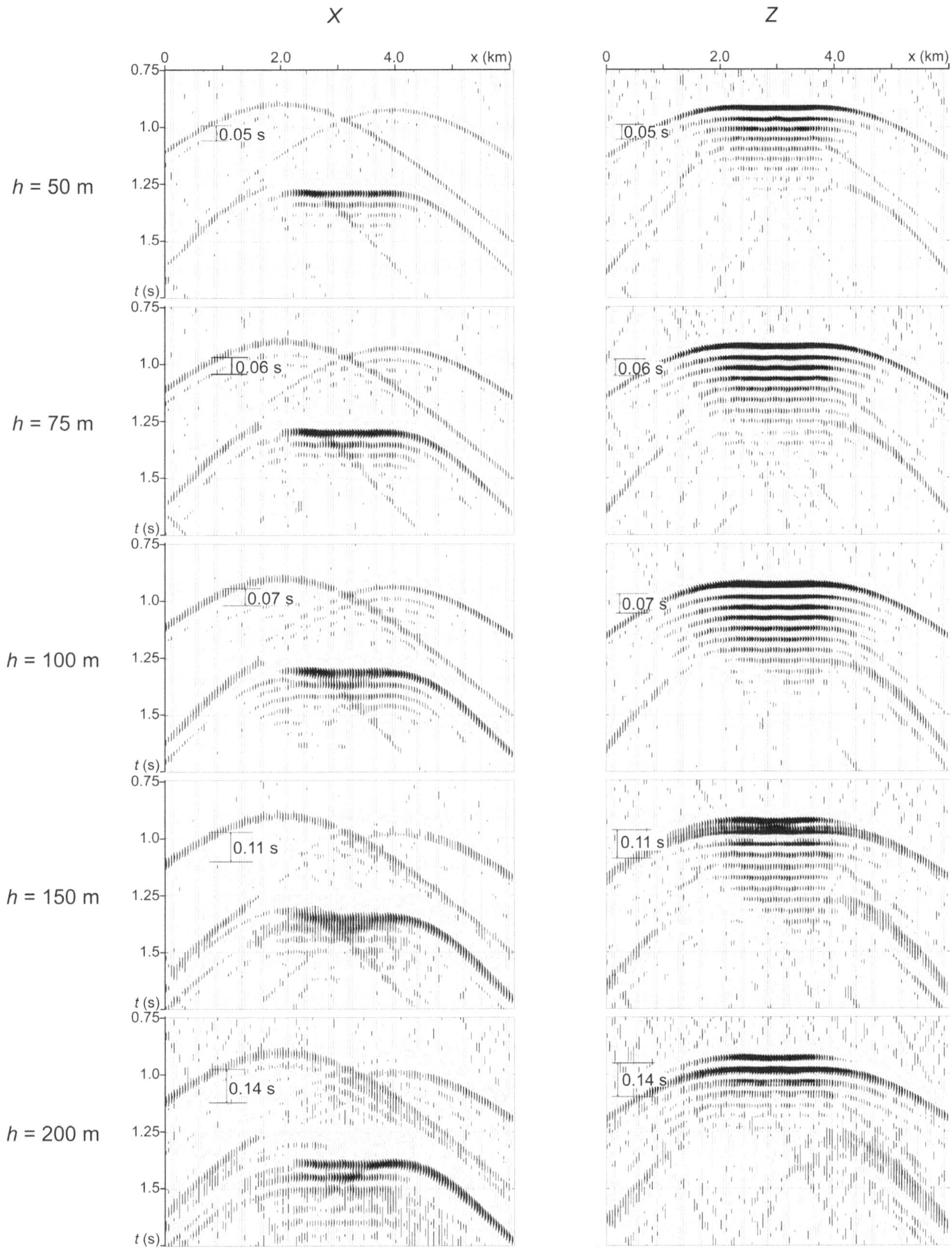

Figure 6.21. *X*- and *Z*-components of common-source gathers showing the dependence of the wave response from a 2000 m wide macrofracture cluster on varying fracture height *h*. The fractures are filled with gas, deviate by 5° from the vertical, and have 100 m spacing.

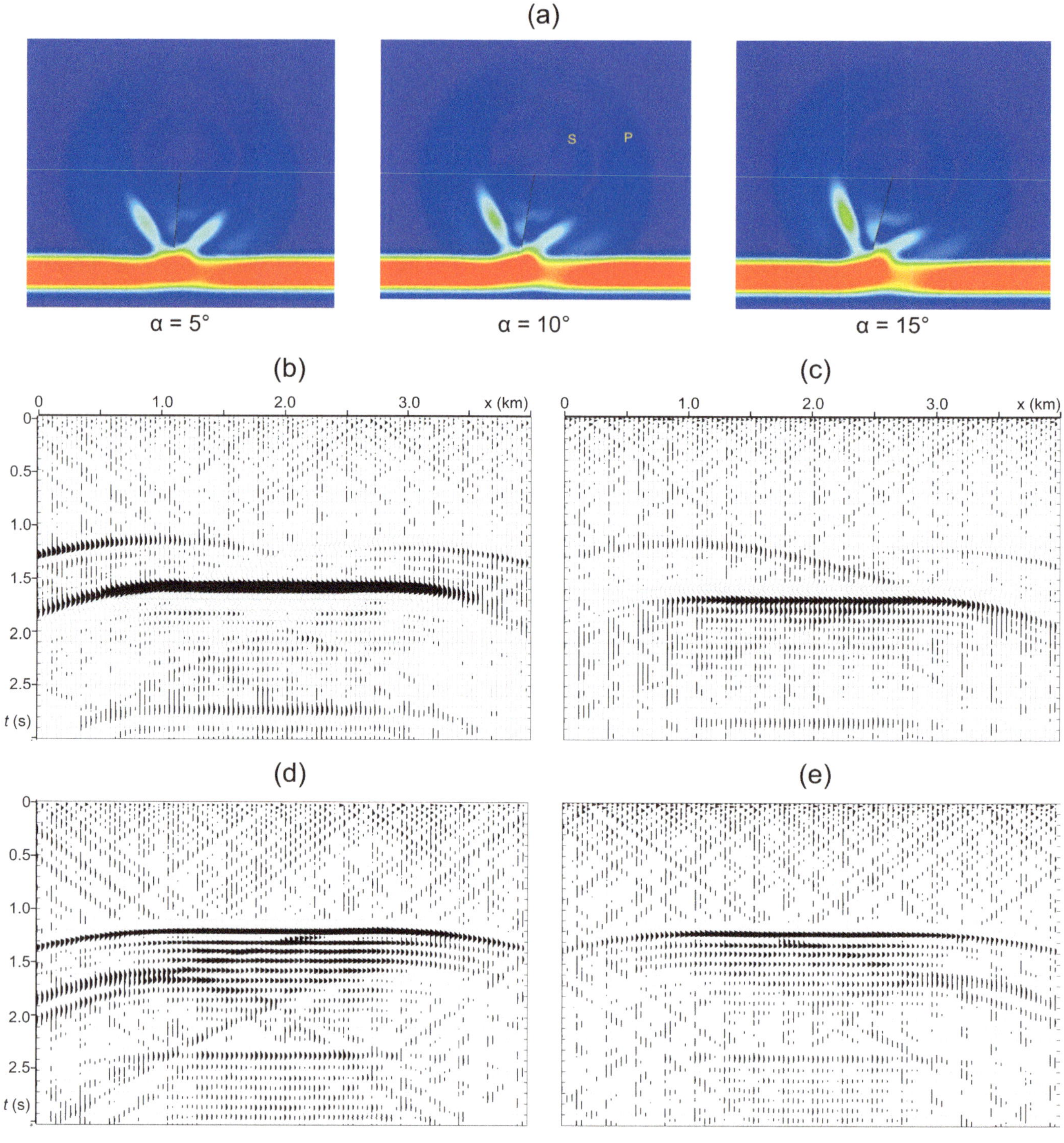

Figure 6.22. (a) Responses from fluid-filled macrofractures with different deviations from the vertical; (b) and (c) *X*-components of responses for fracture deviations of 15° and –15°, respectively; and (d) and (e) *Z*-components of responses for fracture deviations of 15° and –15°, respectively, with lines at the bottom to indicate the cluster position.

the fixed center of both fracture clusters. In Figure 6.22b and 6.22c, this phenomenon is expressed as a considerable displacement of the wavefront to the right and left, respectively. One more manifestation of the asymmetry is observed in Figure 6.22d and 6.22e, where, due to opposite signs of fracture deviations, the vertical components of the *Dps** wavefronts are recorded on opposite sides of the cluster center. Asymmetry in the scattered compressional *Dpp** wavefronts related to the fracture inclinations is less pronounced.

It is obvious that the asymmetry in wavefronts scattered by a cluster is caused by the asymmetry in converted and compressional diffractions from each macrofracture deviated from the vertical (Figure 6.9).

Macrofracture infill type.—The effects of macrofracture infill type—fluid or gas—are illustrated by comparing corresponding source gathers shown in Figures 6.12 and 6.13, 6.15 and 6.16, 6.18 and 6.19, and 6.20 and 6.21. From comparative examination of these Figures and amplitudes (in relative units) in Tables 6.3 and 6.5, we can draw the following conclusions.

1) When the fractures of a cluster are filled with fluid, the front of scattered converted waves on the *X*-component is recorded as a two- or three-cycle horizontal event that coincides with the cluster in terms of position and length. In this case, the background noise is minimal.
2) The response from a cluster of gas-filled macrofractures is much more complex in wave composition than that from a cluster of fluid-filled macrofractures. This manifests itself in the presence of hyperbolic fragments of diffractions from macrofractures at the cluster edges (on both the *X*-component and *Z*-component) and in a substantially greater number of cycles in the principal scattered compressional wavefront.
3) When macrofractures are filled with gas, maximum amplitude is inherent to the multicycle scattered compressional wavefront. This amplitude is 3 times greater than that of the strongest cycle of the scattered converted wavefront when fractures are filled with fluid.
4) When macrofractures are filled with gas, the scattered compressional wavefront on the *Z*-component is substantially more sensitive to variations in fracture parameters, such as height and spacing, than is the scattered converted wavefront on the *X*-component. Note the latter is the principal response when fractures are filled with fluid.
5) In layered media, reflections from near-horizontal boundaries are a major obstacle to identifying scattered compressional wavefronts as indicators of the presence of gas-filled macrofracture clusters. When fractures are filled with fluid, the scattered converted wavefront dominates on the *X*-component. In our idealized case with a plane-wavefront energy source, any reflections from horizontal boundaries are absent; in realistic conditions, however, even if present, they would be rather weak.

6.2.4. Conclusions from Section 6.2

1) When a plane compressional wave impinges on a cluster of parallel near-vertical macrofractures, multicycle fronts of scattered compressional and converted waves are formed due to a constructive interference of diffractions from the tips of the macrofractures.
2) We have verified that, when the reflecting boundaries are near-horizontal and the energy source is a plane wavefront, a scattered converted wavefront can be recorded on the *X*-component as the principal response from a cluster of near-vertical fluid-filled macrofractures.
3) There are differences in the character of responses depending on the macrofracture infill—fluid or gas. When fractures are filled with fluid, they best manifest themselves in a scattered converted wavefront on the horizontal component; when filled with gas, they best appear as a multicycle front of scattered compressional waves on the vertical component.
4) A typical slight deviation of macrofractures from the vertical leads to asymmetry in the structure of scattered wavefields between the opposite sides of a source location.
5) The high-amplitude region in a scattered converted wavefront may be indicative of the location and width of a corresponding macrofracture cluster.

6.3. Assessment of stability in forming a scattered converted wavefront from a system of macrofractures under real conditions

Sections 6.1 and 6.2 present models for numerically simulating responses from macrofracture clusters with idealized characteristics: strict rectilinearity, constant spacing, and constant deviation from the vertical for all of the fractures in a cluster.

In reality, near-vertical macrofractures may be curvilinear, rough, and stepped, which is probably due to the micro- and macro-layering of the sedimentary rocks having natural differences in geomechanical properties and, therefore, in resistance to tectonic stresses. Also, the macrofracture spacing in a cluster may vary within quite broad ranges and the parallelism of the fractures may be disturbed in places, so their dips may vary. Such factors have been confirmed by examination of macrofractures in outcrops (Dorofeyeva, 1986).

Table 6.6. Energy characteristics of the wave response for different levels of variations in fracture spacing (in percent of the nominal value of 100 m).

Variations	0%	10%	20%	30%	40%	60%
E_1	7.3096	6.8554	6.7881	5.8856	6.0736	5.7110
E_2	5.2952	7.8137	7.3925	9.7161	9.2123	10.5192
$(E_{1,0}-E_1)/E_{1,0}$	0.0000	0.0629	0.0714	0.195	0.1693	0.218
E_1/E_2	1.3804	0.8773	0.9182	0.6058	0.6593	0.5429
E_{1ps}/E_{1pp}	20.4567	18.8362	20.2425	13.4326	15.1559	13.5977

Note: E_1 = total energy of the first three cycles over a cluster; E_2 = total energy of the rest of a response, regarded as noise; $E_{1,0}$ = value of E_1 when variations are zero; E_{1ps} = total energy of a scattered converted wavefront on the X-component; and E_{1pp} = total energy of a scattered compressional wavefront on the Z-component.

Table 6.7. Energy characteristics of the wave response for different levels of variations in fracture deviation from the vertical (in percent of the nominal value of 10°).

Variations	0%	10%	20%	30%	40%	60%
E_1	7.3096	7.5443	7.2062	7.5856	6.7926	6.2855
E_2	5.2952	5.1228	5.1208	5.4197	6.0252	7.1472
$(E_{1,0}-E_1)/E_{1,0}$	0.0000	0.0677	0.0970	0.1228	0.1298	0.1410
E_1/E_2	1.3804	1.4727	1.4072	1.3996	1.1274	0.8794
E_{1ps}/E_{1pp}	20.4567	27.2788	26.3113	23.9289	19.9826	18.6872

Note: For designations, see Table 6.6.

to 30%, the wave response practically does not change. At larger levels of variations (up to 40% to 60%), the waves having steep traveltime slopes, which we consider to be noise, become slightly stronger. Plots showing these dependencies are presented in Figure 6.27.

The quantitative estimates support the inference from the visual inspection. With the tested levels of variations in the fracture deviation from the vertical, ranging from 0 to 60%, the scattered converted wavefront energy E_1 decreases by 14%, while the signal-to-noise ratio E_1/E_2 decreases by 36%. Variations up to 30% do not cause any noticeable energy changes.

In real conditions, variations in both fracture spacing and fracture deviation from the vertical are observed. The combined impact of variations in both parameters is assessed for identical levels of their variation (in percentage of corresponding base values).

Figure 6.26 displays both components of common-source gathers for macrofracture clusters with different levels of random variations in both parameters. Table 6.8 provides energy characteristics of the scattered converted wavefront and steeply dipping noise. Plots showing these dependencies are presented in Figure 6.27.

The combined impact of variations in the inclination and spacing of macrofractures certainly reduces the possibility of detecting scattered wavefronts. However, as long as the scatter level is less than 30%, some 80% of the useful signal energy is retained, while the signal-to-noise ratio decreases to 50% or less.

Figure 6.27 shows graphs of energy characteristics of the wave response for different levels of variations in fracture spacing, fracture deviation from the vertical, and in both when they occur simultaneously. It is apparent that, for the investigated levels of variations, the scattered converted wavefront energy decreases to 30% or less, whereas the noise energy is about doubled. Accordingly, the signal-to-noise ratio reduces threefold.

Therefore, the scattered converted wavefronts from clusters of unidirectional macrofractures may be regarded as being sufficiently insensitive to moderate variations in the inclination and spacing of fractures.

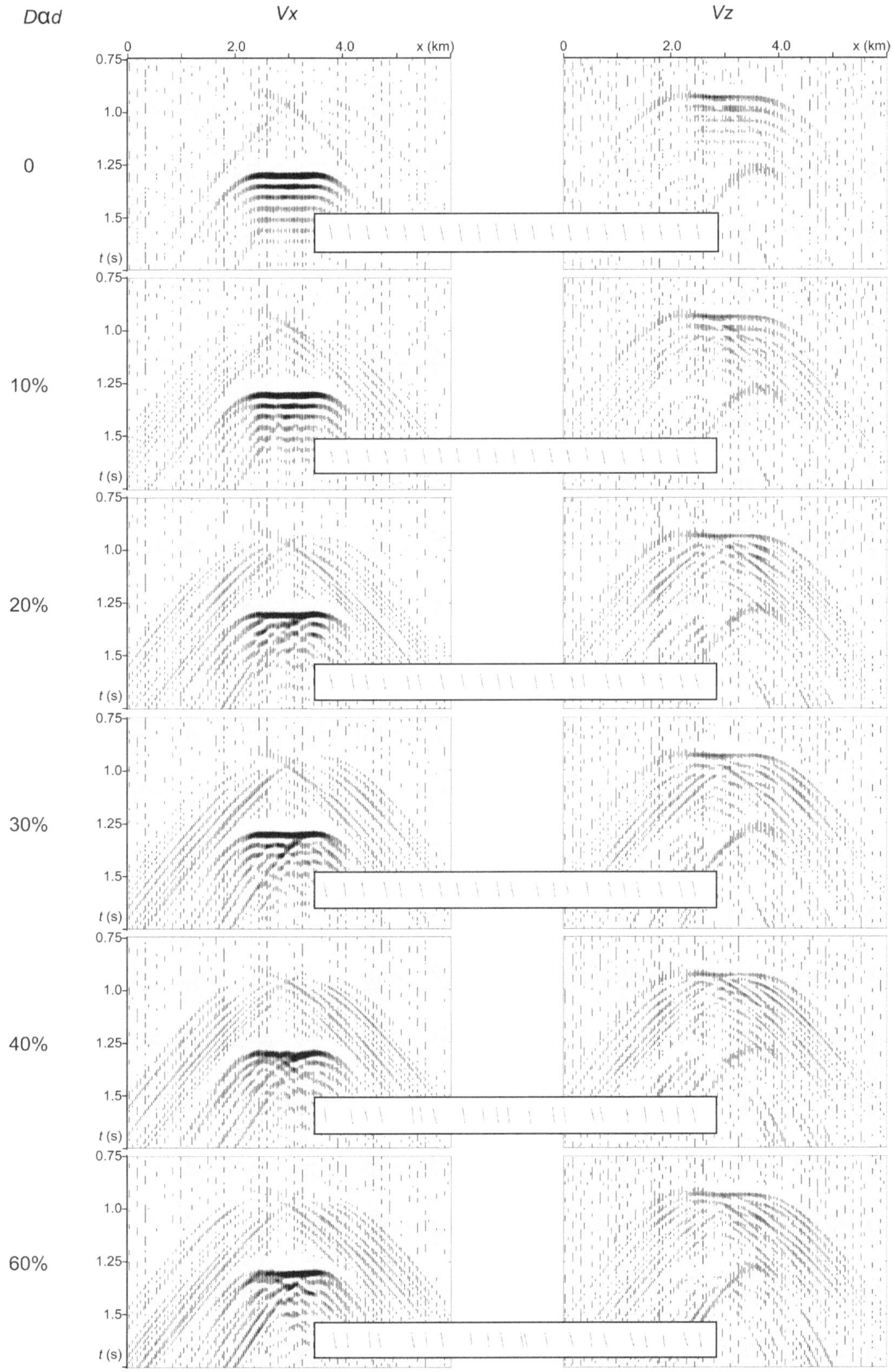

Figure 6.26. *X*- and *Z*-component of common-source gathers for clusters with different levels of random variations in both fracture spacing (in percent of the nominal value of 100 m) and fracture deviation from the vertical (in percent of the nominal value of 10°). Percentages for spacing and deviation are identical. In the rectangles, randomly positioned and randomly inclined fractures are shown.

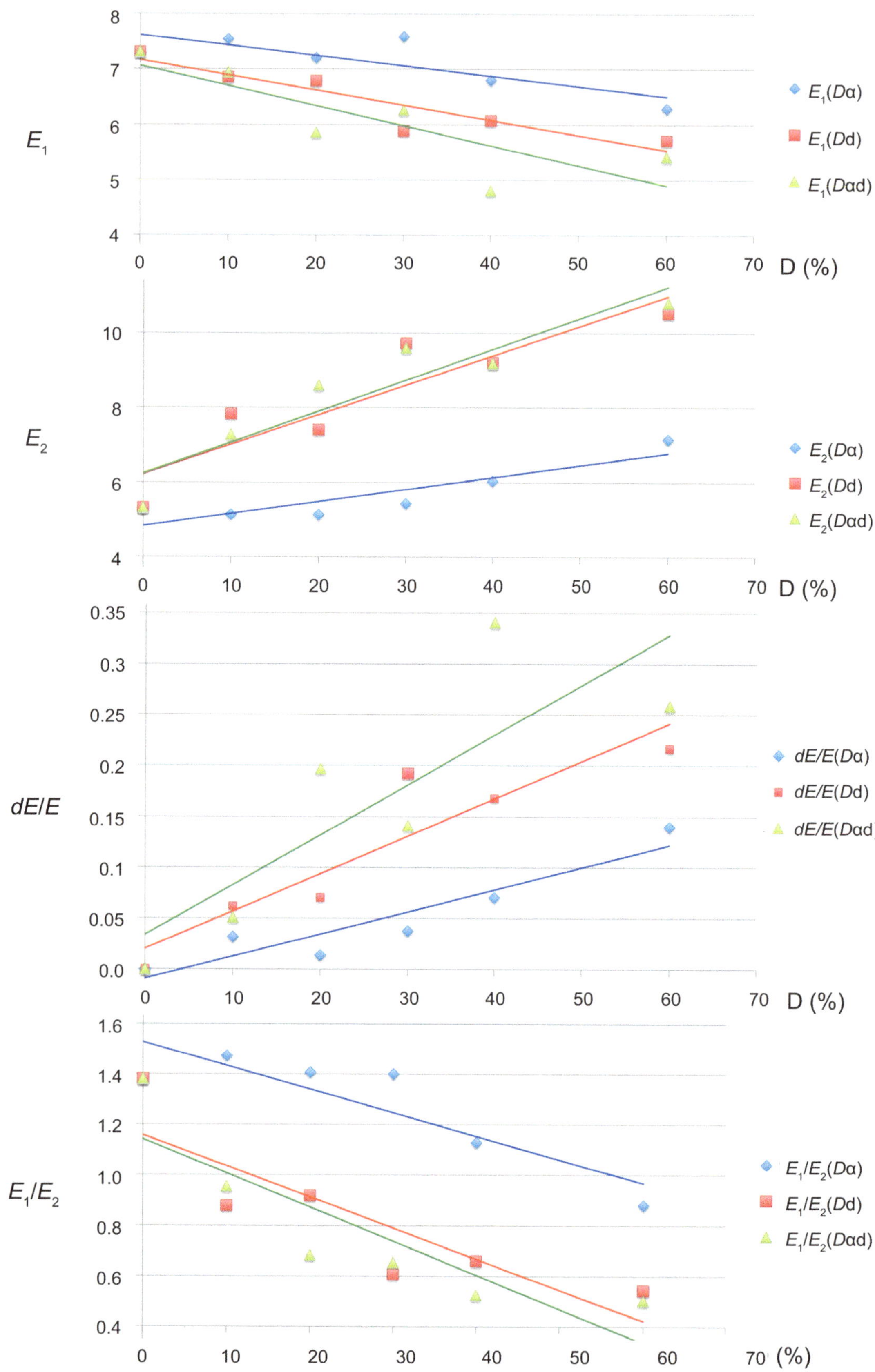

Figure 6.27. Energy characteristics of the wave response for different levels of variations in fracture spacing, fracture deviation from the vertical, and in both when they occur simultaneously. The levels are percentages of corresponding base values.

Table 6.8. Energy characteristics of the wave response for different levels of variations in fracture spacing and fracture deviation from the vertical that occur simultaneously.

Variations	0%	10% and 10%	20% and 20%	30% and 30%	40% and 40%	60% and 60%
E_1	7.3096	6.9389	5.8555	6.2574	4.8004	5.4114
E_2	5.2952	7.2658	8.5799	9.5814	9.1720	10.7829
$(E_{1,0}-E_1)/E_{1,0}$	0.0000	0.050	0.1994	0.1436	0.3431	0.2597
E_1/E_2	1.3804	0.9550	0.6825	0.6531	0.5234	0.5019
E_{1ps}/E_{1pp}	20.4567	19.2118	17.6578	16.5778	12.6953	13.1658

Note: For designations, see Table 6.6.

6.3.3. Determination of a scattered converted wavefront from a macrofracture cluster on stacked time sections when a point energy source is used. Macrofracture cluster identifiers

Based on the investigations discussed, we have established that a plane horizontal wave impinging on a cluster of near-vertical macrofractures forms a front of scattered converted waves that is readily identifiable on the *X*-component. In seismic exploration, however, point sources of the vertical force or expansion center type primarily are used. In order to determine whether the commonly used point source produces similar results, we have performed comparative computations for plane-wavefront source and point source.

A two-layer medium is chosen as a model. The upper layer is 1900 m thick and has the parameters $Vp = 4500$ m/s, $Vs = 2500$ m/s, and $\rho = 2500$ kg/m^3. The lower layer, with parameters $Vp = 5000$ m/s, $Vs = 2600$ m/s, and $\rho = 2650$ kg/m^3, contains a cluster of 21 fluid-filled fractures at a depth of 2000 to 2100 m. The fractures deviate by 10° from the vertical. Data are recorded with a 50 m spacing and a split spread 4000 m wide for a single macrofracture and 6000 m wide for a macrofracture cluster.

For a single macrofracture, the *X*- and *Z*-components of source gathers obtained with a plane-wavefront source and a point source are presented in Figures 6.28a and 6.28b, respectively. The latter pair is NMO-corrected with $Vp = 4500$ m/s.

The vertical component with a point source exhibits a strong, horizontally aligned compressional PP-reflection, as with a plane-wavefront source. The converted reflection is barely discernible.

On the horizontal component, the compressional PP-reflection at 0.85 s and converted PS-reflection at 1.15 s are almost flattened and have a polarity reversal and relatively low amplitudes over the fracture. The converted *Dps* diffraction appears at 1.26 s at the spread center as a hyperbolic fragment without any polarity reversal over the fracture. Its amplitude is approximately half that of the converted PS-reflection and more than an order of magnitude lower than that of the PP-reflection on the *Z*-component.

With a point source, two essential changes occur in the character of the seismic response from the cluster of macrofractures (Figure 6.28d).

1) On the *X*-component, the amplitudes of the compressional and converted reflections remain the same as when using a plane-wavefront source. However, at 1.26 s, which corresponds to the traveltime of the *Dps* diffraction from the single macrofracture, a wave of amplitude that is about 10 times greater than that of the above wave is observed. Such an effect also is noted in Table 6.3. This wave is identified with the front of scattered converted *Dps** waves formed by the superposition of diffractions at the tips of the cluster macrofractures. The wave becomes comparable in amplitude to the compressional PP-reflection on the *Z*-component. The high amplitudes of the *Dps** response provide an understanding of the cluster as a group source of seismic waves with great directivity. The group of 21 sources, which are the tips of macrofractures, provides up to tenfold amplification of the useful signal.
2) As distinct from the continuous diffraction from the single fracture on the *X*-component, the front of scattered converted *Dps** waves from the macrofracture cluster has a polarity reversal at the source location. Also, asymmetry, i.e., a substantial difference between the *Dps** waves on opposite sides of the spread, is observed. The reason for this is that the *Dps** wave corresponds, in terms of kinematic characteristics, to a converted wave reflected from a hypothetical boundary running through the upper tips

6.4.1. Characteristics of wave responses from macrofracture clusters

It is advisable to compare responses from macrofractures with those from mesofractures in extended beds and local zones. To this end, numerical simulations are performed with models A and B shown in Figure 5.15 (Chapter 5, Subsection 5.2.1).

Figure 6.29 shows common-source gathers with maximum offsets of 4 km, as for model A in Figure 5.15. The gathers contain a response from 100 m high macrofractures embedded in an extended bed at a depth of 2 km with a spacing of 100 m, as well as a reflection from a deeper horizontal boundary.

The *X*-component in Figure 6.29a successively records (with respect to arrival times) the following waves: two compressional *Dpp** waves scattered by the upper and lower tips of macrofractures, a compressional PP-reflection from the horizontal boundary, two converted *Dps** waves scattered by the upper and lower tips of macrofractures, and a converted PS-reflection from the horizontal boundary.

Scattered converted waves dominate in amplitude at midrange offsets of about ±2000 m and markedly exceed the *X*-component of the compressional PP-reflection. The amplitudes of the PP and *Dpp* waves are highest at farthest offsets. All waves have a polarity reversal at the source location.

On the *Z*-component shown in Figure 6.29b, only compressional waves, such as the *Dpp** waves scattered by the upper and lower tips of macrofractures and the PP-reflection from the horizontal boundary, are clearly

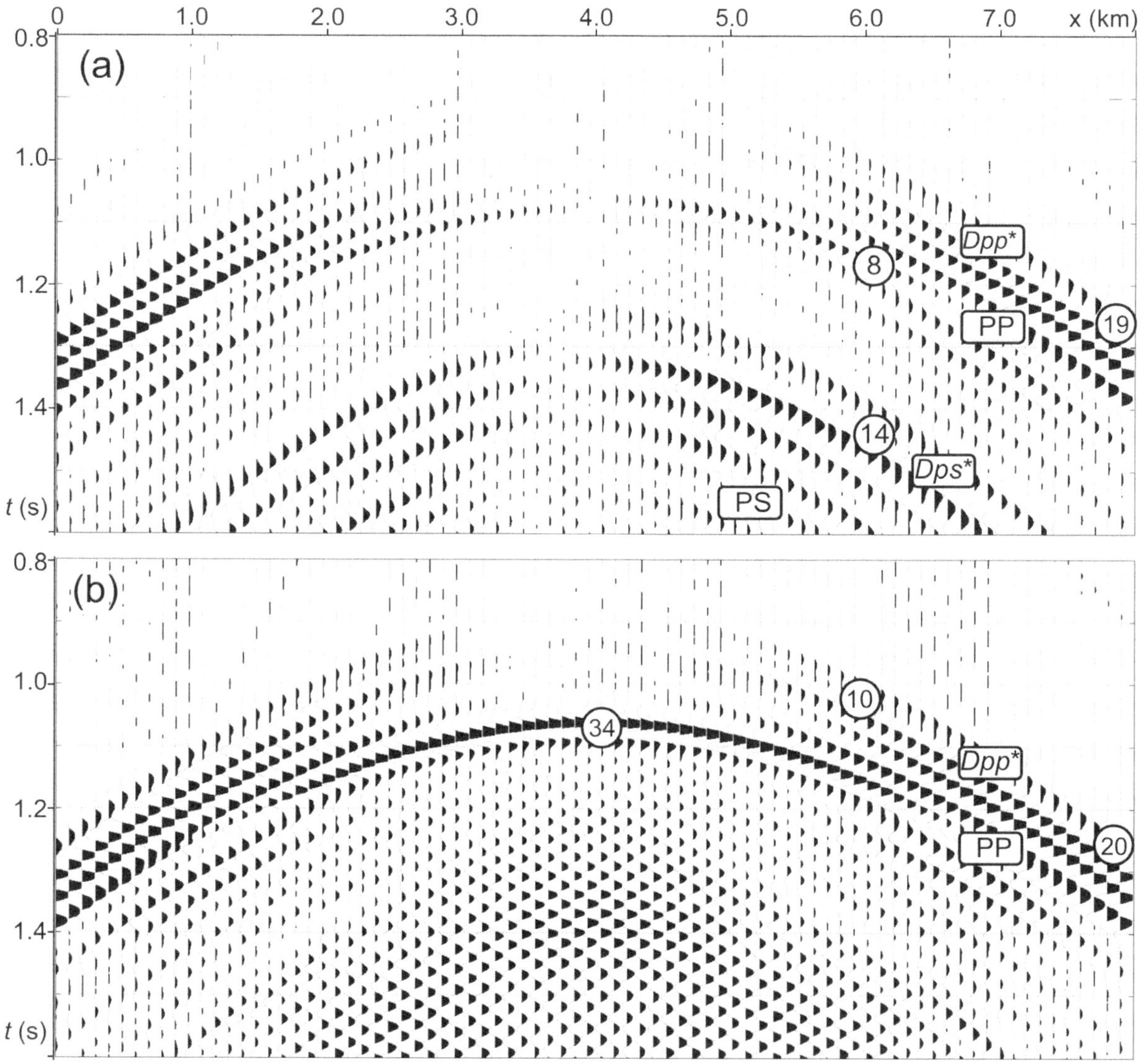

Figure 6.29. (a) *X*-component and (b) *Z*-component of a common-source gather for a macrofractured bed (model A).

identifiable. The amplitude of the PP-reflection substantially exceeds the amplitude of scattered waves at near offsets. The *Dps** wave dominates in amplitude at far offsets and has a multicycle interference character.

The results of numerical simulations for local bed-bounded macrofracture zones (model B in Figure 5.15) are shown in Figure 6.30. On these common-source gathers, the overall character of the wave response resembles that of Figure 6.29. The difference is observable in a better manifestation and a greater concentration of the scattered converted *Dps** wave energy at offsets closer to the source location (±2000 m).

These source gathers show the following.

- The wave response from the macrofracture zone has a wavetrain-like three- to four-cycle character, due to separately recording waves scattered by the upper and lower tips of macrofractures.
- On the *X*-component, the amplitudes of the scattered converted wave are comparable with or greater than those of the PP-reflection.
- On the *Z*-component, the amplitudes of the scattered compressional waves are comparable with and greater than those of the PP-reflection only at farthest offsets.
- The weak character of the *Z*-component after the PP-reflection arrival is apparently attributed to weak converted reflections.

AVO curves of waves from Figures 6.29 and 6.30 are shown in Figure 6.31 for model A and in Figure 6.32 for model B, respectively. For comparison, Figures 6.31 and 6.32 also exhibit AVO curves for the same models but containing mesofractures.

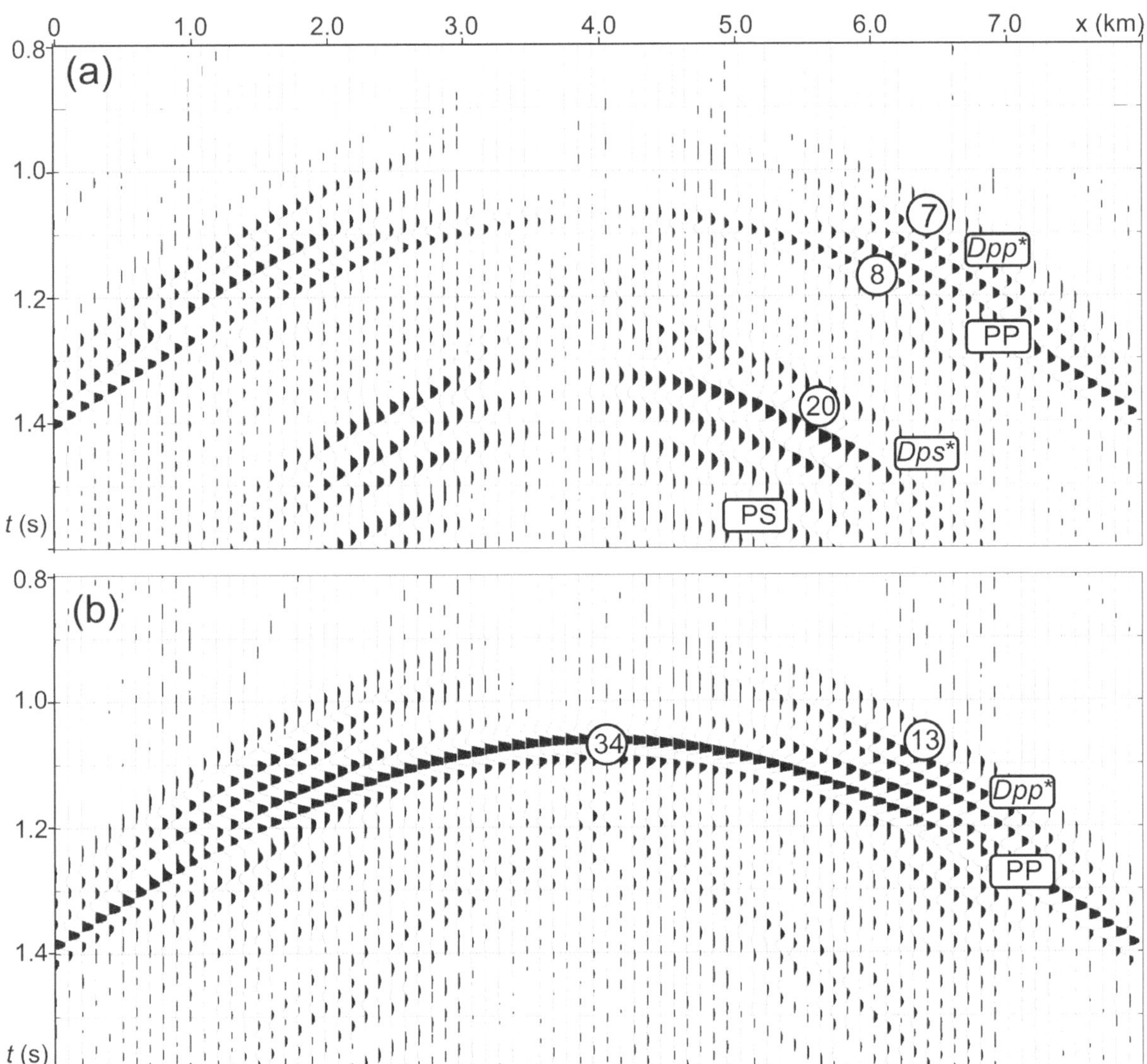

Figure 6.30. (a) *X*-component and (b) *Z*-component of a common-source gather for a local macrofracture zone (model B).

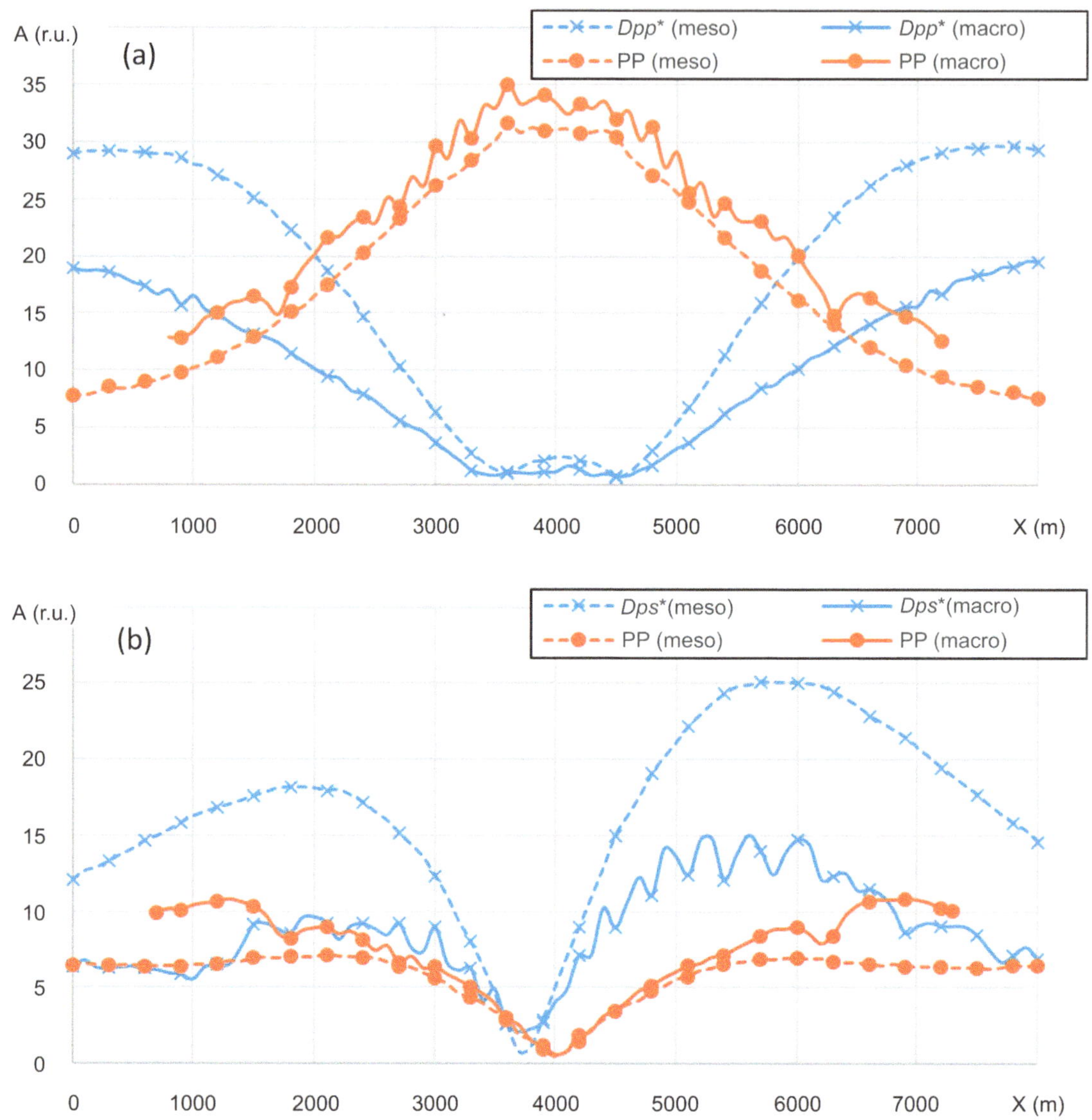

Figure 6.31. AVO curves of the *Dpp** and PP-waves for mesofractures and macrofractures of model A on the (a) *Z*-component and (b) *X*-component.

On the *Z*-component shown in Figure 6.31a, it is apparent that the amplitudes of compressional waves scattered by macrofractures of model A are lower than those of waves scattered by mesofractures of the same model. The amplitude ratio of these waves is about 0.5 to 0.7, while the width of the zone of low amplitudes (less than 10% of the maximum value) increases from 1.5 to 2.5 km when mesofractures are replaced by macrofractures.

The importance of this characteristic for identifying the scattered compressional wave as the ratio of the amplitude of this wave to that of the reflected wave is much lower for macrofractures than for mesofractures. For mesofractures, the ratio is 1.5 to 4.0 at offsets ranging from 1.8 to 4.0 km. For macrofractures, this indicator decreases to 1 to 2, and only at offsets ranging from 3 to 4 km.

The AVO curve and the maximum amplitude of the PP-reflection for macrofractures almost do not differ from those for mesofractures, which is an indication that the comparison is correct. Note high-frequency oscillations of the PP-wave AVO curve on the *X*-component in the case of macrofractures.

It is apparent from Figures 6.31b and 6.32b that the AVO curve of the converted *Dps** wave scattered by macrofractures has amplitudes of 10 to 15 r.u., which are approximately 1.5 times lower than those of the wave scattered by mesofractures (18 to 25 r.u.).

At midrange offsets of –2000 to 2000 m, the *X*-component amplitude of the scattered converted *Dps** wave is determined to be larger than that of the PP-reflection. Both waves undergo a polarity reversal at the

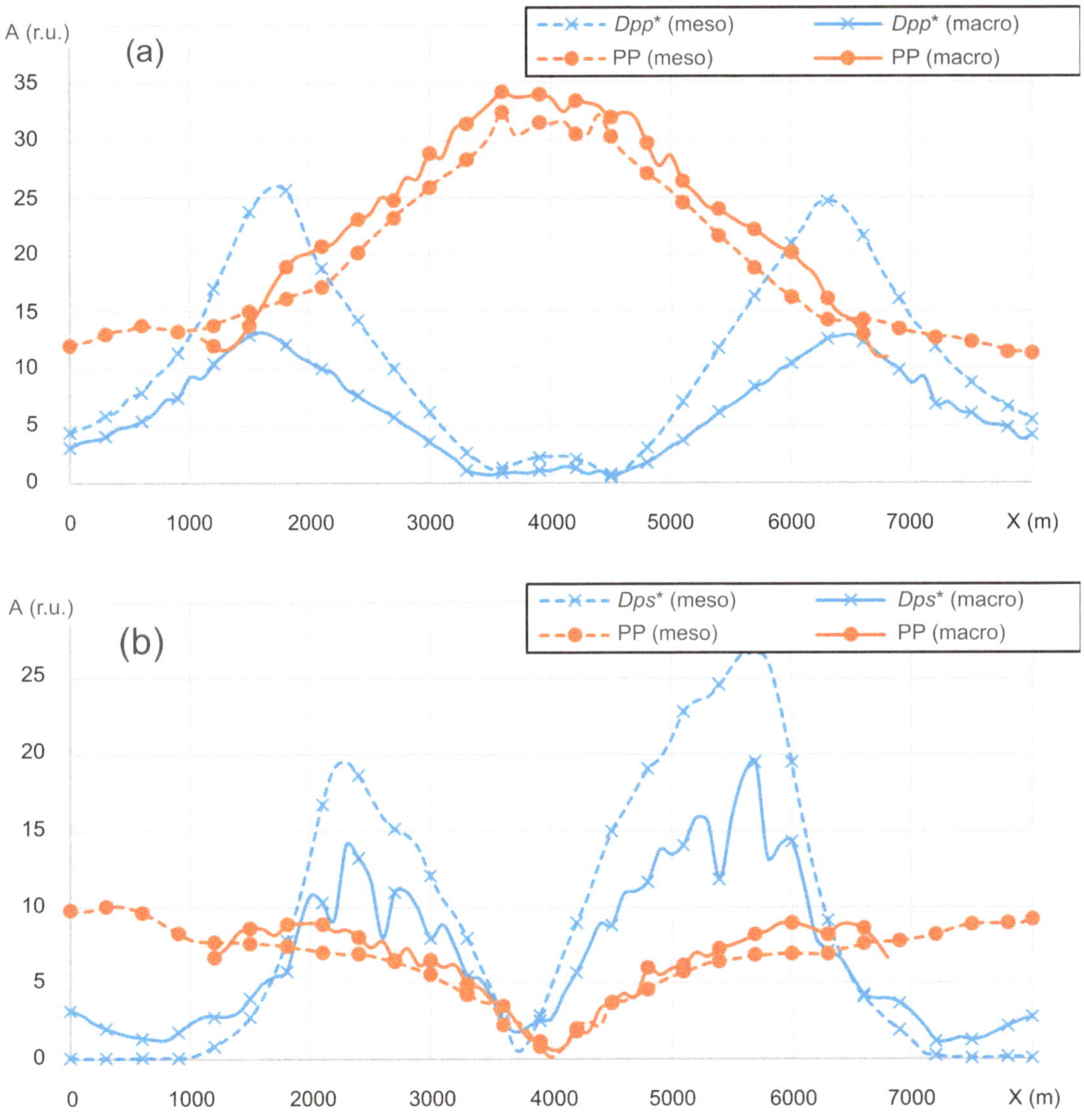

Figure 6.32. AVO curves of the *Dpp** and PP-waves for mesofractures and macrofractures of model B on the (a) *Z*-component and (b) *X*-component.

source location. Moreover, the scattered converted *Dps** wave exhibits a substantial amplitude difference (1.5 to 2 times) between the opposite sides of the source location, as well as a shift of the zero value of the AVO curve away from the source location.

A comparison between responses of macrofractures in models A and B shows the following.

- On the *Z*-component, the compressional wave scattered by macrofractures is approximately 1.5 to 2 times weaker in the region of its maximum amplitudes, i.e., at offsets of 2000 to 3000 m, than the same wave but scattered by mesofractures. For macrofractures, the zone of minimum amplitudes, which are less than 10% of the maximum value, is 1.5 times wider than that for mesofractures.
- On the *X*-component, the converted wave scattered by macrofractures is approximately 1.5 times weaker in the region of its maximum amplitudes, i.e., at offsets of 1000 to 2000 m, than the same wave but scattered by mesofractures. At the same time, it twice exceeds the *X*-component of the PP-reflection from the acoustic boundary and is substantially stronger than the converted PS-wave.
- The converted wave scattered by macrofractures has three to four cycles because it is formed by both the top and bottom fractures, whereas the wave scattered by mesofractures is shorter because it is formed by the top fractures only.

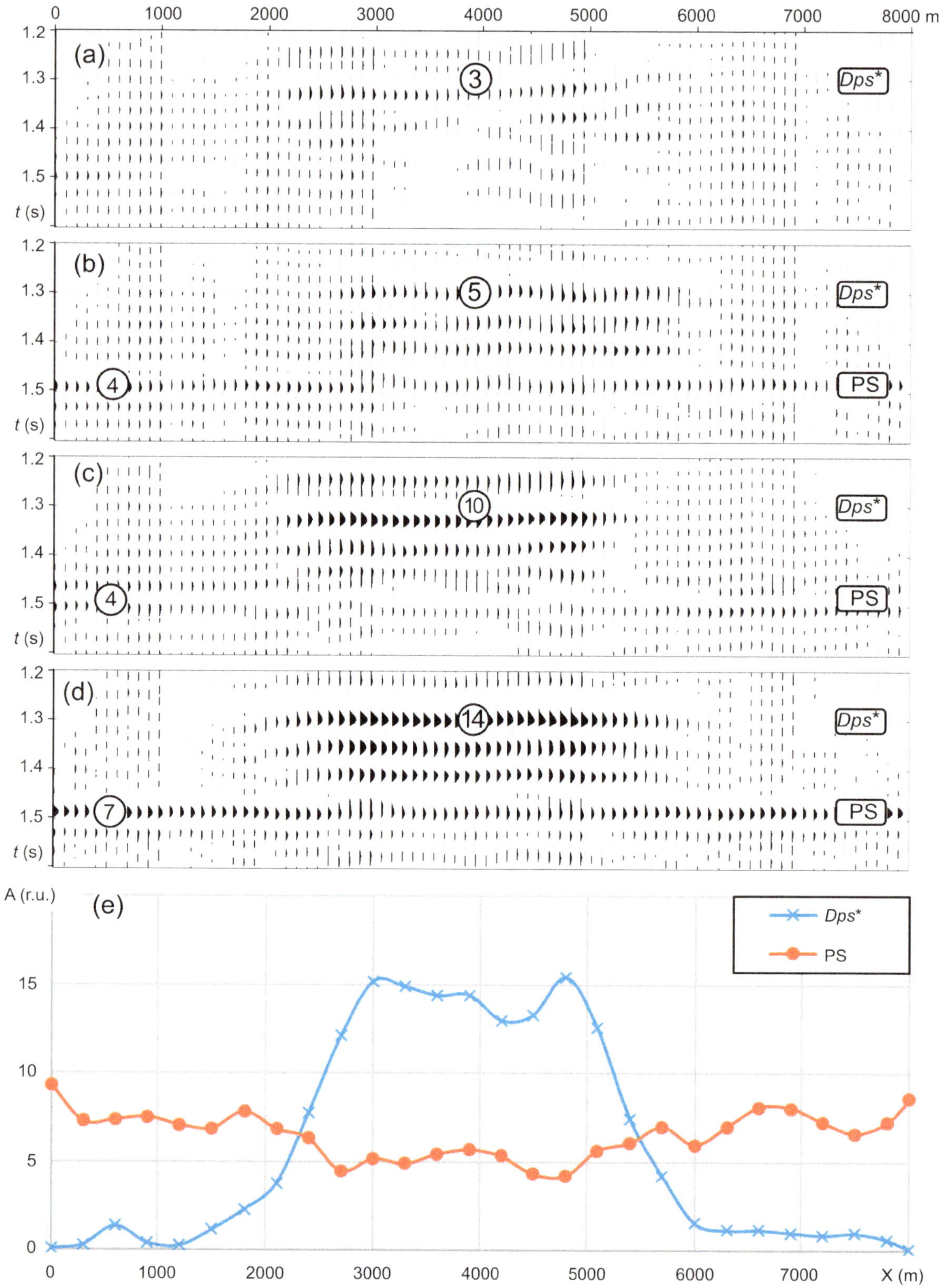

Figure 6.34. Main steps in analyzing CMP seismic data recorded on the *X*-component with the goal of identifying a macrofracture cluster (model B) from scattered converted waves: (a) full stack of all offsets ranging from –2 to 2 km without a polarity correction; (b) partial stack of offsets ranging from –2 to 0 km; (c) partial stack of offsets ranging from 0 to 2 km; (d) full stack of all offsets ranging from –2 to 2 km with a polarity correction; and (e) amplitudes of the scattered converted *Dps** wave and the converted PS-reflection on the full stack with the polarity correction.

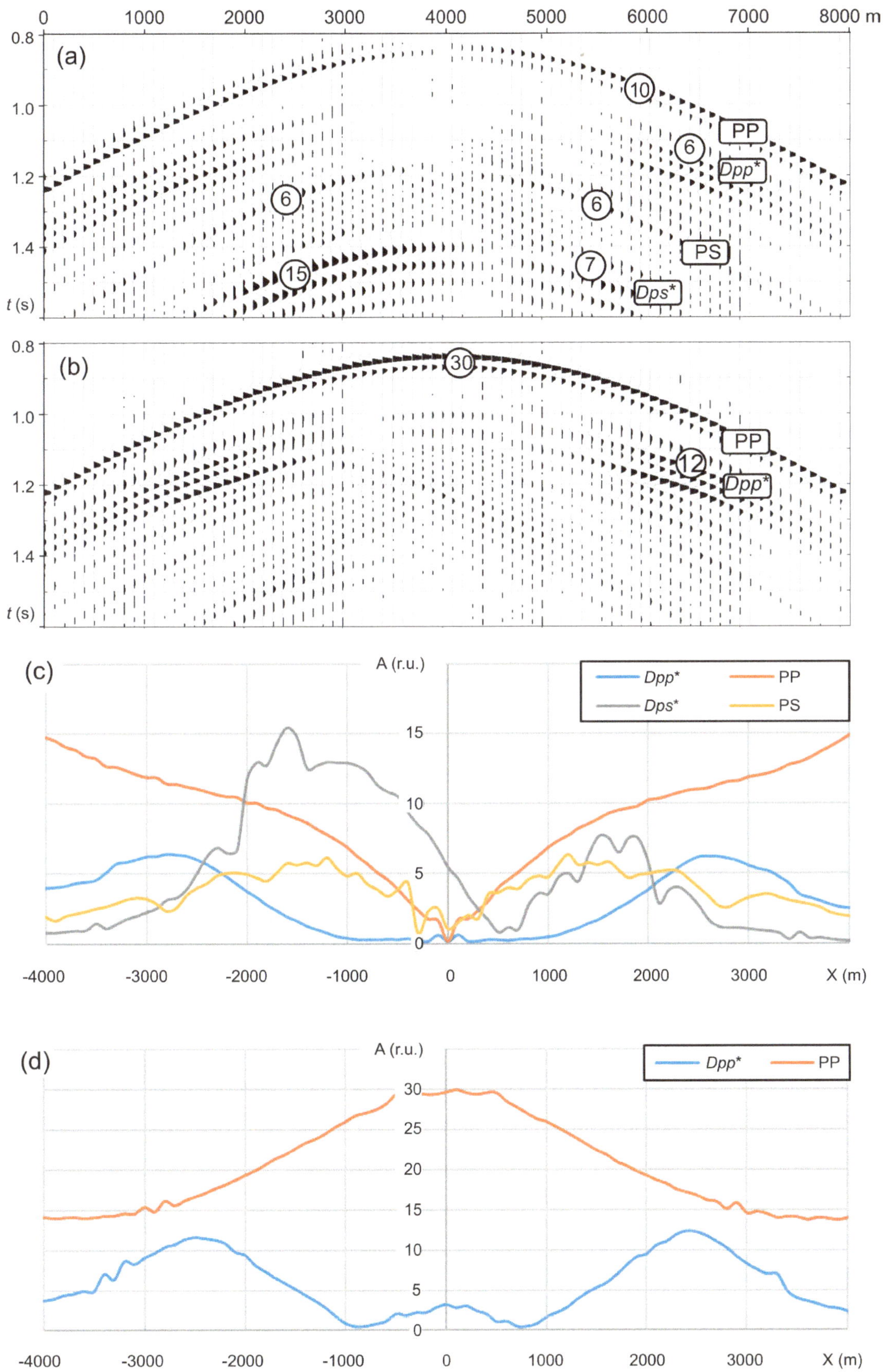

Figure 6.35. (a) X-component and (b) Z-component of a common-source gather for a model with a reflecting boundary above a zone of macrofractures; AVO curves of the (c) X-component and (d) Z-component of different waves.

1.5 times lower than those of the PP-reflection. For the scattered compressional *Dpp** wave, the width of the zone of low amplitudes, which are less than 10% of the maximum value, reduces to ±1200 m as compared with the ±1800 m width of the preceding model (see Figure 6.32b). Overall, the response of scattered waves from the macrofracture system increases in amplitude as compared with the preceding model.

Figure 6.36 illustrates how macrofracture zones (systems) can be identified on the *Z*-component data in the conditions being discussed. The Figure depicts the main steps in recognizing scattered compressional waves using the amplitude ratio of a far-offset stack to a near-offset stack. To this end, we use the common-source gathers whose full CMP stack is shown in Figure 6.36a. Stacking far offsets (Figure 6.36b) reduces markedly the amplitude difference between the scattered *Dpp** wave and the PP-reflection, thus making them comparable. Figure 6.36c allows the amplitudes of these waves on the near-offset stack and far-offset stack to be compared. For the scattered *Dpp** wave, the amplitude ratio of the far-offset stack to the near-offset stack increases about 10 times over the macrofracture zone, thus enabling this feature to be recognized unambiguously.

Figure 6.37 shows how a scattered converted *Dps** wave from the macrofracture zone (system) located below the reflector can be identified on the *X*-component. The Figure provides four types of stack for the *X*-component of data acquired on the same test line: the full stack without a polarity correction, the partial stack of negative offsets, the partial stack of positive offsets, and the full stack with a polarity correction, which are presented in Figure 6.37a through d, respectively.

In Figure 6.37a, 6.37b, and 6.37d, the scattered converted wave is dominant. Here, the best signal-to-noise ratio is obtained via full stacking without a polarity correction, although the signal amplitude is low (Figure 6.37a). A maximum *Dps** amplitude is obtained on the full, polarity-corrected stack (Figure 6.37d).

Factors that reduce the effectiveness of using the amplitude ratio of a far-offset stack to a near-offset stack to detect a macrofracture zone are the same as discussed in Subsection 6.3.1. These factors are: (1) a low amplitude of scattered compressional waves relative to that of reflections, (2) the presence of interference noise at far offsets, and (3) rather small amplitudes of near-vertically propagating waves when the boundaries below the source are near-horizontal.

To overcome the impact of these factors, we may use the method that assesses the AVO gradients of the waves under analysis at midrange offsets, where the amplitudes vary linearly. For the models of geologic media considered above, this gradient is estimated within the 1000 to 2000 m offset interval as the amplitude difference of waves in prestack data. So, for model A of macrofractures (Figure 6.31a), the PP-reflection gradient is negative because the amplitude drops from 29 to 20 r.u., whereas the gradient of the scattered *Dpp** wave is positive because the amplitude rises from 3 to 10 r.u. The gradient of the scattered *Dpp** wave from mesofractures is positive because the amplitude rises from 5 to 20 r.u.

In order to make estimates of the amplitude ratio of a far-offset stack to a near-offset stack more stable, offsets with near-zero amplitudes (up to 500 to 700 m for the models under discussion) should be excluded from the near-offset stack. Note that there is no problem with near-zero amplitudes when the medium has high velocity contrasts and therefore a reflecting boundary is present at the top of the fractured zone.

6.4.3. Conclusions from Section 6.4

1) Analysis of AVO curves for scattered compressional waves, scattered converted waves, and reflections recorded at long offsets shows that the positions of their intervals of maximum amplitudes in relation to the source location are rather different. For the model condition in Figure 5.15, on the *Z*-component, compressional reflections have maximum amplitudes in the vicinity of the source location, whereas compressional diffractions have them at far offsets of 2.5 to 3.5 km; for scattered converted waves on the *X*-component, maximum amplitudes are observed at midrange offsets of 1 to 2 km.
2) The response from macrofracture systems is overall 1.5 to 2 times weaker than that from analogous mesofracture systems.
3) On the *Z*-component, the scattered compressional wave from a macrofracture zone can be identified by an amplitude ratio of a far-offset stack to a near-offset stack. According to (2), however, reliability of their identification will be somewhat lower than that of the identification of scattered compressional waves from mesofractures under the same conditions. In such situations, or in the presence of noise at far offsets, the second recognition method using the AVO gradient at midrange offsets should be employed to assess the type of compressional wave. The sign of the gradient is supposed to be positive for a scattered wave, and negative for a reflection.

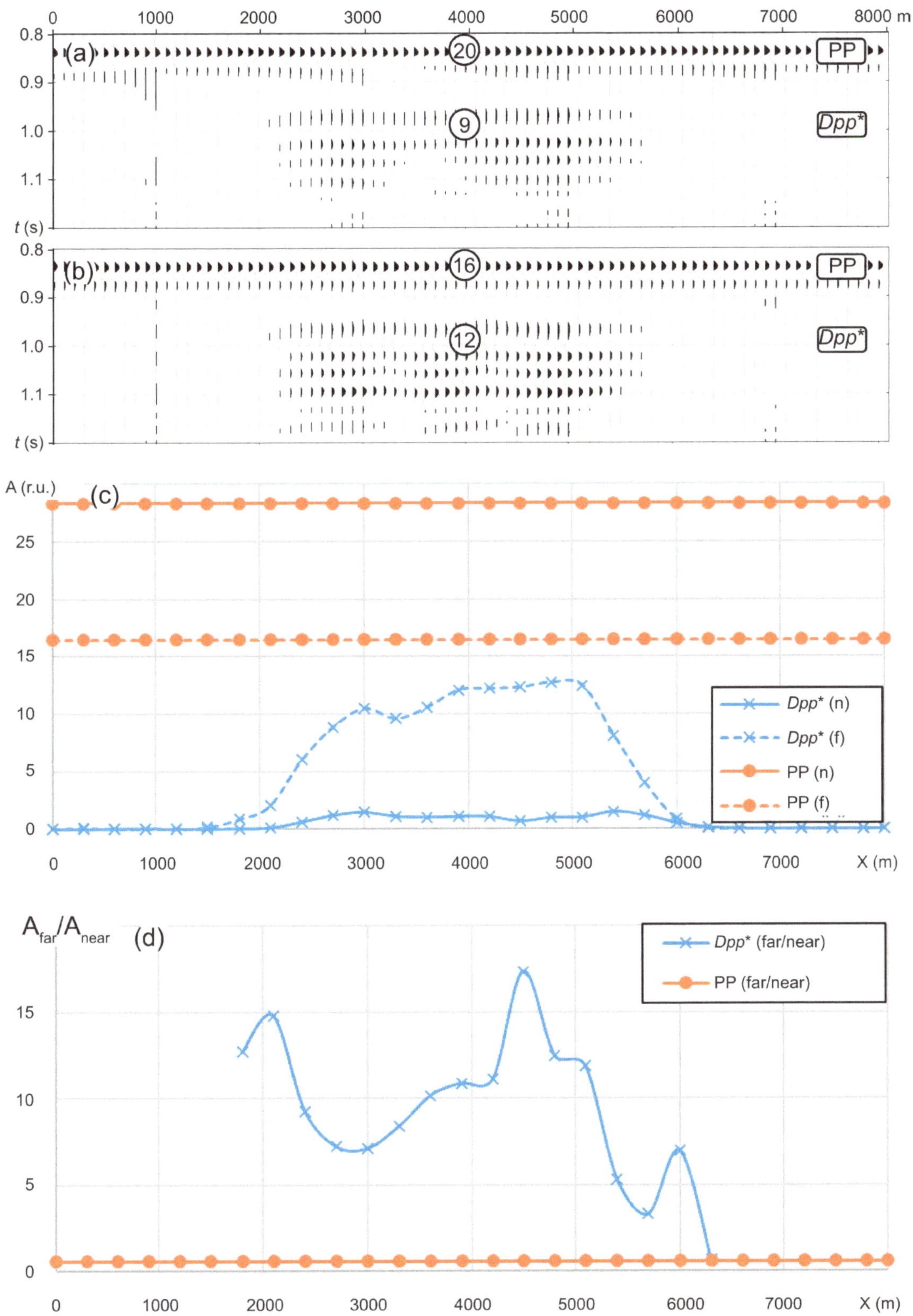

Figure 6.36. Main steps in analyzing CMP seismic data recorded on the *Z*-component with the goal of identifying a macrofracture cluster located under a reflecting boundary from scattered compressional waves: (a) full stack of all offsets ranging from –4 to 4 km; (b) far-offset stack of offsets ranging from –3 to –2 km and 2 to 3 km; (c) amplitudes of the scattered *Dpp** wave and the reflected PP-wave on the near-offset stack of offsets ranging from –1 to 1 km and on the far-offset stack; and (d) amplitude ratio of the far-offset stack to the near-offset stack, which is considered to be a recognition parameter for scattered waves.

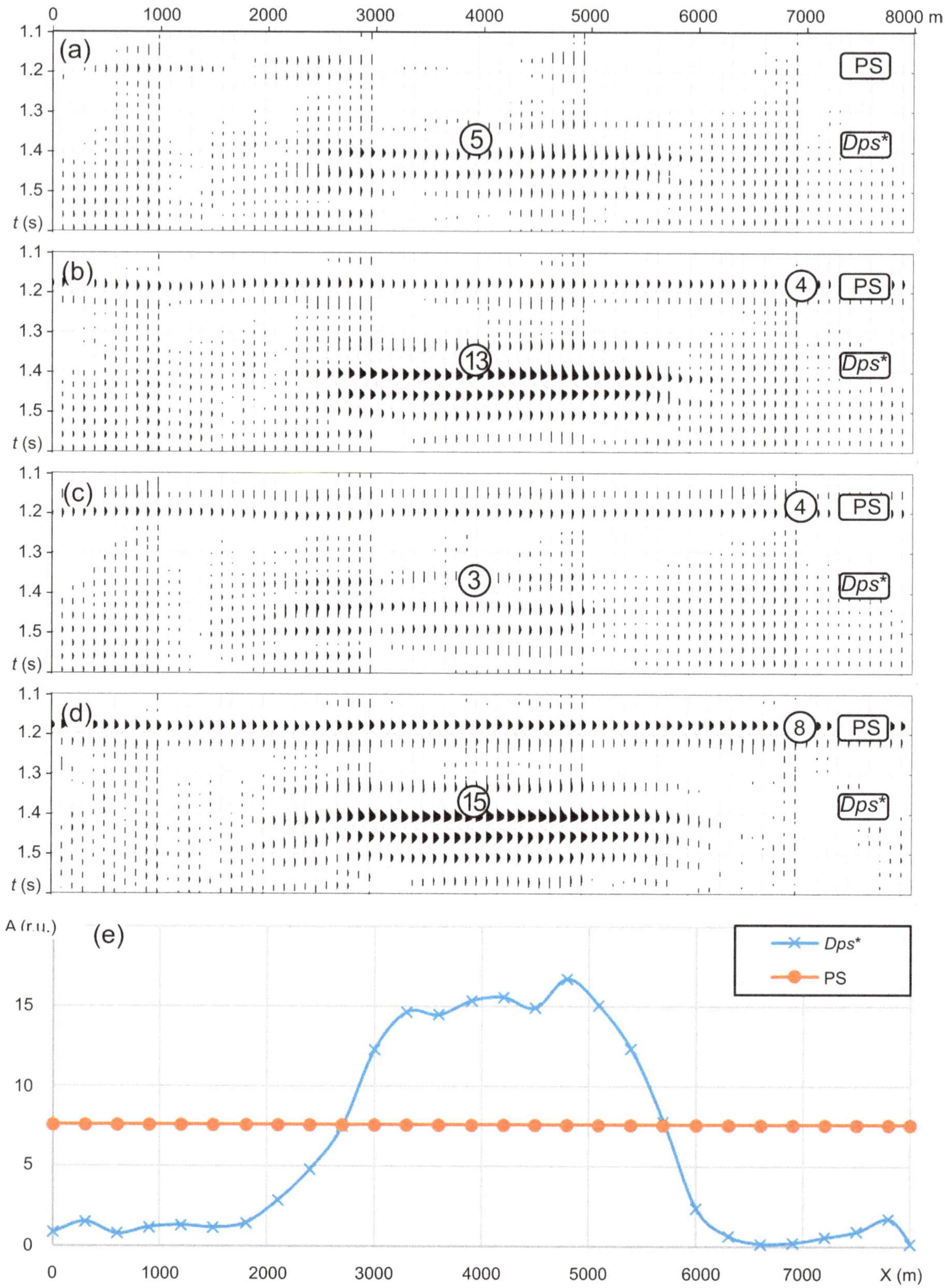

Figure 6.37. Main steps in analyzing CMP seismic data recorded on the *X*-component with the goal of identifying a macrofracture cluster located under a reflecting boundary from scattered converted waves: (a) full stack of all offsets ranging from –2 to 2 km without a polarity correction; (b) partial stack of offsets ranging from –2 to 0 km; (c) partial stack of offsets ranging from 0 to 2 km; (d) full stack of all offsets ranging from –2 to 2 km with a polarity correction; and (e) amplitudes of the scattered converted *Dps** wave and the converted PS-reflection on the full stack with the polarity correction.

The effect of the amplitude attenuation of a wave reflected beneath a macrofracture zone is an important indicator that may be used to identify the zone.

4) At offsets ranging from –2000 to 2000 m, corresponding to the depth of fracturing, the *X*-component amplitudes of a scattered converted wave are comparable with or greater than those of compressional reflections, and substantially exceed those of converted reflections.
5) An important feature of the scattered converted wave is a substantial difference in its structure and its amplitude between opposite sides of a receiver spread. This asymmetry is accompanied by a polarity reversal and a horizontal shift of the zero value of the AVO curve away from the spread center. All of these effects are caused by a deviation of fractures from the vertical.
6) Effective methods of using the *X*-component to identify a converted wave scattered by a macrofracture system include obtaining CMP stacks separately for opposite sides of spreads. Then these stacks are summed with and without a polarity correction. The latter approach considerably attenuates all reflections.
7) Principal indicators on the *X*-component that can be used to predict the presence of a bed-bounded fluid-filled macrofracture zone in a target interval include:
   - high amplitudes of the scattered converted wavefront,
   - a substantial difference in the structure and amplitude of the scattered waves on opposite sides of a receiver spread, even if the deviation of fractures from the vertical is minor, and
   - a horizontal shift of the zero value of the AVO curve away from the source location.

6.5. Conclusions from Chapter 6

1) Using finite-difference simulation, we have obtained a result concerning the polarity structure of a wave diffracted at a near-vertical fluid-filled macrofracture. This result is significantly different from the generally accepted result. We have shown the absence of a polarity reversal on the *X*-component and its presence on the *Z*-component of this wave.
2) We have established that information regarding the location and structure of a system (cluster) of near-vertical unidirectional fluid-filled macrofractures comparable with the wavelength and typical of hydrocarbon reservoirs in tight rocks is contained in scattered compressional waves and scattered converted waves. The former primarily are recorded at long offsets, whereas the latter are recorded at midrange offsets which are shorter than the cluster depth. Both wave types are formed as a result of interference of diffractions from the upper and lower tips of macrofractures.
3) The kinematic characteristics of scattered compressional and scattered converted waves from a macrofracture system correspond to hyperbolas of compressional and converted reflections from a hypothetical boundary running through the tips of macrofractures. Any scattered converted wave has a polarity reversal at the source location, as does any converted reflection.
4) The response from a bed with macrofractures differs from that from a bed with mesofractures by having lower (1.5 to 2 times) amplitudes and by the presence of scattered waves of both types from the lower tips of macrofractures.
5) We find that the energy characteristics of scattered waves depend significantly on the type of fracture infill. When the fractures are filled with fluid, the amplitude of the scattered converted wavefront is substantially higher than that of the scattered compressional wavefront. Conversely, when the fractures are filled with gas, the amplitude of the scattered converted wavefront is lower than that of the scattered compressional wavefront.
6) The *X*-component of the converted wavefront scattered by a fracture zone is insensitive to changes of fracture spacing within a rather wide range. In addition, it demonstrates low sensitivity to random fluctuations (up to 30% to 40% of the average value) of both fracture spacing and fracture deviation from the vertical in a fracture cluster.
7) The degree of asymmetry in amplitudes of converted waves scattered by macrofracture systems differs significantly from that of scattered compressional waves. The compressional waves are practically symmetric. For the scattered converted waves, the difference in amplitude between opposite sides of a source reaches 20% to 30%.
8) When multifold acquisition systems with long offsets are employed, identical methods are applicable for detecting both macrofracture and mesofracture systems. To identify a scattered compressional wave, the amplitude ratio of a far-offset stack to a near-offset stack can be complemented with assessing the AVO gradient at midrange offsets. For the scattered wave, the gradient is supposed to be positive.
9) Effective methods of using the *X*-component to identify a converted wave scattered by a macrofracture system include obtaining CMP stacks from midrange offsets separately for opposite sides of spreads.

Then these stacks are summed with and without a polarity correction. The latter approach considerably attenuates all reflections.

6.6. Recommendations regarding identification of fractured reservoirs in a more general formulation (based on the results discussed in Chapters 5 and 6)

Our investigations have demonstrated an important property of scattered waves for identifying fluid-filled fracturing, such as a positive sign of their AVO gradient. Applying this diagnostic tool automatically filters out all waves with a negative gradient.

Among all waves with a positive gradient, those that are of interest are characterized by a significant AVO gradient for the compressional and converted waves, a high degree of asymmetry in traveltime and amplitude for the converted waves, a horizontal shift of the zero value on the AVO curve (when the boundaries are near-horizontal), and azimuthal anisotropy of velocities.

The fractured beds or zones identified from seismic data, if any, also should be supported by well data acquired with scanners, cross-dipole acoustic tools, etc.

Chapter 7

Macrofractures in 3D Space: The 3D response of a system of near-vertical aligned macrofractures

Modern seismic investigations of hydrocarbon reservoirs are able to provide 3D images and volumetric estimates of properties of the subsurface. Fractured reservoirs typically contain systems (or clusters) of near-vertical macrofractures that form a network of fluid conduits. Detecting these fluid conduits and mapping the details of their structures require analysis of 3D wave responses from the systems of macrofractures. Numerical 3D simulation that employs the grid-characteristic method discussed in Part One is an efficient tool for this purpose.

7.1. Features of the 3D simulation technique

Let us consider in detail how the geometry of fractures in a medium is specified. Prior to automatic computations, in order to set the left- and right-side faces of each fracture, corresponding nodes are determined. A Cartesian coordinate system is used to specify the fracture orientation. The nodes related to the left-side face of a fracture are supposed to have lower coordinates than those related to the right-side face of the fracture. There is one-to-one correspondence between the fracture nodes: each node of the left-side wall has a corresponding node on the right-side face. When a parallelepiped grid is used (Golubev et al., 2013), only three orientations are possible, i.e., along each coordinate plane. Computed values at the nodes related to fracture boundaries depend on the type of infill, gas or fluid. In the case of gas-filled fractures, both fracture faces are considered to be free, that is, the internal fracture pressure is assumed to be much lower than the pressure in the host rock. In the case of fluid-filled fractures, the values at the wall nodes are computed so that the normal stress passes freely and a free slippage is ensured along the wall. The formulae below describe how the boundary and contact conditions on the surfaces of fractures are specified.

A surface of a gas-filled fracture is set as a free boundary:

$$
\begin{aligned}
T_{xx}n_x + T_{xy}n_y + T_{xz}n_z &= 0, \\
T_{xy}n_x + T_{yy}n_y + T_{yz}n_z &= 0, \\
T_{xz}n_x + T_{yz}n_y + T_{zz}n_z &= 0,
\end{aligned}
\tag{7.1}
$$

where T_{xx}, T_{xy}, T_{xz}, T_{yy}, T_{yz}, and T_{zz} are components of a stress tensor, whereas n_x, n_y, and n_z are components of the normal to the fracture surface at a reference point.

A surface of a fluid-filled fracture is accompanied by a free-slip boundary condition:

$$
\begin{aligned}
&v_{1x}n_x + v_{1y}n_y + v_{1z}n_z = v_{2x}n_x + v_{2y}n_y + v_{2z}n_z \\
&f_{1x}^n = -f_{2x}^n, \quad f_{1y}^n = -f_{2y}^n, \quad f_{1z}^n = -f_{2z}^n, \\
&f_{1x}^\tau = f_{2x}^\tau = 0, \quad f_{1y}^\tau = f_{2y}^\tau = 0, \quad f_{1z}^\tau = f_{2z}^\tau = 0,
\end{aligned}
\tag{7.2}
$$

where v_{1x}, v_{1y}, and v_{1z} are velocity components on one side of the surface, while v_{2x}, v_{2y}, and v_{2z} are velocity components on the other side of the surface; f_{1x}^n, f_{1y}^n, and f_{1z}^n are components of a normal force on one side of the surface, while f_{2x}^n, f_{2y}^n, and f_{2z}^n are components of a normal force on the other side of the surface; f_{1x}^T, f_{1y}^T, and f_{1z}^T are components of a tangential force on one side of the surface, while f_{2x}^T, f_{2y}^T, and f_{2z}^T are components of a tangential force on the other side of the surface.

This approach allows us to avoid using a computational grid inside the fracture and replace a computation-intensive solution of equations that describe a gas or liquid state with a correctly selected boundary condition. According to the numerical scheme used, the fracture specified in this manner is equivalent to a real fracture with an opening of less than 1 m positioned exactly between adjacent nodes of a computational grid.

To specify inclined 3D fractures, we use unstructured tetrahedral grids in some computations. The grids are designed using the open-access finalized TetGen™ library (http://wias-berlin.de/software/index.jsp?id=TetGen&lang=1). Here, an integration domain with inclusions—fractures—is defined first. Then, the domain is filled evenly with tetrahedra of predefined minimum volume that satisfy the Delaunay condition. (The Delaunay condition is met when the circumspheres of all tetrahedrons contain no other vertices inside). After this operation is completed, the points at the boundaries of the domain and fractures are optimized for subsequent computations. Next, the grid is divided into parts for use in computing systems with distributed memory. Each grid block that consists of coordinates of nodes, tetrahedra, triangles, and contacting pairs of triangles from the opposite sides of fractures is saved to a binary file and used by an individual computational node when the computational module is in operation. Sharing the integration domain among the computing system nodes at the initial stage helps to minimize the amount of data being transferred in the computational process, thus ensuring high computing speed and desired scale. Figure 7.1 shows examples of the parallelepiped and tetrahedral computational grids.

All nodal values in the domain are computed as follows:

1) Compute the nodal values of the first group using the selected computational scheme.
2) Correct the nodal values of the second group using an absorbing boundary condition and a free boundary condition.
3) Correct the nodal values of the third group according to the selected fracture infill.

Numerical solutions for such huge computational grids, containing millions and hundreds of millions of nodes, require substantial computational resources. To perform numerical simulation in a reasonable time, a parallel computing algorithm based on spatial decomposition is used. Each calculator is given a certain part of the domain in which it performs all of the calculations. At the boundaries between the areas belonging to different calculators, data values are exchanged to ensure that the resulting merged data is seamless. The CPU must be used instead of a calculator (or a core even though this will produce less accurate results). Figure 7.2 illustrates an exchange between border cells for a rectangular 2D grid as a visual aid and as an example of distribution of a computation domain among calculators. The coordinate-wise splitting is helpful in that, at each step, only the cells located along one of the axes participate in exchanging. Solid and dotted arrows denote exchanges at different splitting steps.

In the parallelepiped 3D grid used, the smallest cell size is $10 \times 10 \times 5$ m, and the number of computational nodes is more than 600 million for the base model.

The minimum size of a tetrahedron edge in unstructured tetrahedral grids used is 25 m, and the number of computational nodes is of the order of 200 million. Wave volumes for the model are computed for both gas-filled and fluid-filled macrofractures. The data for gas-filled macrofractures are used primarily for testing purposes.

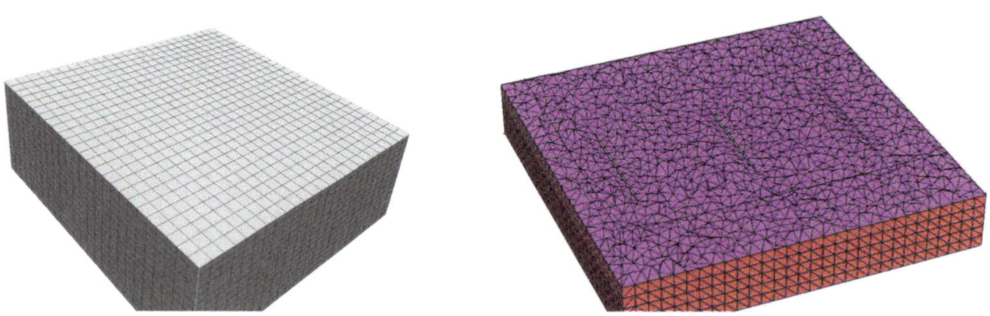

Figure 7.1. Parallelepiped (left) and tetrahedral (right) computational grids used in numerical 3D simulations.

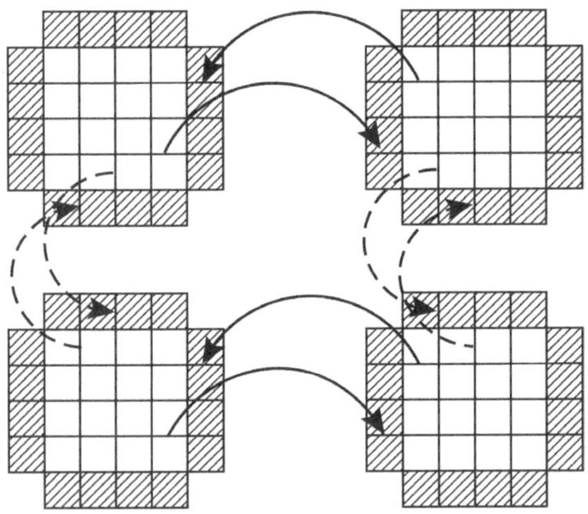

Figure 7.2. Schematic of a parallel computing algorithm based on spatial decomposition, showing how data are exchanged between border cells.

7.2. A 3D model of a macrofracture system and wave volume computation

The integration domain for investigation is $10 \times 10 \times 3$ km in a half space. The parameters of the medium are chosen to be close to those of carbonate rocks: compressional wave velocity $Vp = 4500$ m/s, shear wave velocity $Vs = 2250$ m/s, and density $\rho = 2500$ kg/m^3. In the middle of the medium at a depth of 2000 m, a system of 31 parallel vertical macrofractures occupying an area of 3×3 km is located. The fractures are 3 km long and 100 m high with 100 m spacing (Figure 7.3). Parameters of the medium containing the macrofracture system are identical to those of the host medium.

Seismic energy is generated at the earth's surface. The seismic signal is a 30 Hz rectangular wavelet with a length of 150 m. A plane wavefront is used as energy source under the assumption that a subsequent normal moveout (NMO) correction transforms the wavefield to a form close to that obtained using a point source. Because vertical fluid-filled macrofractures do not respond to a horizontal plane wavefront, a gently dipping (5°) plane wavefront is employed, with only one exception. The aperture of macrofractures, which ranges from a few fractions of a millimeter to a few millimeters in real conditions (i.e., is 5 orders of magnitude less than the seismic wavelength), is specified by a contact condition (i.e., is supposed to be minimal). Both gas-filled and fluid-filled fractures are investigated. For fractures filled with gas, a free boundary condition, as for the earth's surface, is specified. For fractures filled with fluid, a free-slip boundary condition is specified because an ideal fluid transfers only the normal stress at a boundary.

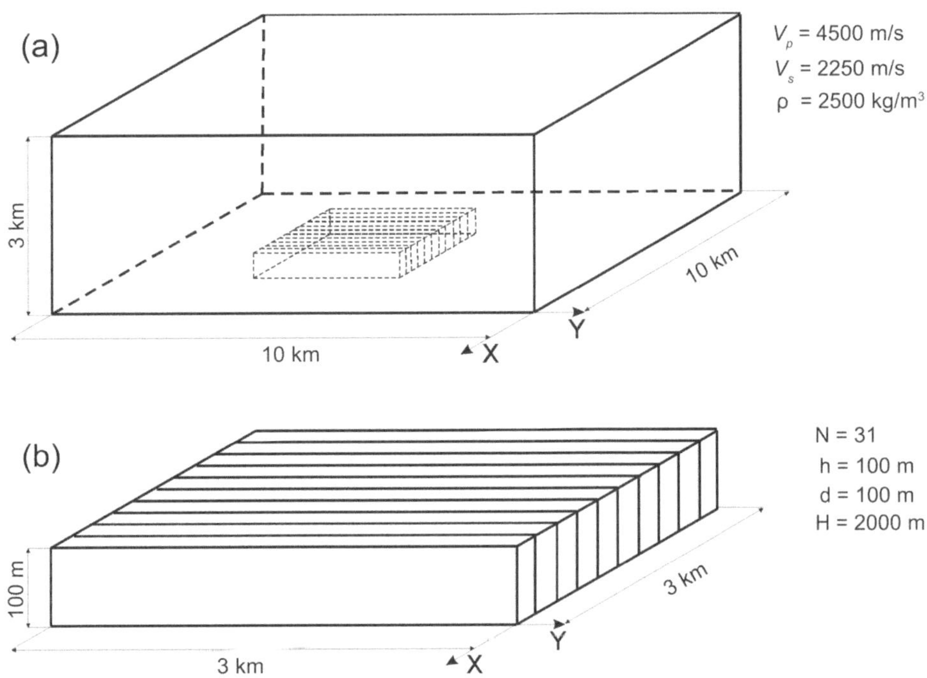

Figure 7.3. (a) Part of a half space containing a cluster of parallel vertical macrofractures, and (b) a 3D model of the cluster.

Figure 7.4. Part of a wave volume.

It should be noted that the widely used term "filled with gas" should be replaced by "empty" because, at the depths of the study of actual macrocracks, gas turns into fluid. Considering, however, that the mathematics for both is very similar, the term "filled with gas" also can be used as a close approximation to "empty."

Figure 7.4 shows the central part of the wavefield volume obtained at 700 ms after the start of the computation for a cluster of macrofractures filled with gas. It is located at a depth of 1000 to 2900 m and has identical sizes of 2000 m along both horizontal axes.

To analyze the properties of 3D wave responses, we use a series of snapshots and common-source gathers recorded on the earth's surface.

7.3. Common-source gathers of the 3D response from a single macrofracture in a 3D half space

As the basis of interpretation of 3D wave responses from clusters of macrofractures, first we examine the character of the 3D response from a single macrofracture.

Figure 7.5 shows orthogonal cross sections of the 3D response from a single empty (gas-filled) vertical macrofracture. The cross sections are normal to (Figure 7.5a) and parallel to (Figure 7.5b) the fracture plane.

The vertical component of the source gather shown at the top of Figure 7.5a contains two diffractions, compressional and converted. No polarity reversal is observed on either diffraction. However, although the amplitudes of the compressional wave are rather high along its entire front, the amplitudes of the converted diffraction decrease to almost zero over the macrofracture.

The horizontal *X*-component of the source gather shown in the middle of Figure 7.5a illustrates that the two diffractions have a polarity reversal over the macrofracture, with the amplitudes of the compressional wave being remarkably reduced at this place.

Because the horizontal *Y*-component of the source gather has zero amplitude, the bottom portion of Figure 7.5a is used to present a 3D image of diffraction from the macrofracture as a hyperbolic cylinder along the *Y* axis.

At the top of Figure 7.5b, the vertical component of the response cross section along the macrofracture plane shows a horizontal event, which is the compressional diffraction coincident with the macrofracture position. On the horizontal *X*-component shown in the middle of Figure 7.5b, a very weak horizontal event is observed, which is the converted diffraction. On the horizontal *Y*-component shown at the bottom of Figure 7.5b, two compressional waves diffracted at the macrofracture tips and having *Y*-components are visible.

Figure 7.6 demonstrates analogous cross sections for a macrofracture filled with fluid. The response here is many times weaker than that of the gas-filled fracture.

The vertical component of the source gather at the top of Figure7.6a shows the response as barely discernible compressional and converted diffractions.

At the top of Figure 7.6b, the vertical component of the response cross section along the macrofracture plane exhibits zero level of useful signals. Because the horizontal *Y*-components of both cross sections also are zero, they are not shown in this picture. Only on the horizontal *X*-components, shown at the bottom of Figures 7.6a and 7.6b, are wave responses clearly visible. At the bottom of Figure 7.6a, the diffraction has no polarity reversal over the macrofracture.

An event that is the converted diffraction from the macrofracture is shown at the bottom of Figure 7.6b. At the bottom of Figure 7.6a, the *X*-component of the converted wave diffracted at the vertical fluid-filled macrofracture has no polarity reversal. This is consistent with the similar effect mentioned in Chapter 6, where it is accompanied by a polarity reversal on the *Z*-component.

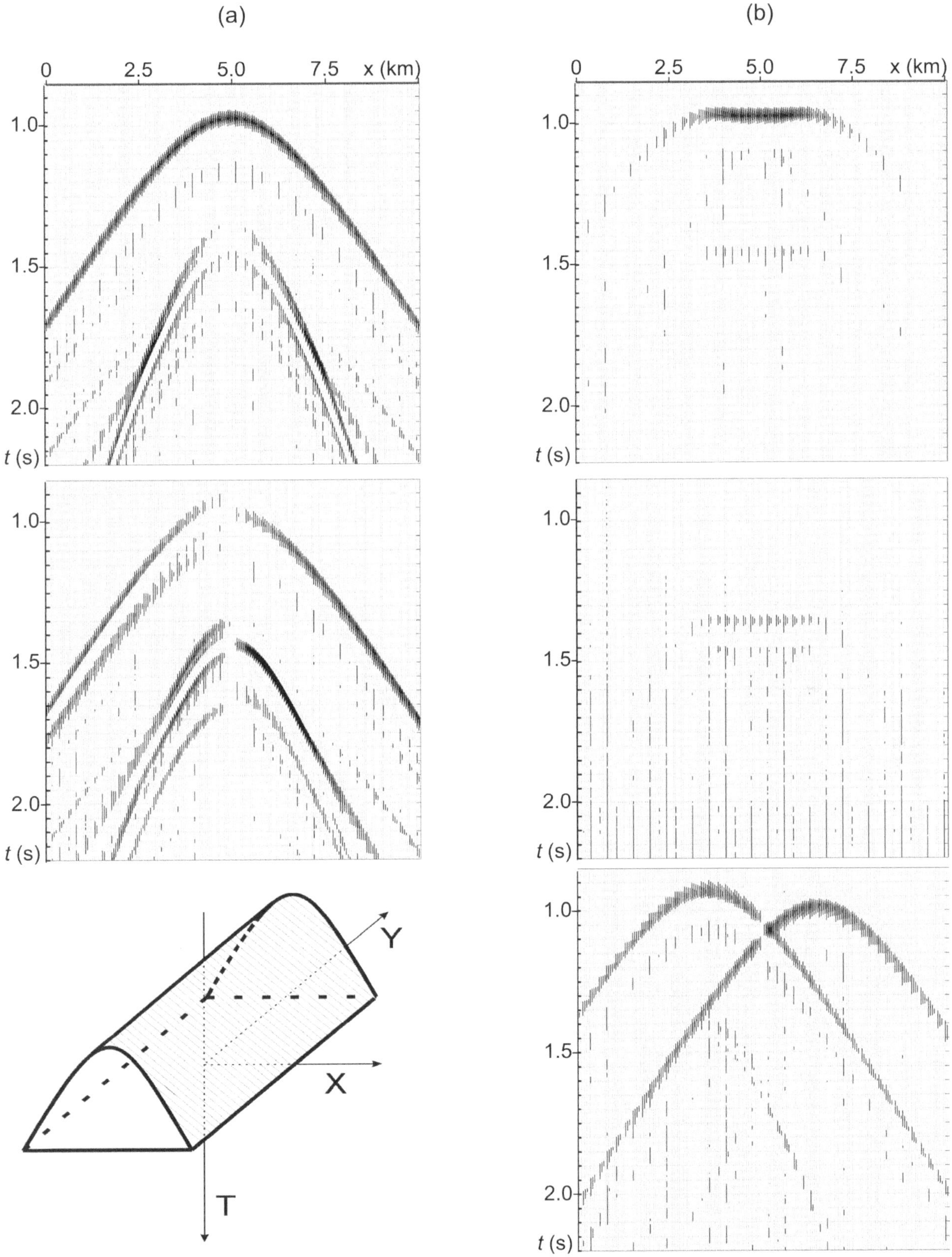

Figure 7.5. Cross sections of the 3D wave response from a single empty vertical macrofracture: (a) orthogonal to and (b) parallel to the fracture plane. Each cross section contains the vertical component (top), the horizontal *X*-component (middle), and the horizontal *Y*-component (bottom). Because the horizontal *Y*-component of the response has zero amplitude, the bottom portion of Figure 7.5a presents a 3D image of diffraction from the macrofracture as a hyperbolic cylinder along the *Y* axis.

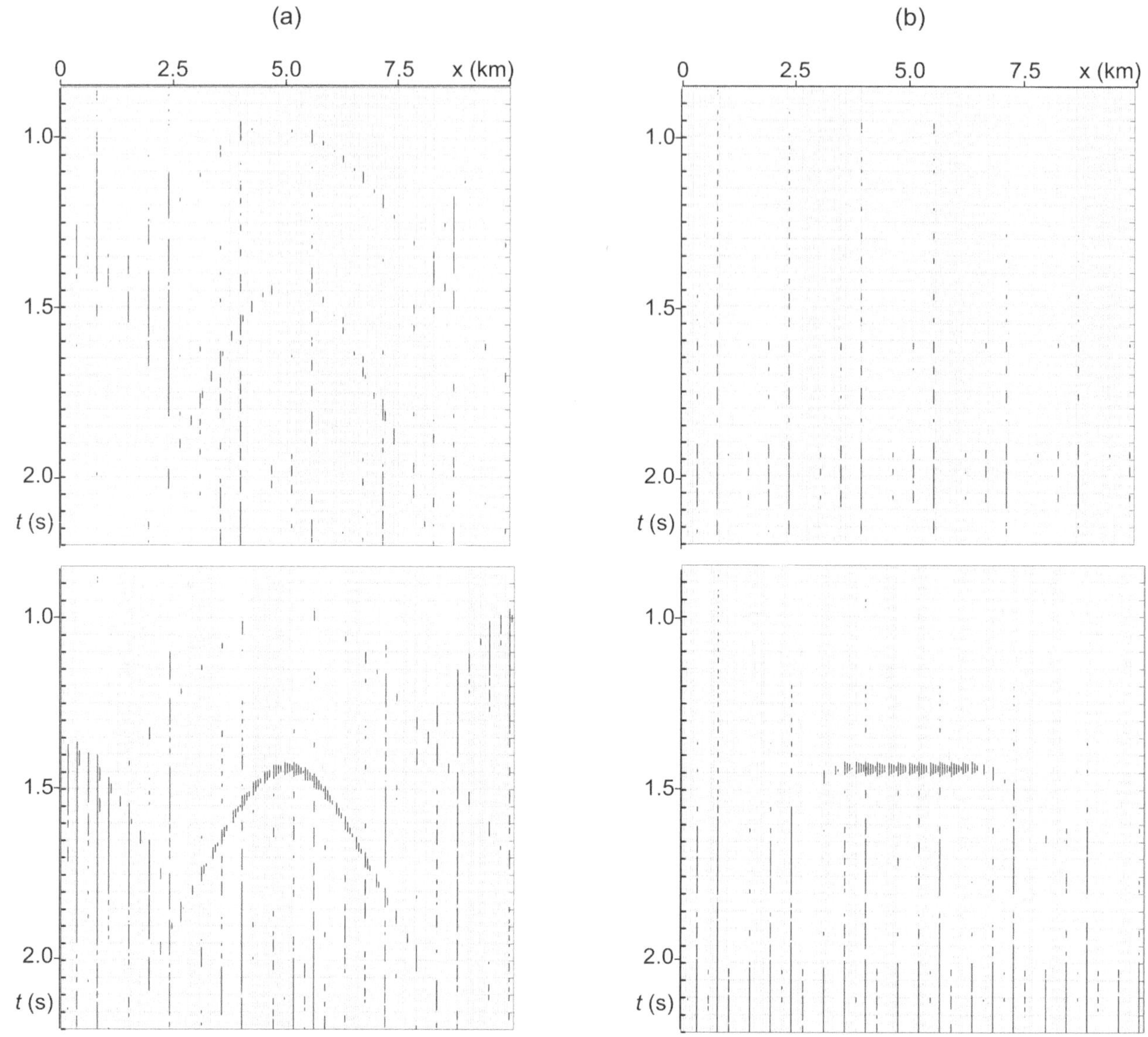

Figure 7.6. Cross sections of the 3D wave response from a single vertical macrofracture filled with fluid: (a) orthogonal to and (b) parallel to the fracture plane. Each cross section contains the vertical component (top) and the horizontal *X*-component (bottom).

7.4. Characteristics of the 3D response from a cluster of macrofractures on snapshots

Snapshots displayed in Figure 7.7 show how the seismic wave response from a system of macrofractures filled with gas develops over time. Figures 7.7a and 7.7b contain snapshots in the central *OXZ*-plane and central *OYZ*-plane, which are perpendicular and parallel to the planes of fractures, respectively, illustrating the formation and the upward and downward movement of several straight-line wavefronts located over and beneath the fracture cluster, as well as the development of concentric fronts of diffracted waves from the edge macrofractures of the cluster.

To demonstrate the nature of formation of components of the seismic response from a cluster of fractures filled with gas and fluid, snapshots in the central *OXZ*-plane and central *OYZ*-plane at 850 ms are shown in Figures 7.8 and 7.9, respectively. These snapshots are used to estimate the amplitude and direction of particle motion.

Tables 7.1 and 7.2 summarize the amplitudes of identified waves in relative units and as percentages of incident wave amplitude of 0.05 r.u.

Figures 7.8 and 7.9 show that, after the incident compressional wavefront has passed through the cluster

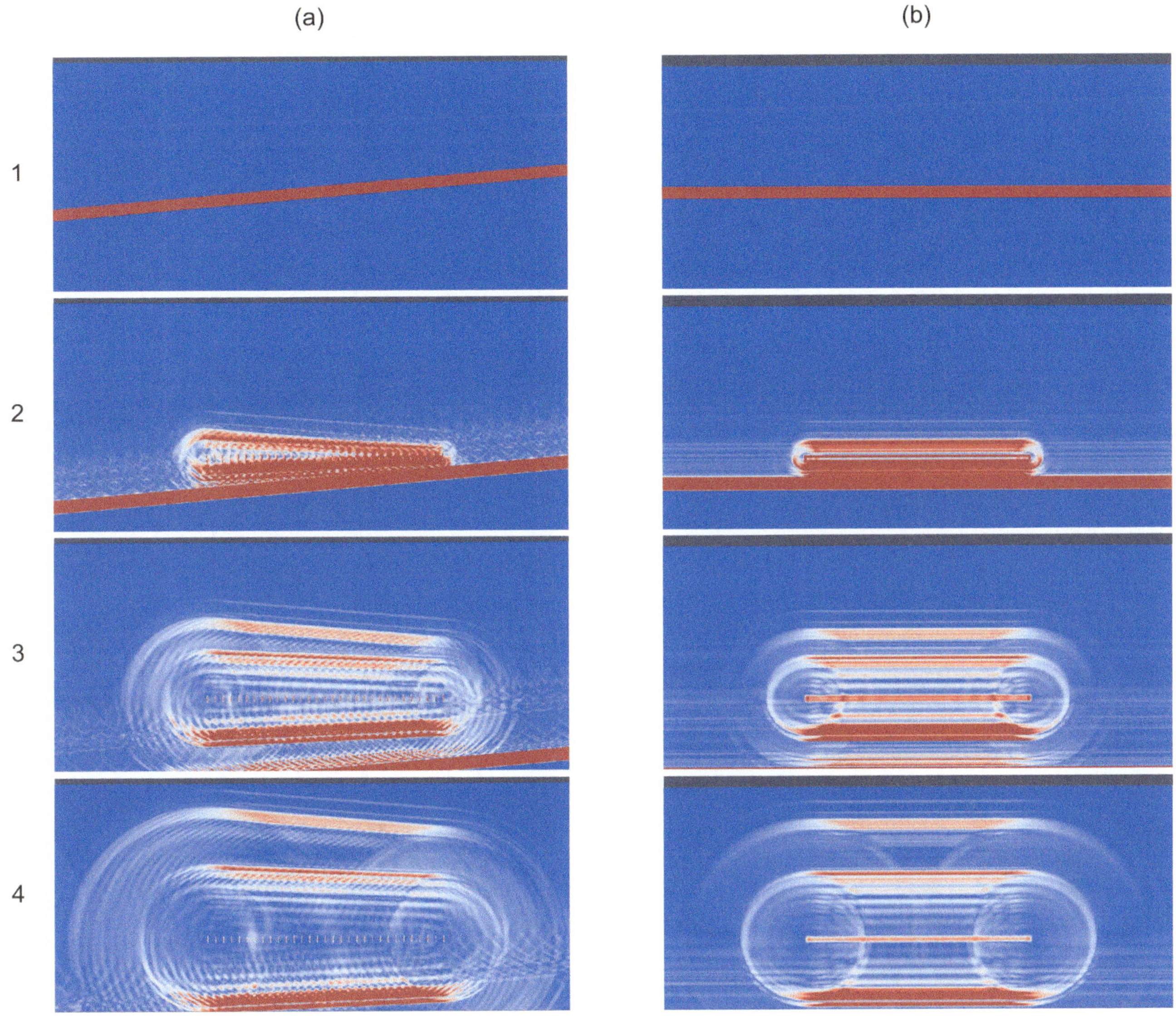

Figure 7.7. Snapshots of the absolute value of particle velocity showing how the seismic wave response from a system of macrofractures filled with gas develops over time in the (a) central *OXZ*-plane, which is perpendicular to the planes of fractures, and (b) central *OYZ*-plane, which is parallel to the planes of fractures; 1 = 400 ms, 2 = 550 ms, 3 = 700 ms, and 4 = 850 ms.

of macrofractures, the fronts of scattered compressional and scattered shear waves, resulting from the interference of diffractions from the macrofracture tips, propagate both upward and downward.

The main upgoing elements of the response are the fronts of scattered compressional (3) and scattered converted (4 and 5) waves. When fractures are filled with gas, the waves have approximately equal amplitudes (see Table 7.1 and Figure 7.8). The amplitude of the downgoing transmitted scattered converted wavefront (11) is substantially higher (by 1.5 to 2 times) than that of the upgoing compressional and shear wavefronts (3 and 5). Figure 7.8c shows the wave character as vectors. It is obvious that all shear particle motions correspond to the SV type and are in the *OXZ*-plane. The main shear wavefronts (4, 5, and 11) are followed by the periodical, gradually decreasing in amplitude upgoing (6 and 7) and downgoing (9 and 10) wavefronts of the same (SV) type. They apparently are associated with intra-cluster reflections. Any such events following the compressional wavefront (3) are almost unidentifiable. The empty macrofractures of the cluster themselves are a strong source of shear (SV) particle motion (8).

Figure 7.9 presents snapshots for the same cluster but with the macrofractures filled with fluid rather than

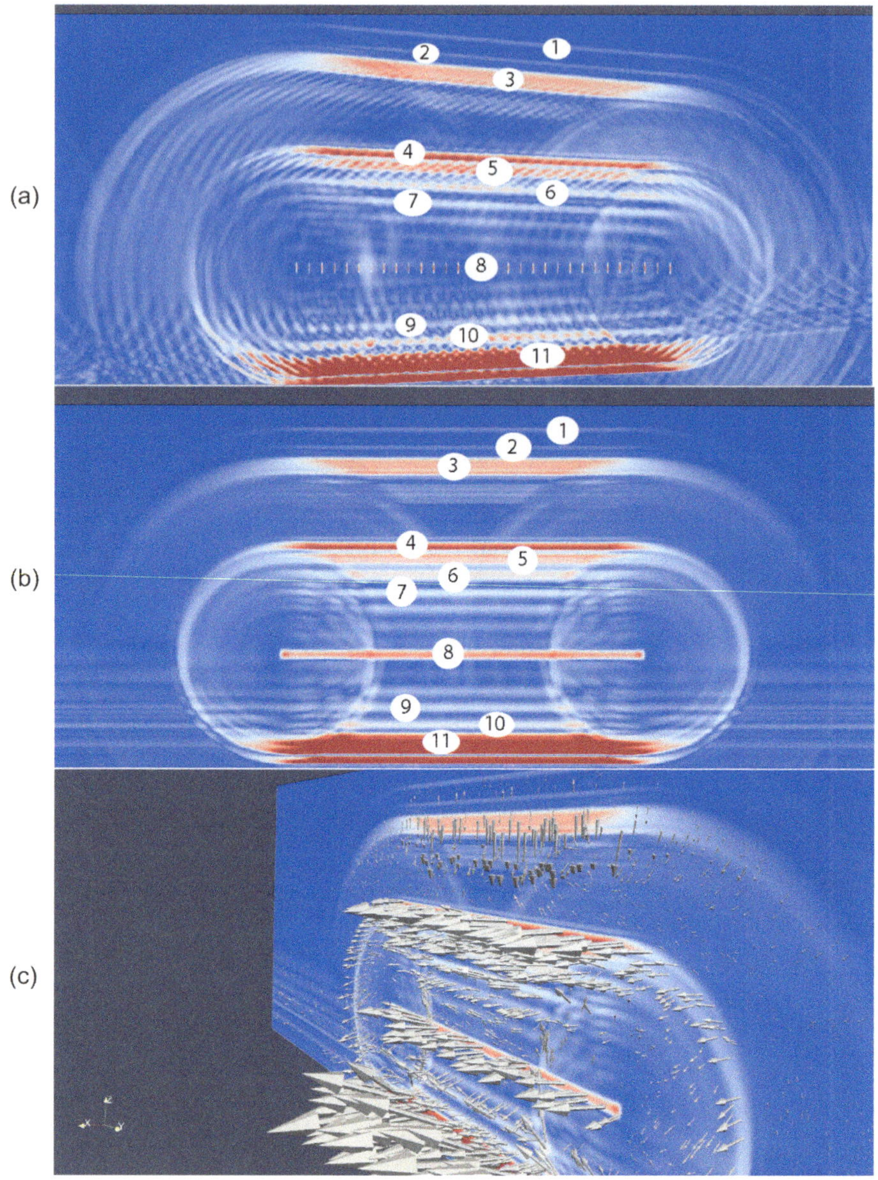

Figure 7.8. Snapshots of the absolute value of particle velocity in the (a) central *OXZ*-plane, which is perpendicular to the planes of macrofractures, (b) central *OYZ*-plane, which is parallel to the planes of macrofractures, and (c) central *OYZ*-plane with vectors indicating the amplitude and direction of particle motion. The snapshots are related to 830 ms, and the macrofractures are filled with gas. Here, 1 and 2 = weak cycles preceding the scattered compressional wavefront; 3 = scattered compressional wavefront; 4 and 5 = cycles of the scattered converted wavefront with shear SV particle motion; 6 and 7 = cycles of the periodical tail with SV particle motion induced by wavefronts 4 and 5; 8 = waves from macrofractures; 9 and 10 = cycles of the periodical tail with SV particle motion induced by the transmitted scattered converted wavefront; and 11 = transmitted scattered converted wavefront.

Table 7.1. Amplitudes (in relative units and as percentage) of waves from a cluster of macrofractures filled with gas (Figure 7.8).

Wave	1	2	3	4	5	6	7	8	9	10	11
Amplitude in relative unts	2	2	10	14	11	8	6	10	6	7	26
Amplitude in percentage of that of the incident P-wave 0.05	0.4	0.4	2.0	2.8	2.2	1.6	1.2	2.0	1.2	1.4	5.2

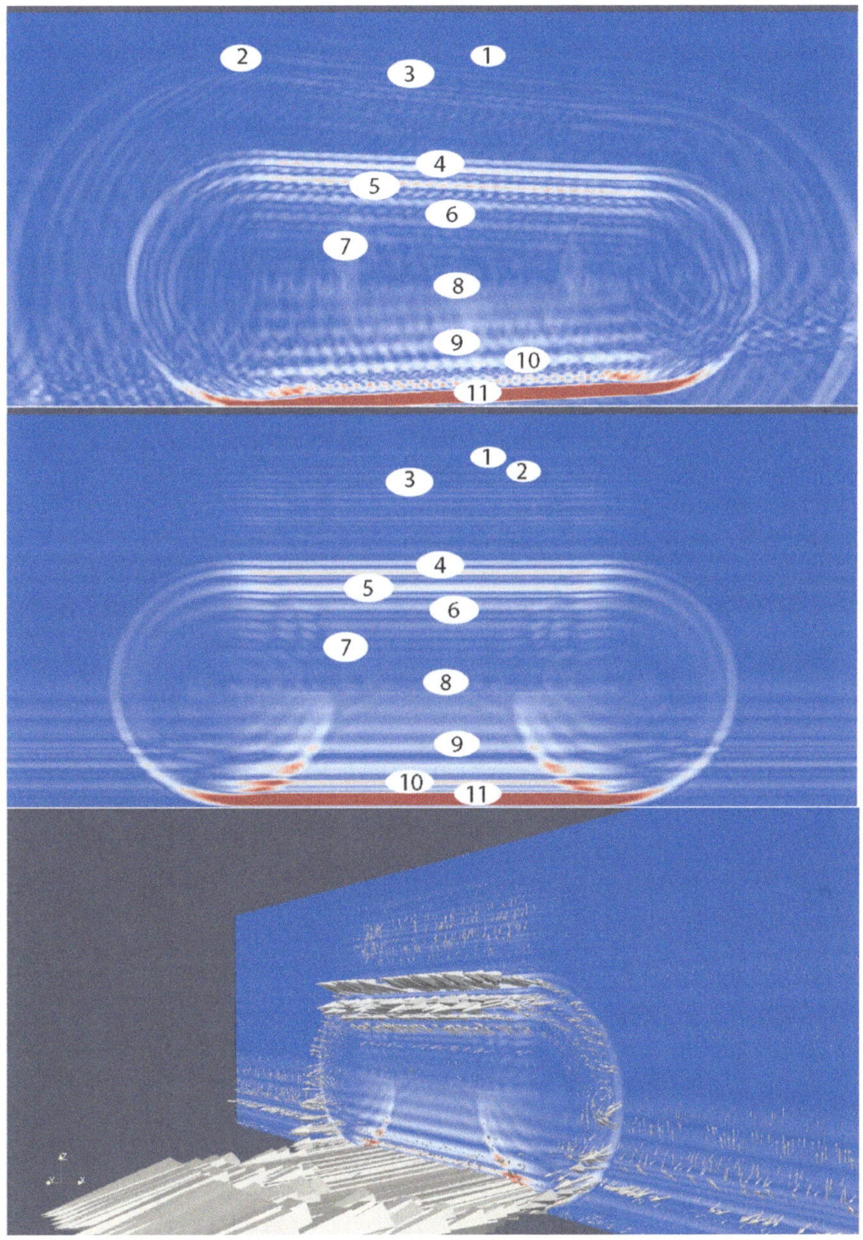

Figure 7.9. Snapshots of the absolute value of particle velocity in the (a) central *OXZ*-plane, which is perpendicular to the planes of macrofractures, (b) central *OYZ*-plane, which is parallel to the planes of macrofractures, and (c) central *OYZ*-plane with vectors indicating the amplitude and direction of particle motion. The snapshots are related to 830 ms, and the macrofractures are filled with fluid. For wave designations, refer to Figure 7.8.

Table 7.2. Amplitudes (in relative units and as percentage) of waves from a cluster of macrofractures filled with fluid (Figure 7.9).

Wave	1	2	3	4	5	6	7	8	9	10	11
Amplitude in relative units	0.5	1	1.2	8	7.4	4	2	1	5	8.8	40
Amplitude in percentage of that of the incident P-wave 0.05	0.1	0.2	0.24	1.6	1.48	0.8	0.4	0.2	1.0	1.76	8.0

7.6. Common-source gathers of the 3D response from a cluster of macrofractures filled with gas

For a detailed investigation of the response from a cluster of macrofractures, we use three orthogonal components of common-source gathers recorded on the earth's surface. This allows all of the different elements of the response to be identified—compressional PP-waves, shear SV-waves, and, probably, shear SH-waves.

Similar to Figure 7.11, the 3D wave response of a cluster of vertical gas-filled macrofractures to a plane-wavefront energy source inclined by 5° is illustrated in Figure 7.13. It is represented by cross sections in the *OXZ*-plane, which is orthogonal to fractures, and in the *OYZ*-plane, which is parallel to fractures. From these sections, we can identify the following main components.

In the 0.89 to 1.02 s interval on the vertical component shown at the top of Figure 7.13, the ***first wavetrain*** is visible. Denoted as 1, it is formed by the scattered compressional wavefront and other main waves. The first cycle of the wavefront, which has an amplitude of 11.64 r.u. and a period of 80 ms, dominates in amplitude. It is followed by two to three cycles with amplitudes ranging from 5.24 to 6.43 r.u. and with a much lower prevailing period of 30 ms. The horizontal part of the wavefront is consistent with the cluster width. Outside the cluster, the front weakens rapidly and turns into the hyperbolic waves diffracted at the edges of the cluster. These waves are denoted as 2. The structures of the front on the orthogonal cross sections are almost the same. In the 1.1 to 1.5 s time interval on both cross sections, a rather weak multicycle wavetrain of 50 ms period is observed as a tail of the scattered compressional wavefront. It probably is generated by intra-cluster reflections of compressional waves. At 1.85 s, the first multiple reflection of the scattered compressional wavefront is observed.

The ***second wavetrain*** is formed by the scattered converted wavefront with the SV particle motion and is denoted as 4. It appears as a strong low-frequency event with an amplitude of 3.09 r.u. in the 1.35 to 1.40 s interval on the *X*-components of both orthogonal cross sections in the middle of Figure 7.13. Outside of the intensive, straight-line segment, which has a length equal to the cluster width, the front turns into intensive multicycle edge hyperbolas, denoted as 4. At the cluster ends shown in the middle of Figure 7.13a, the hyperbolas have opposite signs. In the middle of Figure 7.13b, i.e., on the *X*-component of the cross section parallel to fractures, the edge waves are much shorter and weaker. With a 150 ms time delay, the principal single-cycle wavefront is followed by an approximately 50 ms long train of periodic events. It is denoted as 6, also has SV particle motion, and is generated by intra-cluster reflections of converted waves. On the *Y*-component shown at the bottom of Figure 7.13, this wavefront is practically invisible.

The ***third multicycle wavetrain***, denoted as 8, is most pronounced in the 1.50 to 1.90 s interval on the vertical component and the horizontal *X*-component of the cross section parallel to fractures, which is shown at the top and in the middle of Figure 7.13b, respectively.

The converted wavefront, denoted as 4, is missing at the top and middle of Figure 7.13b. Here, the events of the third wavetrain, denoted as 8, are partially superimposed on multicycle trains of compressional and converted waves, denoted as 3 and 4, respectively. The width of the events in this wavetrain is close to the cluster width, while the amplitude of the first five to six phases is adequately pronounced.

In the time interval of the third wavetrain, horizontal events are missing on the vertical component of the cross section orthogonal to fractures, shown at the top of Figure 7.13a. Instead, a wavefield with a staggered structure is observed, which is formed by the steep flanks of converted waves diffracted at the tips of the macrofractures. These diffractions have a substantial vertical component with amplitude ranging from 2.0 to 3.3 r.u.

At 1.6 s on the cross section orthogonal to fractures, shown at the bottom of Figure 7.13a, a multicycle scattered converted wavetrain with rather low amplitude of 0.3 r.u. is observed, denoted as 7. In structure, this wavetrain resembles both the scattered converted wavefront observed at 1.35 s, which has SV particle motion and is denoted as 4, and its multicycle tail on the horizontal *X*-component. The nature of this wavetrain is associated with the front of SV-waves diffracted at the ends of the cluster parallel to the *X* axis.

At 0.9 s on the horizontal *Y*-component of the cross section parallel to fractures, shown at the bottom of Figure 7.13b, compressional waves diffracted at the tips of the macrofractures are observable. These waves are denoted as 2 and apparently have a horizontal *Y*-component. There may be alternative explanations for the nature of these waves, however.

When a horizontal plane wave impinges on vertical macrofractures, these listed characteristics of the response on the vertical component and horizontal *Y*-component remain, which is clearly visible at the top and bottom of Figure 7.14.

It is important to note that a scattered converted wavefront, denoted as 4, and observed at 1.35 s on snap-

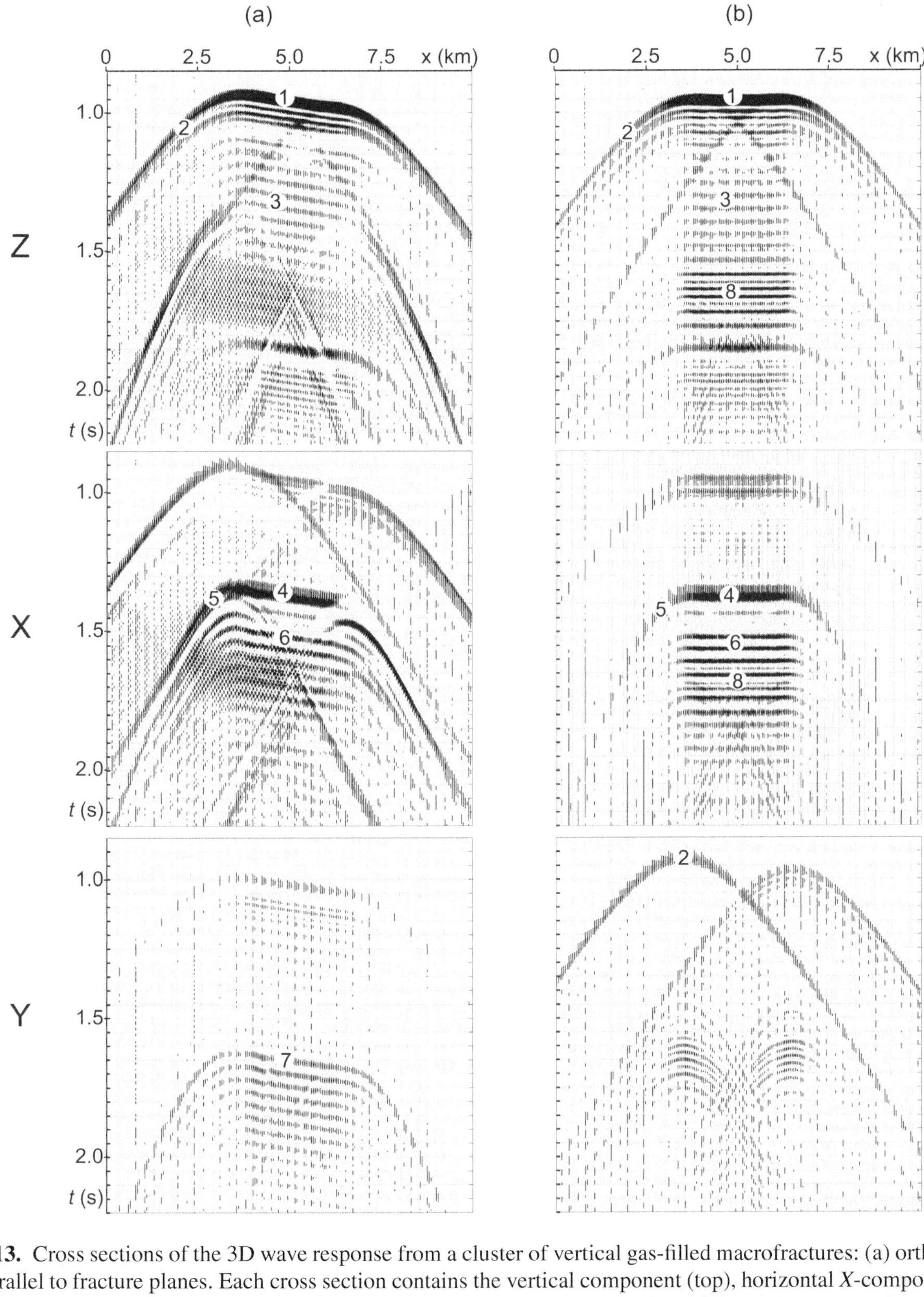

Figure 7.13. Cross sections of the 3D wave response from a cluster of vertical gas-filled macrofractures: (a) orthogonal to and (b) parallel to fracture planes. Each cross section contains the vertical component (top), horizontal *X*-component (middle), and horizontal *Y*-component (bottom). A plane-wave energy source inclined by 5° is used: 1 = scattered compressional wavefront; 2 = compressional diffractions from the edges of the cluster; 3 = intra-cluster reflections of compressional waves; 4 = scattered converted wavefront with SV particle motion; 5 = converted diffractions from the edges of the cluster; 6 = intra-cluster reflections of converted waves; 7 = probable front of SH-waves; and 8 = side cross-sections of steep flanks of compressional and converted diffraction hyperbolas.

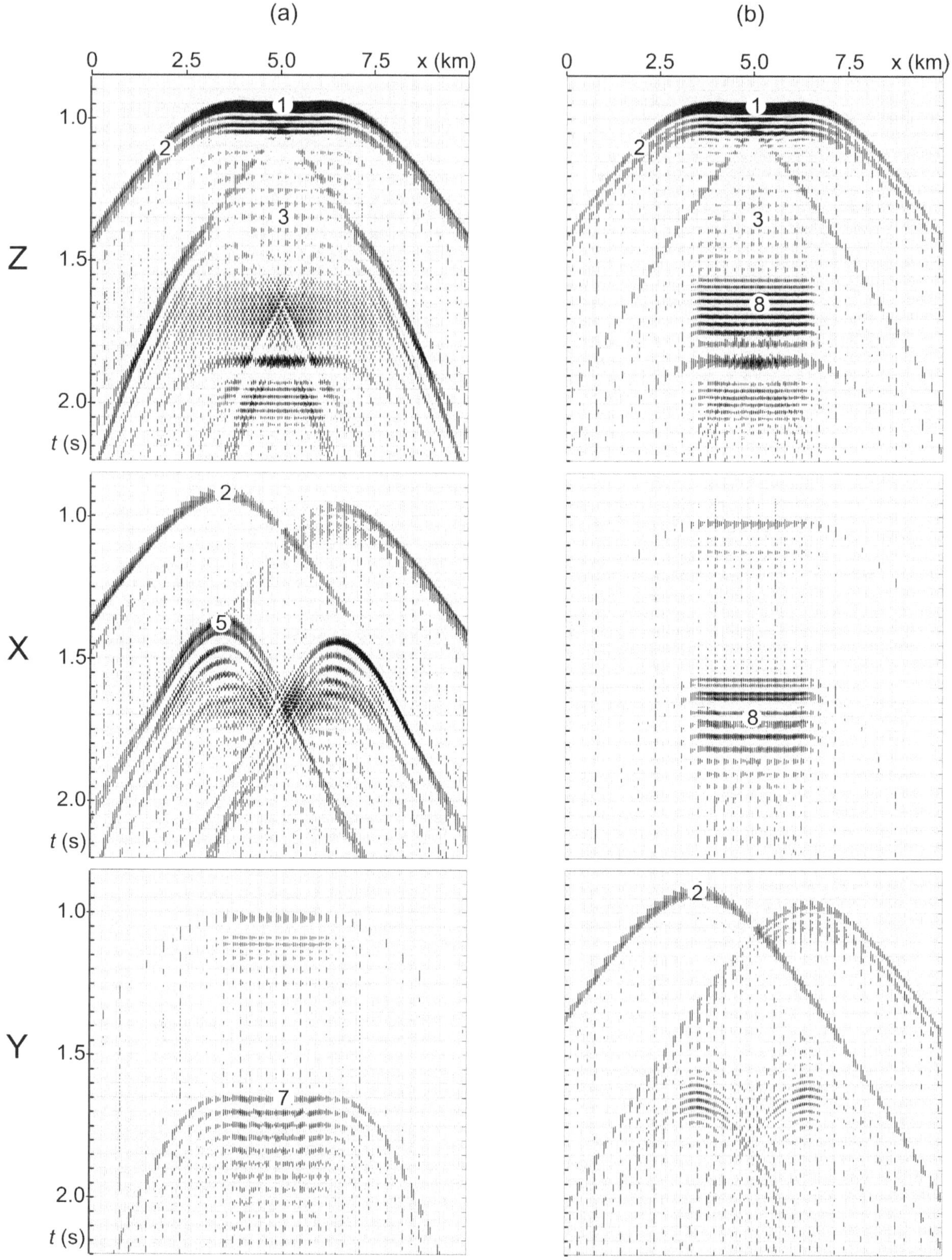

Figure 7.14. Cross sections of the 3D wave response from a cluster of vertical gas-filled macrofractures (a) orthogonal to and (b) parallel to the trend of fractures. Each cross section contains the vertical component (top), horizontal *X*-component (middle), and horizontal *Y*-component (bottom). A horizontal plane-wavefront energy source is used. For wave designations, refer to Figure 7.13.

shot aX in Figure 7.13 is totally missing from snapshot aX in Figure 7.14.

On snapshot aX in Figure 7.14, individual compressional *Dpp* diffractions, denoted as 2, and converted *Dps* diffractions, denoted as 5, from the cluster edge fractures are dominant. On snapshot bX in Figure 7.14, events, denoted as 8, are only present; they are side cross-sections of steep flanks of the compressional and converted diffraction hyperbolas.

7.7. Difference in wave composition between the cross sections of the 3D response parallel and orthogonal to fractures

Depending on the type of macrofracture infill, gas or fluid, the wave compositions of seismic responses are significantly different along and across the macrofracture planes. The difference is especially great when a horizontal plane wave impinges on vertical gas-filled macrofractures (see Figure 7.14). Differences between the two orthogonal cross sections are visible on the vertical component at the top of Figure 7.14, on the horizontal *X*-component in the middle of Figure 7.14, and on the horizontal *Y*-component at the bottom of Figure 7.14. The differences are especially significant on both horizontal components. As shown in Figures 7.13 and 7.14 for both types of the plane-wavefront energy source, the difference on the vertical component is minimal, whereas the difference on the horizontal *Y*-component is maximal. On the horizontal *X*-component, the difference between Figures 7.13 and 7.14 is much less, because of a strong front of scattered converted waves present on both orthogonal cross sections.

As shown in Figures 7.11 and 7.12 for macrofractures filled with fluid, the difference between waves on the orthogonal cross sections is minor.

In order to quantitatively describe the degree of similarity or difference between waves on the orthogonal cross sections, Table 7.4 summarizes the amplitudes of scattered compressional and scattered converted wavefronts for the two clusters of fluid-filled fractures. The amplitudes are given as percentages of the amplitude of the incident wavefront taken to be equal to 0.05 r.u. For both cluster models, the difference between the horizontal *X*-components of the scattered converted wavefronts *Dps* on the orthogonal cross sections is insignificant and does not exceed 6% to 7%. The amplitudes of the other wave elements are also quite close. So, we can assert that there is no practical difference in the wave composition between the orthogonal cross sections of the 3D response from the cluster of macrofractures filled with fluid.

7.8. Character of the amplitude anomaly induced by a cluster of fluid-filled macrofractures on 3C-3D data

The applicability of using amplitude anomalies of scattered converted wavefronts for detecting clusters of fluid-filled near-vertical macrofractures is shown in Leviant et al. (2012) based on an analysis of results of numerical 2D modeling.

Figure 7.12 shows how the front of converted waves scattered by a cluster of parallel fluid-filled macrofractures manifests itself on the horizontal *X*-component of the cross sections of the 3D response that are perpendicular and parallel to the strike of macrofractures.

Several amplitude snapshots are selected from the time interval of recording this wavefront. Those related to times 1.18, 1.23, and 1.28 s are presented in Figures 7.15a–c, respectively. The amplitude anomaly is most pronounced in Figure 7.15b.

Figure 7.15d shows the amplitude anomaly assessed quantitatively, with a tie to the modeling target. The interval between isolines in this figure is chosen to be one-sixth of the maximum amplitude, which is approximately 1% of that of the incident wave. Most of the square

Table 7.4. Amplitudes of the principal response waves as percentages of the amplitude of the plane incident wave on two orthogonal cross sections.

Model	Fracture deviation from the vertical, horizontal plane-wave energy source				Vertical fractures, slightly inclined plane-wave energy source			
Component	*X*		*Z*		*X*		*Z*	
Wave	*Dpp*	*Dps*	*Dpp*	*Dps*	*Dpp*	*Dps*	*Dpp*	*Dps*
***OXZ*-plane**	0.4%	6.4%	1.6%	0.4%	0.0%	5.7%	1.6%	0.2%
***OYZ*-plane**	0.4%	6.0%	1.2%	0.4%	0.0%	5.3%	1.8%	0.2%

Table 8.1. Maximum amplitudes (in relative units) of waves generated by megafractures with different types of infill.

Model type	Fracture tip	Compressional wave				Converted wave			
		X-component		*Z*-component		*X*-component		*Z*-component	
		Amplitude	Polarity reversal	Amplitude	Polarity reversal	Amplitude	Polarity reversal	Amplitude	Polarity reversal
20 fluid-filled macrofractures	Upper Lower	1.00 2.50	No	1.17 3.59	Yes	3.36 9.23	No	1.50 2.93	Yes
20 gas-filled macrofractures	Upper Lower	8.81 11.88	Yes	6.91 11.53	No	8.85 24.89	Yes	6.96 14.86	Yes
Averaged model	Upper Lower	2.48 3.27	Yes	3.20 4.14	No	2.14 2.78	No	1.84 2.63	Yes

that from the megafracture filled with a homogeneous rock with reduced physical parameters.

- The amplitudes of both diffracted waves, *Dpp* and *Dps*, are higher from the lower tips of the megafractures than from the upper tips.
- For all types of infill, the *X*-component of the converted *Dps*-2 diffraction has the highest amplitude of all of the response waves.
- There is no polarity reversal on the *X*-component of the converted waves diffracted at the fluid-filled megafracture and the *Z*-component of the compressional waves diffracted at the gas-filled megafracture.

8.3. Effects of megafracture internal structure on wave response

The effect of the internal structure of a megafracture on the wave response is investigated in the framework of the general base model outlined in the previous section, assuming that the macrofractures within the megafracture are filled with fluid.

Figure 8.3 shows five primary types of megafracture models with different internal structures. The impact of the megafracture's internal structure on the character of the wave response from that fracture is illustrated in Figure 8.4, which presents the *X*- and *Z*-components of common-source gathers for these five model types. They all exhibit identical composition of wave response that, as discussed in Section 8.2, consists of two compressional waves and two converted waves diffracted at the upper and lower tips of the megafracture.

It is worth noting that, for all five models, the *X*-component of the converted diffraction does not have any polarity reversal. Here, the waves diffracted at the lower tips of the megafractures are clearly visible along the entire seismic spread. The *Dps*-2 wave of model 2, with a rhombic network of macrofractures, is the only exception. For models 1 and 2, the *X*-components of the compressional diffractions have zones of decreased

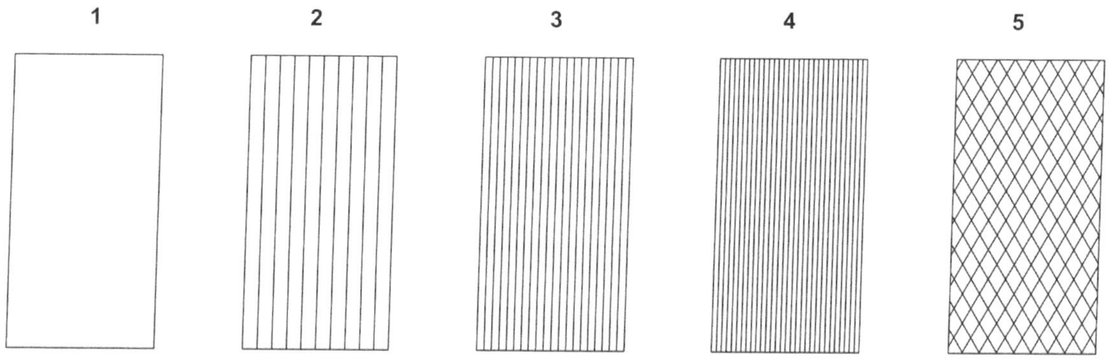

Figure 8.3. Five primary model types for the internal structure of megafractures: 1 = 20 parallel macrofractures having the same parameters as the host rock; 2 = a rhombic network of macrofractures having the same parameters as the host rock; 3 = homogeneous rock with a 20% decrease in physical parameters relative to those of the host rock; 4 = 20 parallel macrofractures with a 20% decrease in physical parameters relative to those of the host rock; and 5 = a rhombic network formed by two systems of macrofractures deviated by 30° and –30° from the vertical and having a 20% decrease in physical parameters relative to those of the host rock.

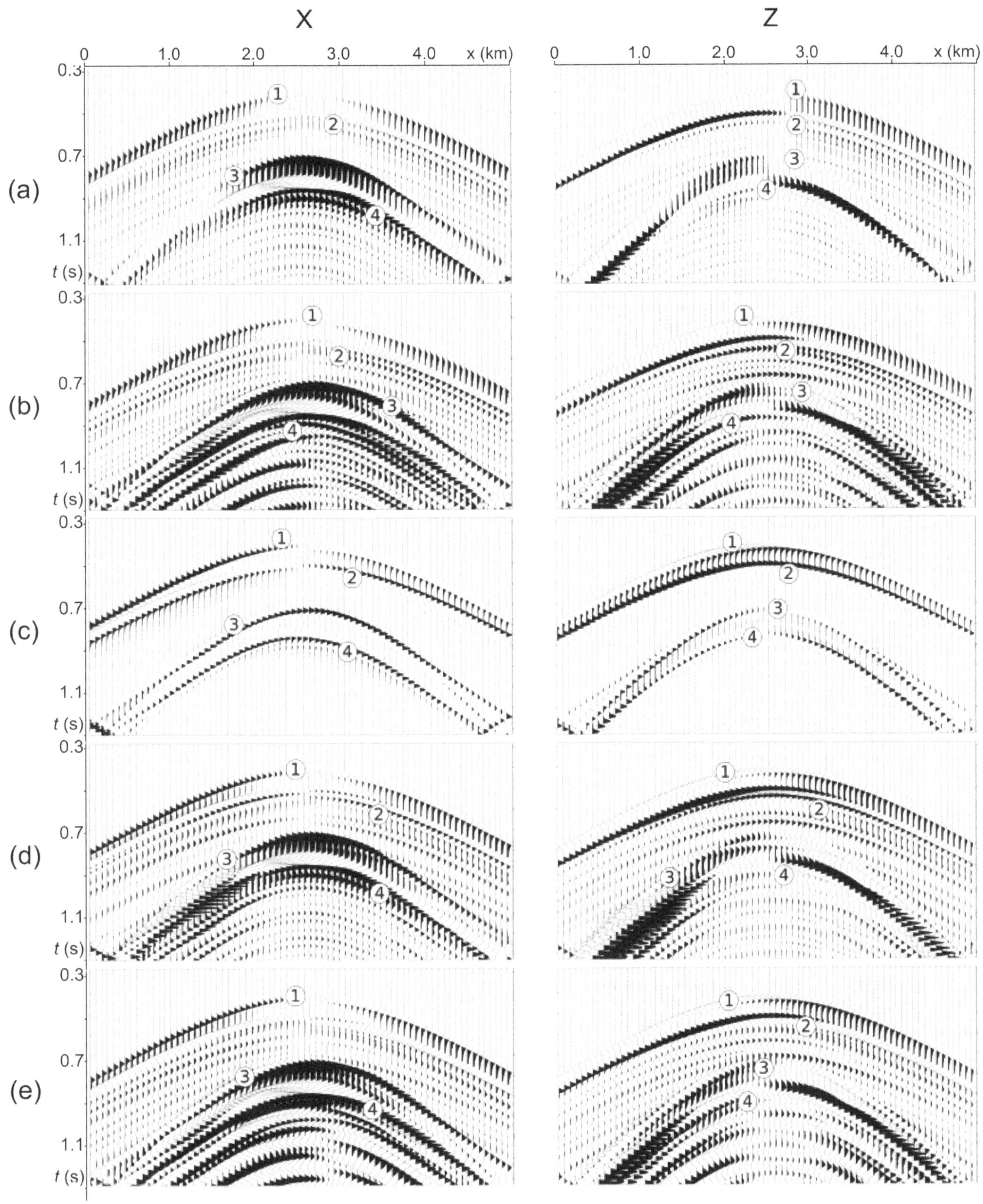

Figure 8.4. *X*- and *Z*-components of common-source gathers for (a) model 1, (b) model 2, (c) model 3, (d) model 4, and (e) model 5 shown in Figure 8.3. Here, 1 and 2 indicate, respectively, compressional waves diffracted at the upper tip (*Dpp*-1) and lower tip (*Dpp*-2) of the megafracture; 3 and 4 indicate, respectively, converted waves diffracted at the upper tip (*Dps*-1) and lower tip (*Dps*-2) of the megafracture.

Table 8.2. Maximum amplitudes (in relative units) of waves generated by fluid-filled megafractures with different internal structures.

Model type	Fracture tip	Compressional wave				Converted wave			
		X-component		*Z*-component		*X*-component		*Z*-component	
		Amplitude	Polarity reversal	Amplitude	Polarity reversal	Amplitude	Polarity reversal	Amplitude	Polarity reversal
Model 1	Upper Lower	1.00 2.50	No	1.17 3.59	Yes	3.36 9.23	No	1.50 2.93	Yes
Model 2	Upper Lower	1.04 2.36	No	1.87 3.82	Yes	3.76 8.93	No	2.24 4.88	Yes
Model 3	Upper Lower	2.48 3.27	Yes	3.20 4.14	No	2.14 2.78	No	1.84 2.63	Yes
Model 4	Upper Lower	1.97 3.81	No	2.62 6.11	Yes	3.46 9.56	No	2.44 11.32	Yes
Model 5	Upper Lower	1.66 3.03	No	2.88 4.85	Yes	3.57 9.83	No	3.74 6.58	Yes

amplitudes over the megafracture but do not undergo a polarity reversal. For the other models, these waves have polarity reversals.

On the *Z*-component, the converted diffractions of all five models have a polarity reversal. The compressional diffractions in the responses of all models also have a polarity reversal except for model 3, i.e., the averaged model.

Models with a relatively simple megafracture structure—models 1 and 3 with macrofractures parallel to the megafracture boundaries—have a simpler form of the wave response than those with a more complex internal structure—models 2, 4, and 5. In model 2, two macrofracture systems form a rhombic network; in models 4 and 5, the megafracture, in addition to internal mesofractures, has a 20% decrease in parameters relative to the host medium. The complexity of the wave response of model 2 manifests itself as a multi-cycle converted wavetrain due to the interference of the *Dps*-1 and *Dps*-2 diffractions. In models 4 and 5, a splitting of diffractions leading to the multi-cycle and complex behavior of the entire wavefield is observed.

From analysis of the common-source gathers shown in Figure 8.4 and data in Table 8.2, the following common features in the responses of fluid-saturated megafractures are apparent.

- On both components, the amplitudes of converted diffractions are higher than those of compressional diffractions, except for model 3. The averaged amplitude ratios of these waves on the *X*-component and *Z*-component are 3.0 and 1.25, respectively. The averaged amplitude ratio of the *Dps* wave on the *X*-component to the *Dpp* wave on the *Z*-component is 2.0.
- The amplitude of the wave response from the lower tips of megafractures is, on average, 2 to 2.5 times higher than that from the upper tips.
- Making the internal structure of a megafracture more complex or making the megafracture's internal parameters different from those of the host rock produce a more complex wave response. In particular, the response may have multiple cycles, be blurred, and have increased amplitudes of diffractions. For instance, the ratio of the amplitude of compressional waves for model 4 to that of model 1 is 2. For converted waves, the amplitude ratio is 1.05 to 1.10. The ratio of the amplitude of compressional waves for model 5 to that of model 2 is 1.5. For converted waves, the amplitude ratio is 1.1 to 1.2.
- Asymmetry of the dynamic properties of the diffractions, especially the converted diffractions, is caused by a deviation of megafractures from the vertical.

8.4. Effects of megafracture height and thickness on the wave response

To analyze the impact of height and thickness, a megafracture model containing 11 parallel macrofractures with a 10° deviation from the vertical and a 2000 m depth of its top is considered. A single horizontal reflection boundary is positioned at a depth of 2800 m. A 7 km long receiver line with 50 m receiver spacing is used. The height and thickness of the megafracture are variable.

Figure 8.5 shows the impact of megafracture height on the seismic response by comparing the *X*- and *Z*-components of common-source gathers acquired for different heights (50, 100, and 200 m).

As megafracture height increases, the character of the response changes only slightly, the change primarily manifesting in the extent to which the waves diffracted at the upper and lower tips of the megafracture are separated in time. When fracture height is 200 m, the compressional waves diffracted at the upper (*Dpp*-1) and lower (*Dpp*-2) tips of the megafracture are fully separated, with a 70 ms interval. In turn, when the height is 100 m, they form a complex three-cycle signal 30 to 35 ms long; when the height is 50 m, they almost merge into a single two-cycle wave.

The converted waves *Dps*-1 and *Dps*-2 diffracted at the upper and lower tips of the same megafracture but

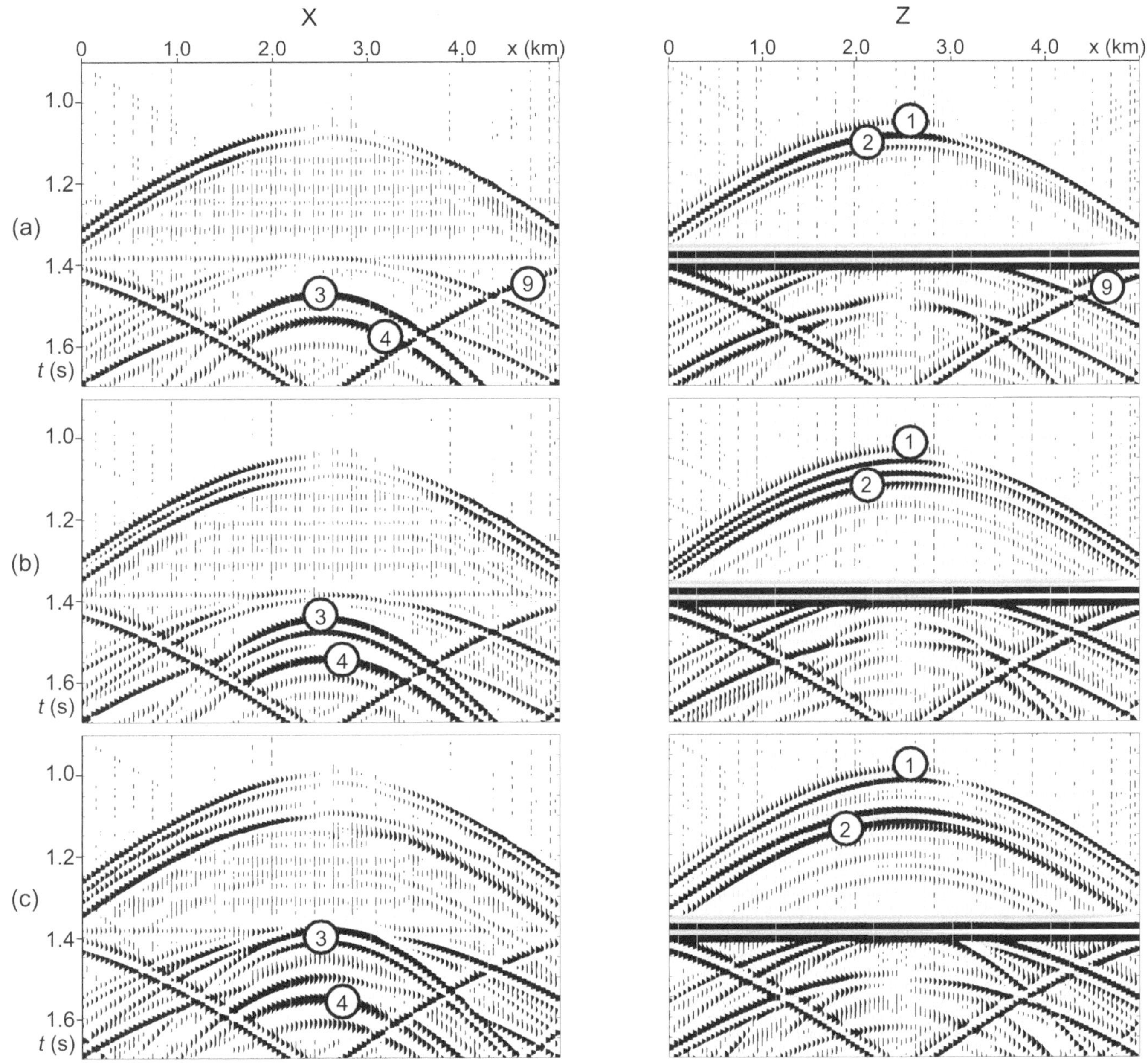

Figure 8.5. *X*- and *Z*-components of common-source gathers for a megafracture with different heights: (a) 50 m, (b) 100 m, and (c) 200 m. The megafracture contains 11 water-filled macrofractures and has a 2000 m depth of its top. Here, 1 and 2 indicate, respectively, compressional waves diffracted at the upper tip (*Dpp*-1) and lower tip (*Dpp*-2) of the megafracture; 3 and 4 indicate, respectively, converted waves diffracted at the upper tip (*Dps*-1) and lower tip (*Dps*-2) of the megafracture.

Table 8.3. Amplitudes (in relative units) of diffractions depending on megafracture height.

Megafracture height (m)	1 – *Dpp*-1 (*Z*)	2 – *Dpp*-2 (*Z*)	3 – *Dps*-1 (*X*)	4 – *Dps*-2 (*X*)
50	2.81	2.81	2.70	1.25
100	2.74	5.89	3.20	4.47
200	2.7	5.1	3.97	4.57

with a reduced velocity of shear waves, which is equal to 2500 m/s, are recorded separately for all of the height values under consideration. The intervals between them are 130, 100, and 65 ms when the heights are 200, 100, and 50 m, respectively. For the heights of 200 and 100 m, when the signals do not interfere, the diffraction amplitudes barely vary (Table 8.3).

Figure 8.6 shows that a sixfold increase in the megafracture thickness, from 5 to 30 m, with the fracture

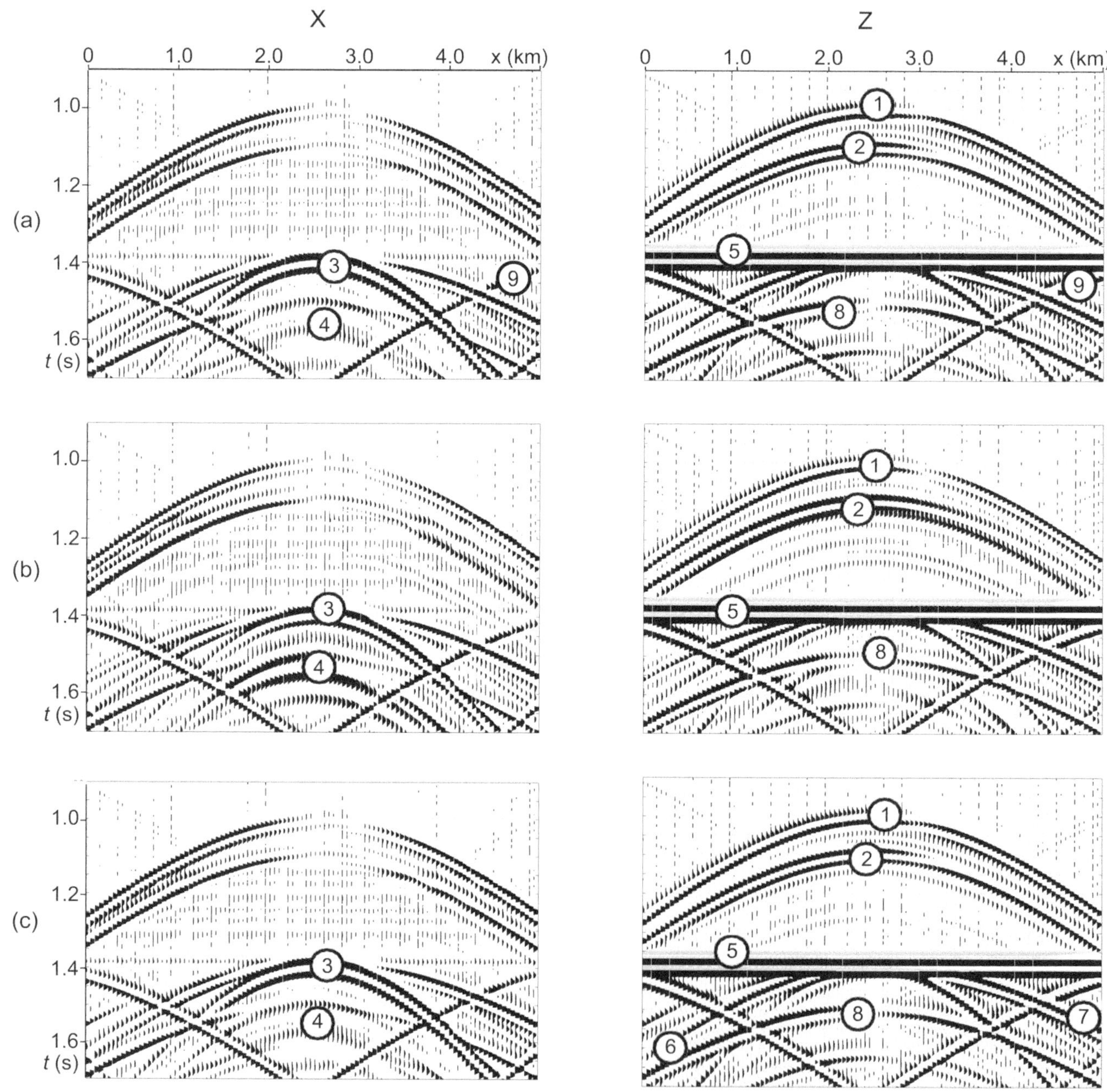

Figure 8.6. *X*- and *Z*-components of common-source gathers for a megafracture with different thicknesses: (a) 30 m, (b) 10 m, and (c) 5 m. The megafracture contains 11 water-filled macrofractures and has a 2000 m depth of its top. Here, 1 and 2 indicate, respectively, compressional waves diffracted at the upper tip (*Dpp*-1) and lower tip (*Dpp*-2) of the megafracture; 3 and 4 indicate, respectively, converted waves diffracted at the upper tip (*Dps*-1) and lower tip (*Dps*-2) of the megafracture.

Table 8.4. Amplitudes (in relative units) of diffractions depending on megafracture thickness.

Megafracture thickness (m)	1 – *Dpp*-1 (*Z*)	2 – *Dpp*-2 (*Z*)	3 – *Dps*-1 (*X*)	6 – *Dpp*-2* (*Z*)
30	4.21	3.66	7.22	1.51
10	2.7	5.1	3.97	1.36
5	1.96	4.86	3.22	1.18

height staying unchanged, also produces little effect. When the megafracture thickness is minimal (5 m), the dominant frequency of the response waves increases somewhat. Table 8.4 summarizes the amplitudes (in relative units) of the response waves depending on the thickness of the megafracture. With a sixfold decrease in the fracture thickness, the signal amplitudes of almost all of the waves decrease by 1.5 to 2 times.

So, it may be stated that the effects of height and thickness of a megafracture on the wave response are insignificant.

8.5. Characteristics of the wave response from a megafracture, using a point energy source and multifold acquisition geometry (long offset)

Identifying megafractures as potentially highly permeable channels for hydrocarbon movement in a reservoir is important for modeling a hydrocarbon transport and designing an optimum regime for both development and production. Numerical modeling with the grid-characteristic method applied to geologic settings and acquisition geometries typical of real seismic exploration is a worthwhile approach to this task.

To simulate wave responses, we use a four-layer subsurface model mimicking a multi-layer carbonate medium with both velocity and density increasing with depth. The model parameters are shown in Table 8.5. The model contains a 200 m high and 10 m thick megafracture with an upper tip which is 100 m below the top of layer 2. The megafracture consists of 10 parallel fluid-filled macrofractures deviated by 10° from the vertical.

The acquisition line is 8 km long with 50 m receiver spacing. The sources are positioned so that one is in the center of the spread and the other two are shifted 1500 m to the left and right of center. The receiver line is normal to the megafracture trend. The record length is 2.4 s and the sampling rate is 2 ms. The point source generates a 30 Hz Berlage wavelet.

Figures 8.7 and 8.8 present, respectively, the *X*- and *Z*-components of common-source gathers computed for the sources located at and outside of the spread center. This method of acquisition is intended to make the seismic response better distinguishable from the megafracture against the background of reflected waves.

Wave designations in the figures are: 1 = direct wave from a source; 2 = Rayleigh wave; 3 = compressional reflection (PP-1) from the first boundary; 4 = compressional reflection (PP-2) from the second boundary; 5 = compressional reflection (PP-3) from the third boundary; 6 = converted reflection (PS-1) from the first boundary; 7 = converted reflection (PS-2) from the second boundary; 8 = converted reflection (PS-3) from the third boundary; 9 = compressional diffraction (*Dpp*-u) from the megafracture's upper tip; 10 = compressional diffraction (*Dpp*-l) from the megafracture's lower tip; 11 = converted diffraction (*Dps*-u) from the megafracture's upper tip; 12 = converted diffraction (*Dps*-l) from the megafracture's lower tip; 13 = shear diffraction (*Dss*-u) from the megafracture's upper tip; 14 = shear reflection (*Dss*-l) from the megafracture's lower tip; 15 = compressional duplex (double bounce) wave (PPP); 16 = converted duplex (double bounce) wave (PPS); 17 = shear reflection (SS-1) from the first boundary; and 18 = shear reflection (SS-2) from the second boundary.

Table 8.5. Parameters of a four-layer medium.

Layer	Layer thickness (m)	Compressional wave velocity, *Vp* (m/s)	Shear wave velocity, *Vs* (m/s)	Density, ρ (kg/m^3)	Reflection coefficient
1	2000	4051	2272	2272	
2	310	4500	2500	2500	0.1
3	500	5000	2750	2750	0.1
4	1500	5601	3111	3000	0.1

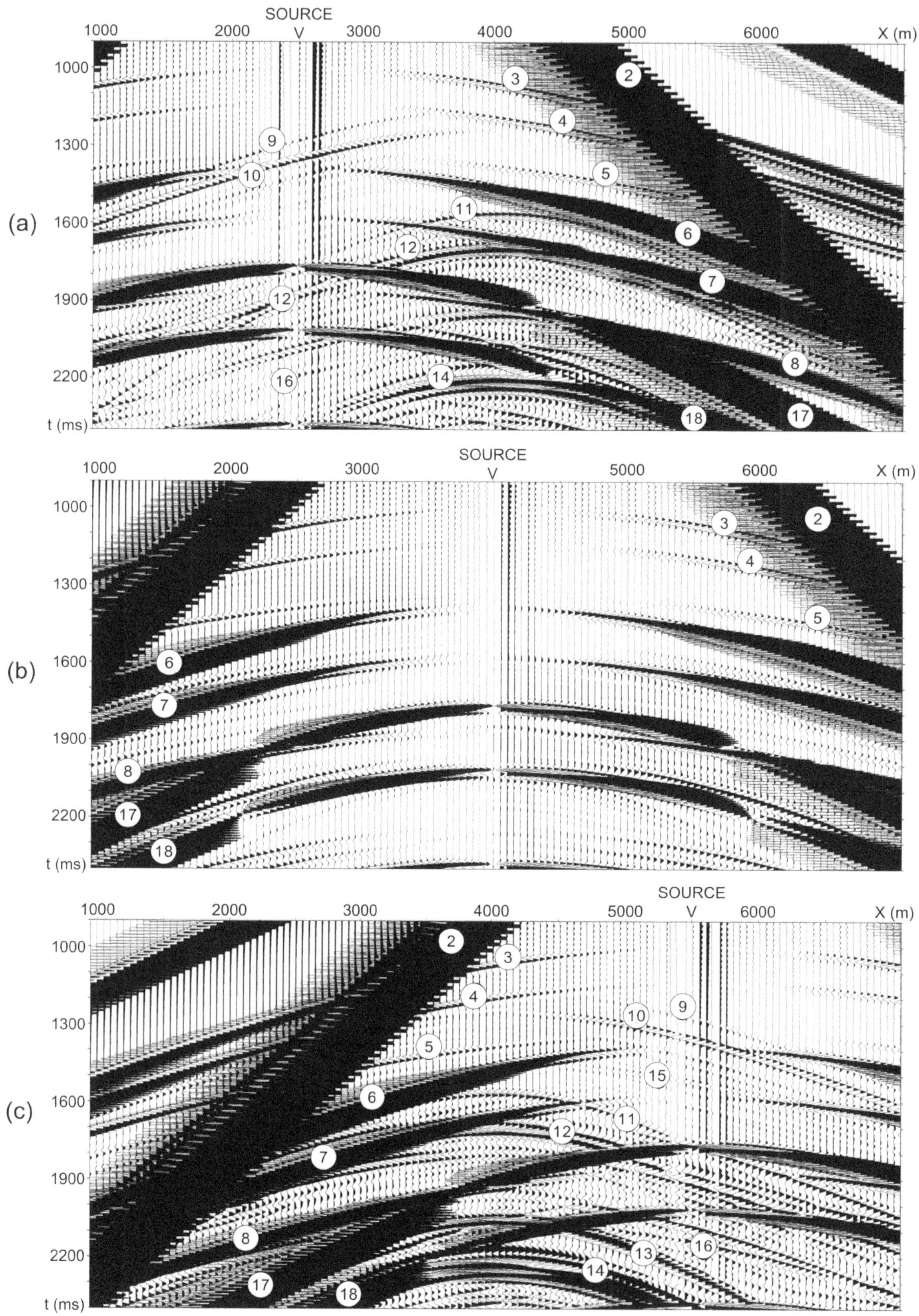

Figure 8.7. *X*-components of common-source gathers for different source locations: (a) 1500 m left of the spread center, (b) in the spread center, and (c) 1500 m right of the spread center.

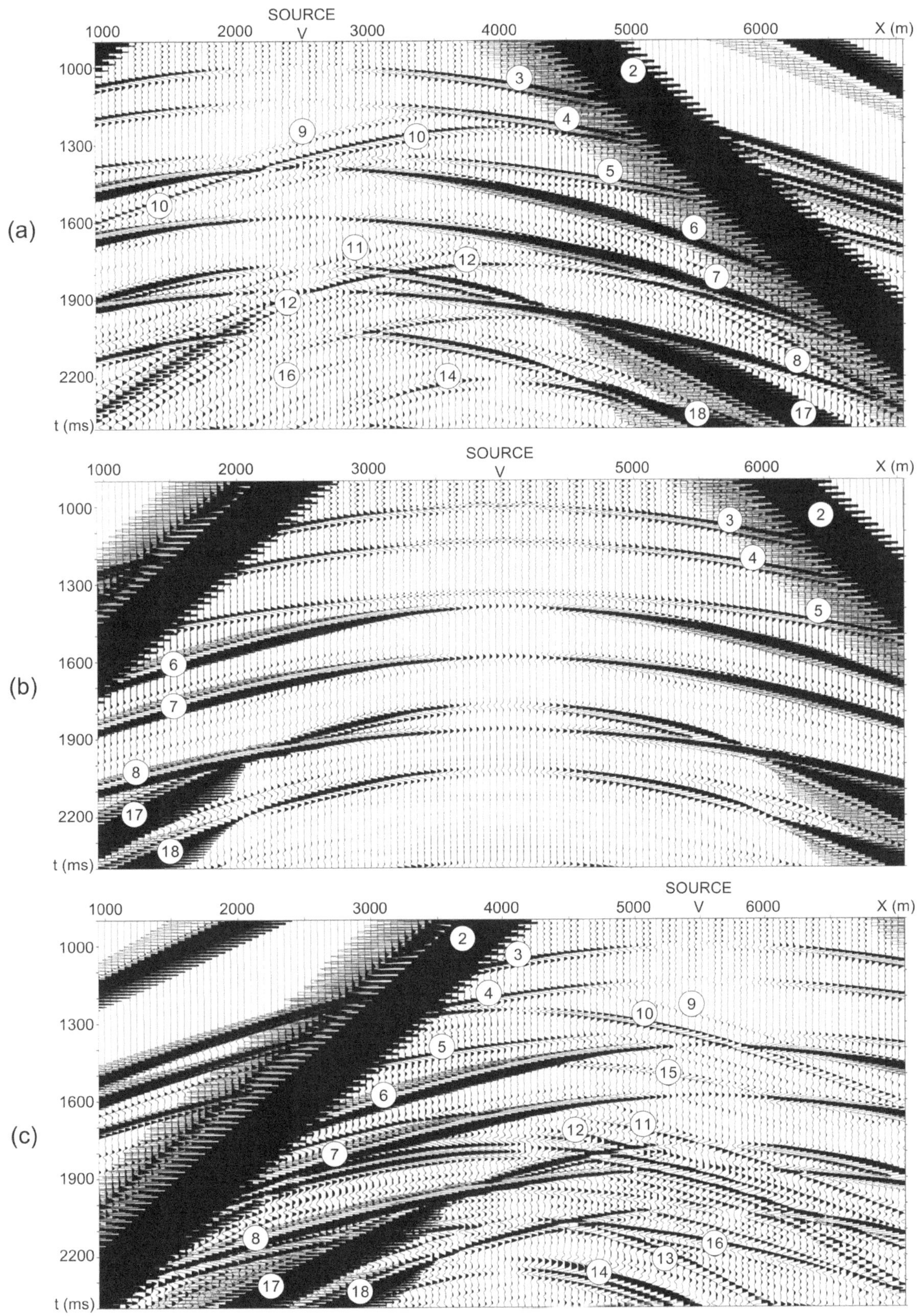

Figure 8.8. *Z*-components of common-source gathers for different source locations: (a) 1500 m left of the spread center, (b) in the spread center, and (c) 1500 m right of the spread center. Same wave designations as used in Figure 8.7.

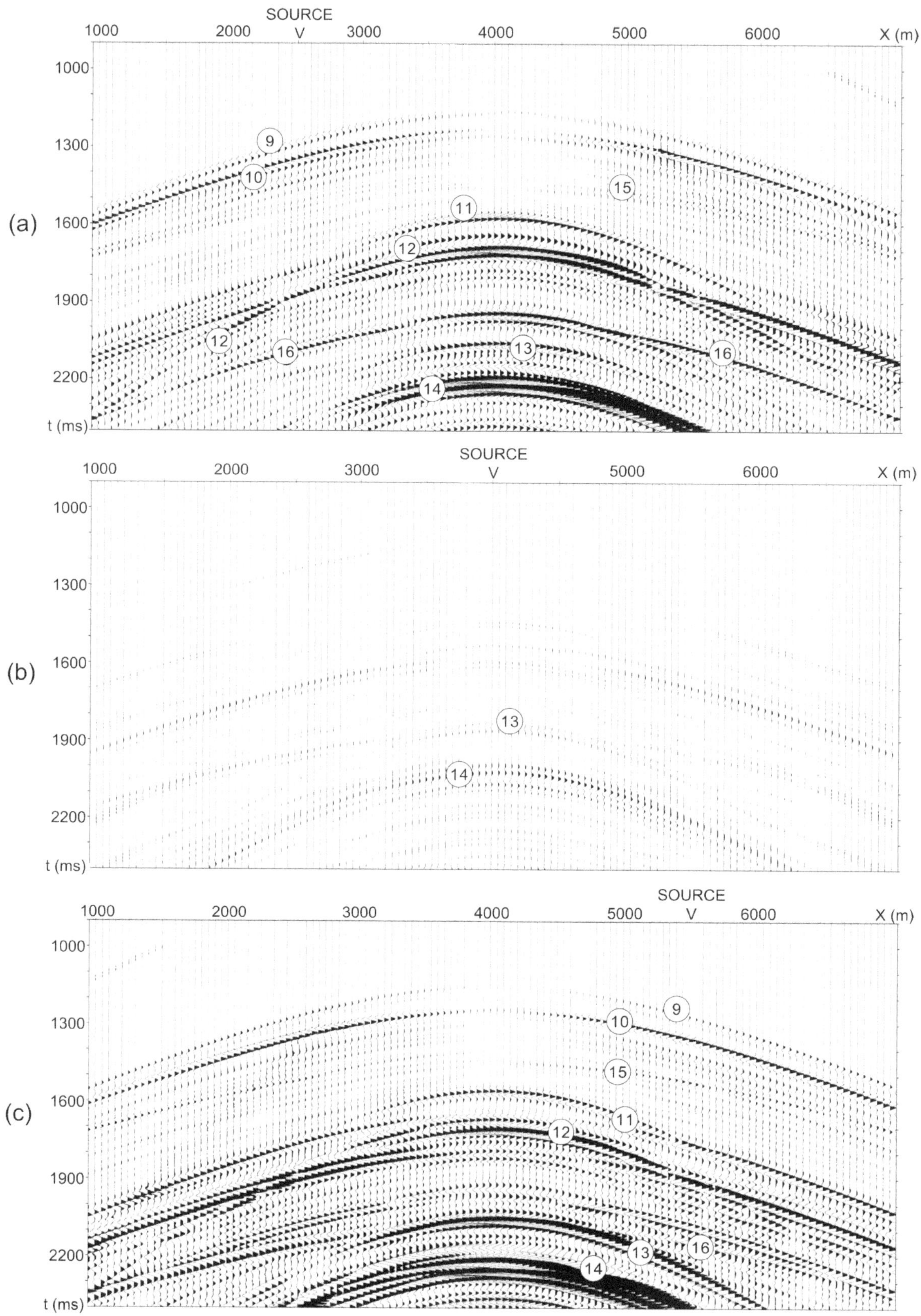

Figure 8.9. *X*-component of the difference between source gathers obtained with and without a megafracture in a layered medium. The source gathers are computed for different source locations: (a) 1500 m left of the spread center, (b) in the spread center, and (c) 1500 m right of the spread center. Same wave designations as used in Figure 8.7.

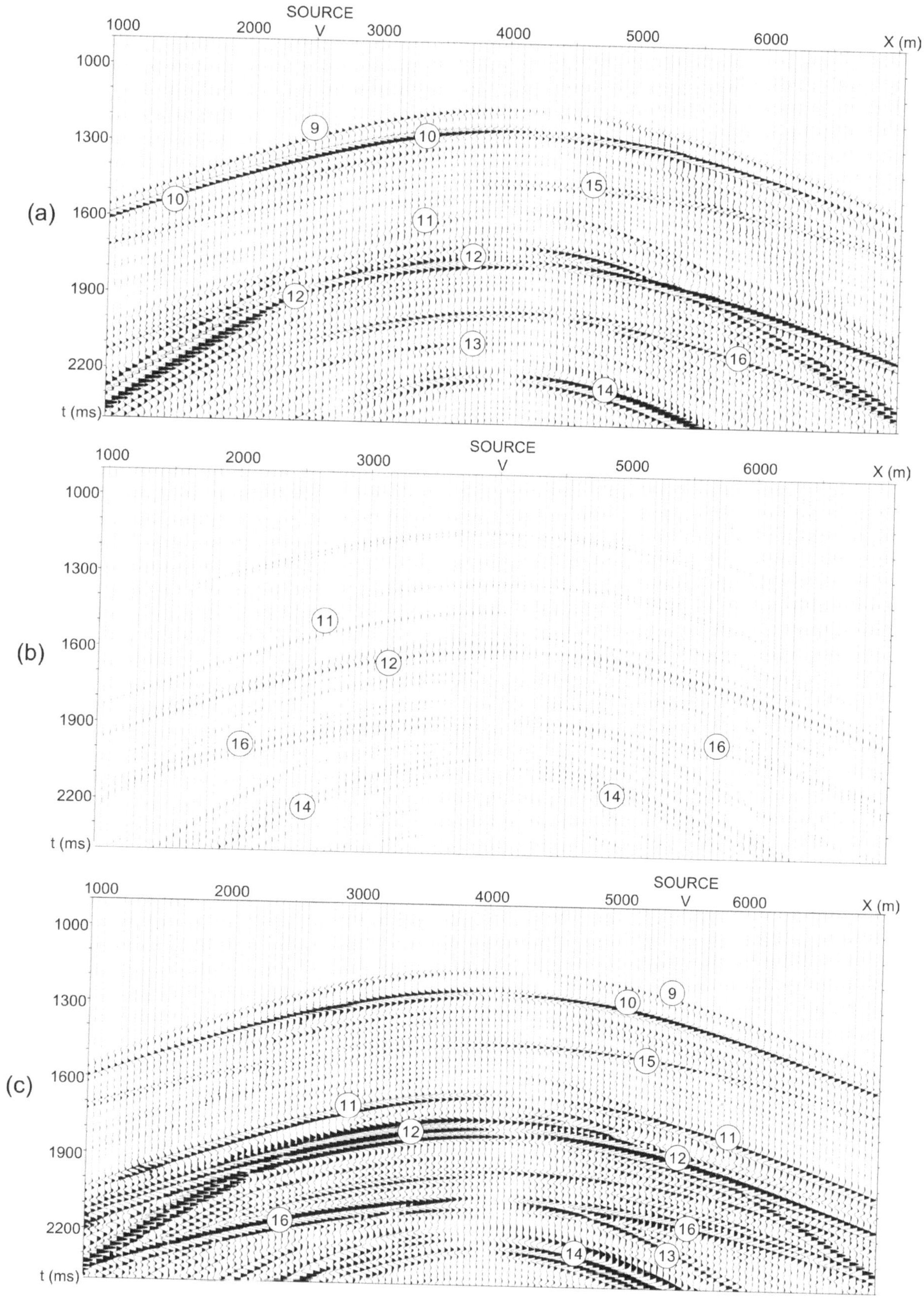

Figure 8.10. *Z*-component of the difference between source gathers obtained with and without a megafracture in a layered medium. The source gathers are computed for different source locations: (a) 1500 m left of the spread center, (b) in the spread center, and (c) 1500 m right of the spread center. Same wave designations as used in Figure 8.7.

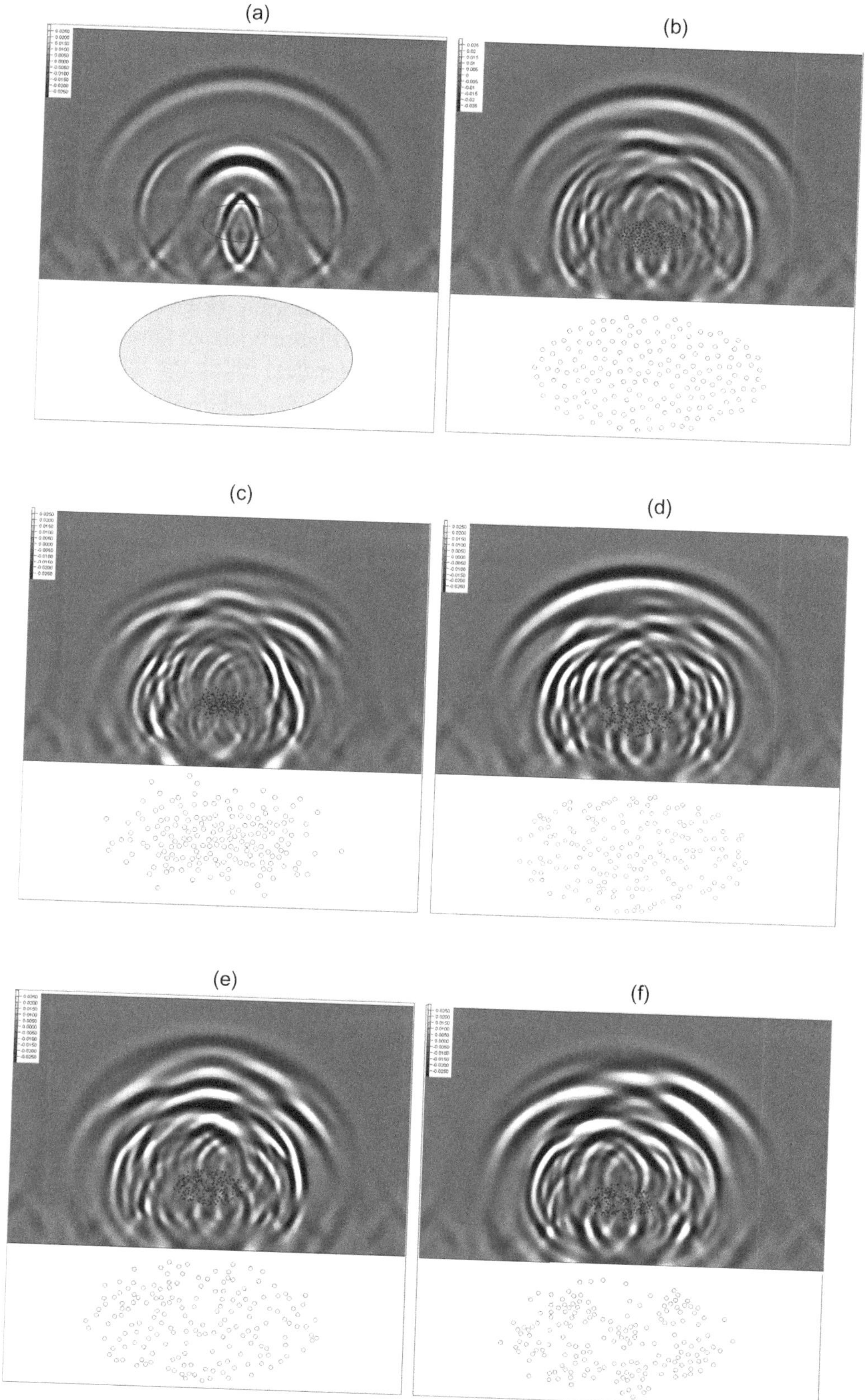

Figure 9.1. Wave response for the models of cavernous elliptical reservoirs containing different types of cavity distribution: (a) continuous medium with no heterogeneities inside the elliptical zone; (b) uniform distribution of cavities; (c) diffuse blurred boundary of the elliptical zone; (d) randomly non-uniform distribution of cavities, with the zone boundary being sharp; (e) moderately non-uniform distribution of cavities; and (f) strongly non-uniform distribution of cavities, with their size being reduced.

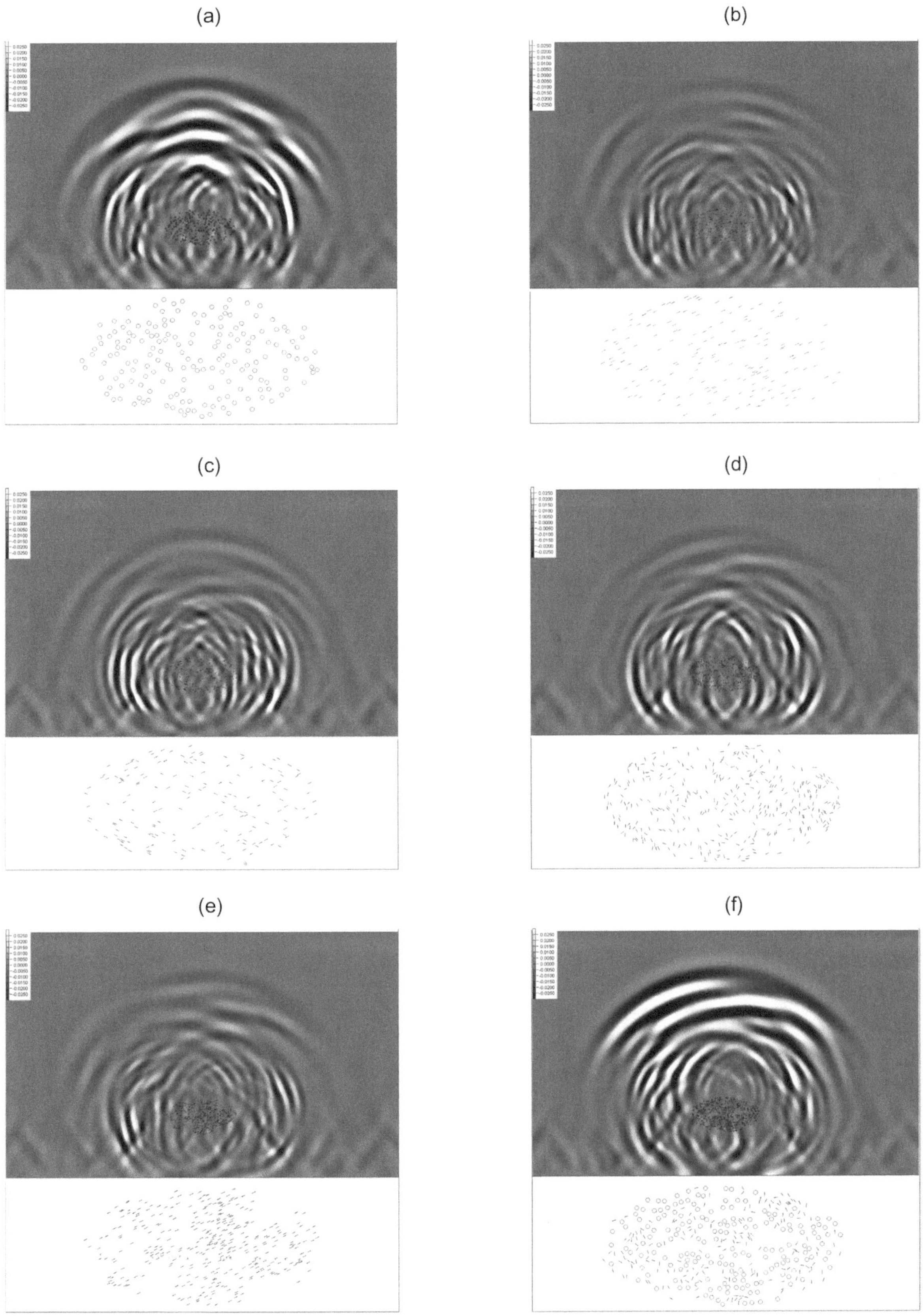

Figure 9.2. Wave response for various types of fractured reservoirs: (a) moderately non-uniform cavity distribution; (b) near-horizontal unidirectional fractures; (c) near-horizontal fractures in two directions; (d) random distribution of fracture directions; (e) double concentration of near-vertical fractures; and (f) non-uniform distribution of fractures and cavities.

concentration), fracture size *d* is 100 cm, and the fracture distribution is moderately non-uniform.

The following conclusions can be drawn from a visual inspection of Figure 9.2:

1) In the overall comparison of cavities of the same size (compare Figure 9.2a with Figure 9.2e and Figure 9.2e with Figure 9.2f), the fractures produce much weaker seismic responses, primarily due to considerable attenuation of PP-waves.
2) For the same size and number of heterogeneities, as compared with the cavernous zones, the fractured zones form much stronger converted PS-waves relative to the compressional PP-waves.
3) Unlike the cavities, a unidirectional fracture field generates an asymmetric seismic response, and the change of fracture orientation influences the direction of the energy peak.
4) With a twofold increase in the density of unidirectional fractures, the energy of the wave response almost doubles.

Observations 1 and 2 can be explained as follows: in the mathematical model, the water-filled fractures are quite transparent to compressional waves propagating strictly parallel or perpendicular to the fracture surface.

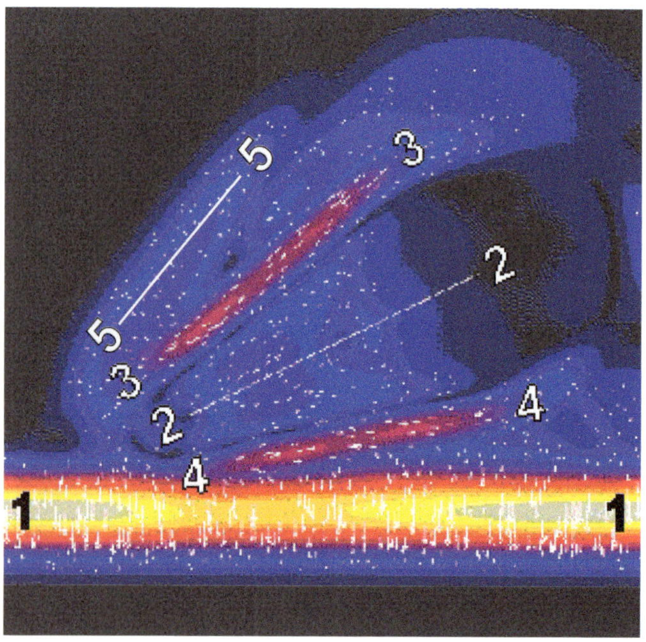

Figure 9.3. Wave responses of fluid-filled fractures: 1–1 = incident plane wavefront; 2–2 = inclined fracture; 3–3 = converted reflection from the fracture; 4–4 = transmitted converted wave; and 5–5 = weak compressional reflected (scattered) wave.

In the general case, the response energy should be smaller than in the case of cavities. The fact that the filled fractures generate stronger converted waves is illustrated in Figure 9.3, which shows the transmission of a plane compressional wave through an inclined fracture.

As can be observed in Figure 9.3, when a wave propagates through a water-filled fracture, a converted wave is mainly scattered back, whereas the compressional wave (5–5) is scattered much less. For this reason, a zone with numerous fractures, whether or not orientated randomly, will generate predominantly converted waves.

9.4. Effects of non-uniform distribution of micro-inhomogeneities on the scattered energy response

Next, we address the dependence of the response on the nature of the cluster. To do so, we examine the relationship between the degree of non-uniformity (degree of clusterization) of the distribution of the inhomogeneities inside the macrozone and the strength of the scattered energy response. We compute models to represent real conditions as much as possible with regard to the minimum size of inhomogeneities and their maximum number. The total porosity is taken to be 5%, which is typical of such reservoirs. Snapshots of the seismic responses from cavities (cycle 2), fractures (cycle 9), and cavernous-fractured reservoirs (cycle 12) (Table 9.1) for the four levels of non-uniformity are depicted in Figure 9.4. The snapshot on each panel is accompanied by a histogram containing information on the distribution of micro-inhomogeneities in the reservoir zone, an estimate of the response energy (at the upper right corner of each snapshot), and a visual snapshot of the distribution of cavities and fractures within the elliptical macrozone.

As the degree of non-uniform distribution of cavities increases (see histograms in the left column of Figure 9.4), the variation in the seismic response character is shown by the following:

- greater fragmentation of the PP- and PS-wavefronts,
- greater number of phases (cycles), and
- general increase in response energy (*E*) (84, 93, 128, 218).

The same tendency can be observed on the four snapshots displayed in Figure 9.4b showing the effect of the degree of non-uniform distribution of fractures. In general, the structure of the wavefield and the relative domination of converted waves are preserved. However, the response

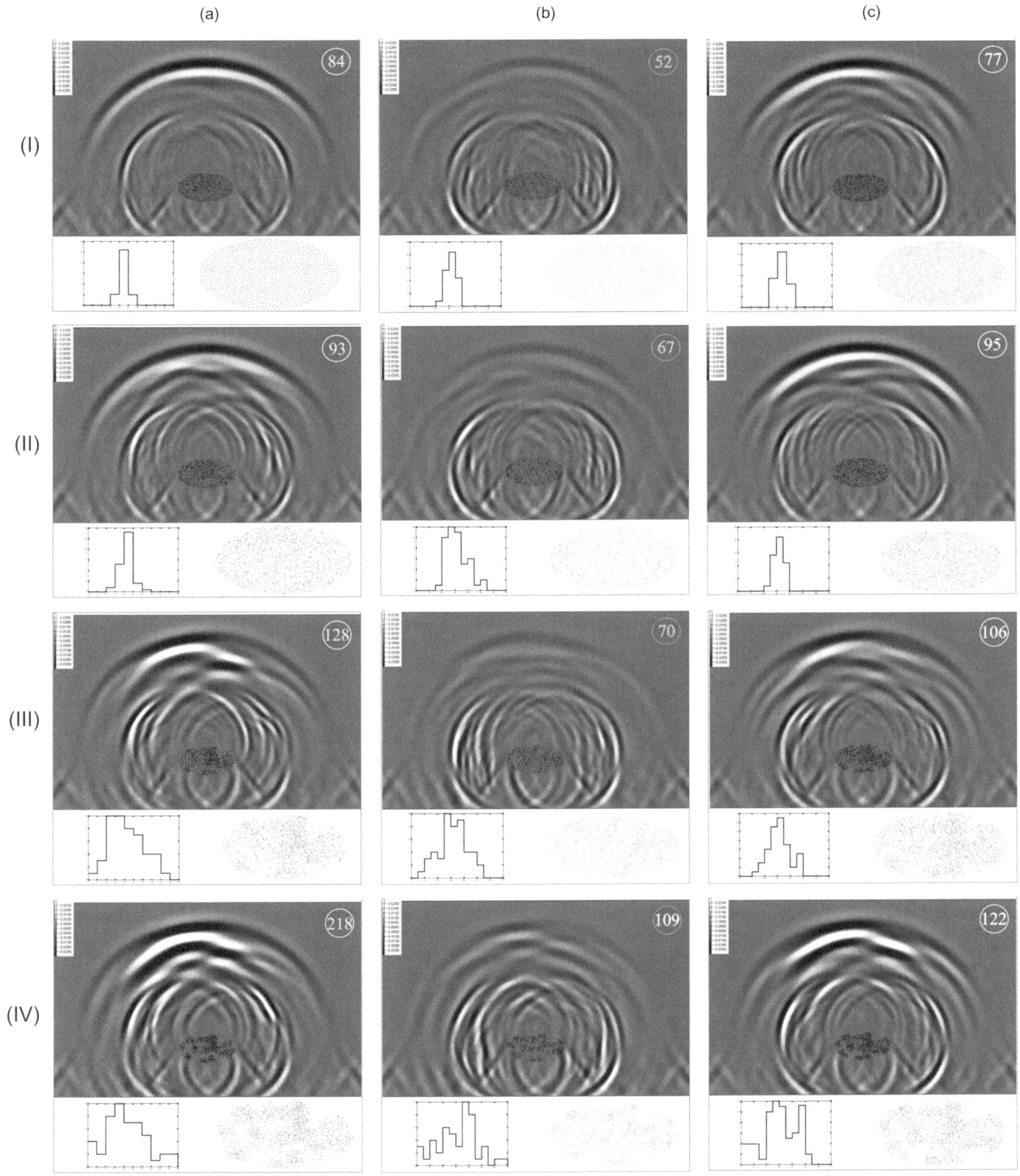

Figure 9.4. Impact of non-uniform distribution of micro-inhomogeneities on the character and strength of the scattered energy response for (a) cavernous, (b) fractured, and (c) combined (cavernous and fractured) reservoirs and different types of inhomogeneity distribution: uniform (I), randomly non-uniform (II), moderately non-uniform (III), and strongly non-uniform (IV). The number in the circle in the upper right corner of each snapshot represents the response energy E.

energy and the degree of interference of both PP- and PS-waves increase with increasing non-uniformity. The same tendency—greater response energy and greater wavefront disintegration—also is observed in Figure 9.4c, which shows the seismic responses for the combined (fractured and cavernous) zones.

9.5. Quantitative estimation of non-uniform distribution of inhomogeneities

We propose the following method to characterize the degree of non-uniform distribution of fractures and cavities in a rock. Let us divide the elliptical region into a number of cells with equal area (Figure 9.5) and calculate the number of centers of cavities and fractures in each cell. The sum of heights of all columns of the histogram shown in Figure 9.6 is always equal to the number of cells. The uniform distribution of inhomogeneities yields a histogram with one or two columns, because the number of objects in each cell is about the same. Greater non-uniformity increases the number of columns. The presence of several separate clusters produces a histogram with two distinct peaks: one near the zero density and one near the average density of objects in the clusters (see Figure 9.6).

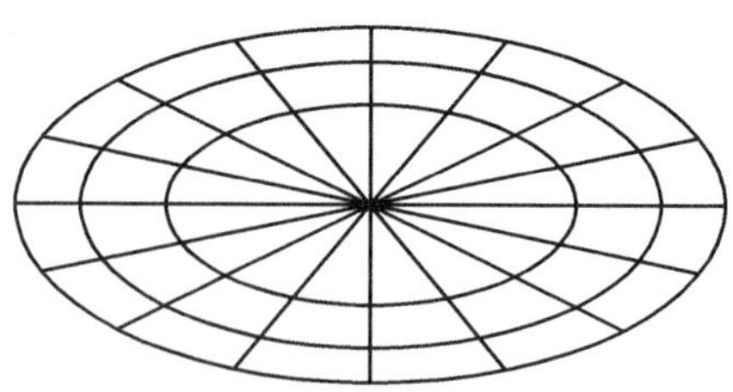

Figure 9.5. Partition of the elliptical region into equiareal cells used to construct histograms.

The histograms provide a visual but qualitative characteristic of the distribution of micro-inhomogeneities in the reservoir zone. However, it is sometimes necessary to analyze the relationship of this characteristic to other parameters for different modeling targets. In this case, only one indicator of non-uniform distribution of inhomogeneities is required. For this purpose, we use entropy (H):

$$H = -\frac{\sum_i p_i \ln p_i}{\ln N}, \tag{9.1}$$

where p_i is the fraction of cavities belonging to the ith cell of the ellipse and N is the total number of cells. The entropy found in this way is zero when all cavities occur in a single cell and 1 when an equal number of cavities are found in all cells, i.e., we have a uniform distribution of micro-inhomogeneities.

9.6. Quantitative estimation of the energy level of the wave response from a micro-inhomogeneous zone

In the course of modeling, the total energy within the integration domain is calculated:

$$W = \frac{1}{2}\int (\mathbf{e} : \mathbf{T} + \rho \vec{v}^2) dV, \tag{9.2}$$

where the first term is the potential energy of an element of the elastic medium (the Clapeyron equation), the second term is its kinetic energy, $\mathbf{e}$ is the strain tensor, and ":" denotes the double scalar product.

Using the relationship between stress $\mathbf{T}$ and strain $\mathbf{e}$ (Hooke's law)

$$\mathbf{T} = \lambda(\mathbf{e} : \mathbf{I})\mathbf{I} + 2\mu\mathbf{e}, \tag{9.3}$$

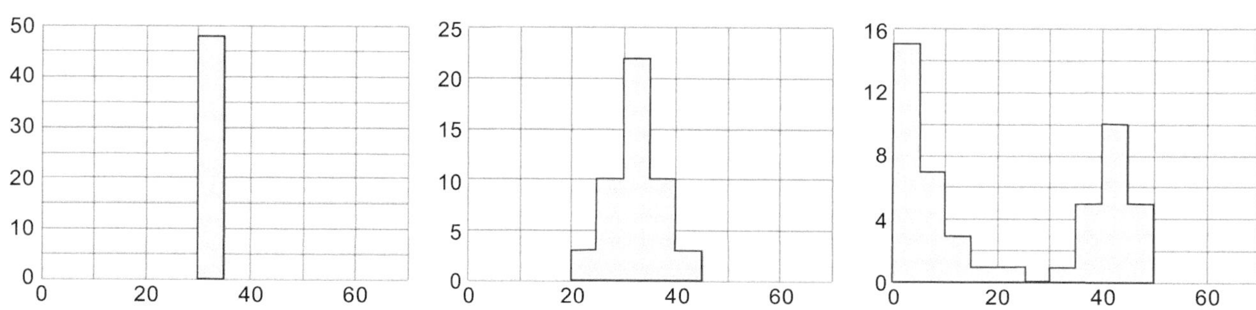

Figure 9.6. Typical histograms (left to right): uniform distribution, moderately non-uniform distribution, and clear presence of clusters.

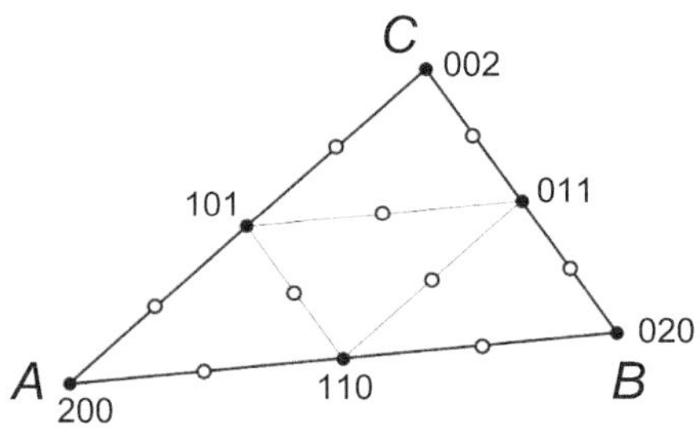

Figure 9.7. Position of auxiliary points in a triangle and their indices.

we can express the potential energy only in terms of stress, which is calculated numerically as

$$\frac{1}{2}\mathbf{e} : \mathbf{T} = \frac{1}{4\mu}\left[(\mathbf{T}:\mathbf{T}) - \frac{\lambda}{d\lambda + 2\mu}(\mathbf{T} : \mathbf{I})^2\right] \qquad (9.4)$$

where $D = 2$ is the space dimension.

The integral over a mesh triangle is calculated under the assumption that the data in it are represented in a piecewise linear manner (see Figure 9.7):

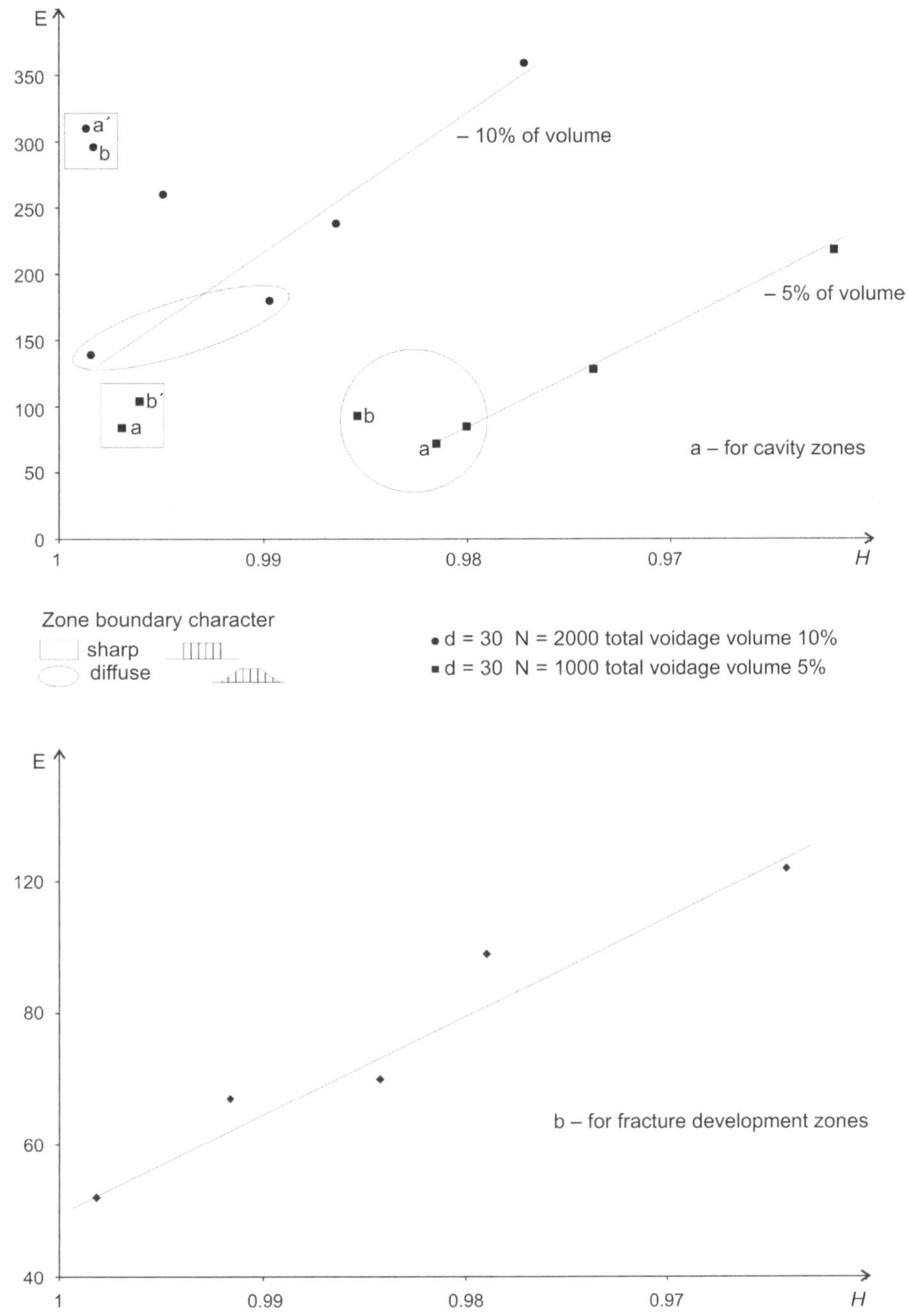

Figure 9.8. Dependence of the energy E of a response from the macrozone on the degree of non-uniform distribution (concentration) (H) for cavity zones (top) and fracture development zones (bottom).

$$\int uds \approx \frac{S}{12}(u_{200} + u_{020} + u_{002} + 3u_{110} + 3u_{011} + 3u_{101}), \quad (9.5)$$

where S is the triangle area and u_{ijk} are the values at its nodes. Note, in this case, the accuracy of integration is the same as that of the 1D trapezium method.

At the end of computation, the energy values characterize signals reflected by the reservoir zone. If the response of each fracture and cavity did not depend on the location of the neighboring inhomogeneities, the residual energy for various distributions of cavities in the host rock would be the same. But, as the computations show, there is a relatively strong dependence of the total energy of the reflected signal on the degree of clusterization of inhomogeneities, which is illustrated in Figure 9.8.

9.7. Assessing the relationship between seismic response energy and degree of non-uniform distribution of inhomogeneities in the reservoir zone

We have used the following characteristics to establish the existence of the relationship:

a) an estimate of the total energy at the time of generating compressional and converted waves reflected from cavernous or fractured zones, and
b) an estimate of entropy as a measure of non-uniform distribution of inhomogeneities within the elliptical reservoir zone.

Figure 9.8 shows the results of numerical simulation characterizing the quantitative relationship between the energy E of the seismic response produced by scattering from a cavernous and/or fractured zone and the degree of non-uniform distribution of micro-inhomogeneities in the macrozone defined via entropy H. It follows from the diagrams that:

- when the distribution of inhomogeneities changes from uniform to strongly non-uniform (with distinct clusters), the total energy of the response of cavities and fractures approximately doubles;
- the energy reflected from a fractured zone is about half that from a cavernous zone (with the same size and number of inhomogeneities); and
- the E–H plot for a macrozone containing fractures and cavities is almost the same as that for cavities.

9.8. Conclusions from Chapter 9

1) The assumption of the cluster nature of the seismic response from zones of diffuse cavities in a massive host rock is confirmed by identifying a unique quantitative relationship between the degree of non-uniform distribution of inhomogeneities and the strength of the seismic response.
2) It is established that, in the zones of diffuse fracturing, the strength of the seismic response increases with the increasing degree of non-uniform distribution of fractures (strengthening of cluster structure), exactly as in the case of cavernous inhomogeneity.
3) The main factor governing the generation of an elastic energy response by a zone of diffuse micro-inhomogeneities with no distinct boundary is the change in the ratio of the size of the micro-inhomogeneity to the wavelength, ranging from about 10^{-5} to cluster sizes comparable with the wavelength.
4) The response of a fractured macrozone is much weaker than that of an analogous cavernous zone. As this occurs, the ratio of compressional PP-waves to shear PS-waves changes in favor of the latter, implying the possibility of predicting the predominant type of inhomogeneities from separate recordings of compressional and shear waves.
5) The proposed methods for quantitative evaluation of non-uniform distribution of cavities and/or fractures using the calculation of the entropy and energy of the seismic response of a scattering feature have made it possible to improve the reliability of the relationship under investigation.

Conclusion

The study of wave responses from fractures, typical of those found in petroleum reservoirs, has been performed numerically using the grid-characteristic method (GCM) and are presented in this text. The mathematics of the method is described in detail in Part One. The method takes into consideration the characteristic properties of hyperbolic-type dynamic systems of equations, which allow the construction of accurate algorithms of integration at the domain boundaries, interfaces inside a model, and fracture surfaces. The method uses explicitly specified boundary conditions on the fracture surfaces, which allows proper account to be taken of fracture inhomogeneities that effectively have almost zero volume but finite surface area, as well as infill that may be gas or fluid.

The advantages of GCM over other methods are that GCM makes it possible to specify discretely every fracture, regardless of the number of fractures, and to specify explicit boundary conditions on the surface of every fracture. The method approximates the aspect ratios of physical fractures equal to 10^{-5} to 10^{-3}. Such an approach enables the wavefields associated with fractures having apertures (opening) of less than 1 mm to be modeled, thus meeting the requirements of real geologic conditions.

Computations, involving the Galerkin method, have shown that the use of an infinitely thin fracture (ITF) model in GCM numerical modeling is appropriate for modeling mesofractures and macrofractures, the major fracture types in hydrocarbon reservoirs, with an actual height to aperture ratio ranging from 1/100,000 to 1/1000.

It has been shown that the anisotropic properties of a fluid-filled fractured medium of the ITF model are consistent with the effective media theory.

Common properties typical of single fractures

Simulating wave responses of near-vertical fluid-filled and gas-filled fractures typical of hydrocarbon reservoirs with the GCM has helped us establish that:

- compressional and converted diffractions radiated from the tips of a near-vertical fracture are the principal elements of the response that carries information about the fracture to the earth's surface;
- compressional diffractions are always stronger than converted waves on the *Z*-component, while converted waves are stronger on the *X*-component;
- in the case of fluid-filled fractures, the converted diffracted wave is strongest among the others on the *X*-component, while, in the case of gas-filled fractures, the compressional diffracted wave is strongest among the others on the *Z*-component;
- an important feature of the seismic response from a fluid-filled fracture is that a polarity reversal in the compressional and converted diffracted waves occurs above the fracture on the *Z*-component rather than the *X*-component, as in reflections from a flat horizontal boundary, and the converted diffracted wave on the *X*-component does not undergo a polarity reversal;

- a fracture plane deviation from the vertical, even if minor (5° to 10°), causes asymmetry of the branches of the converted diffraction relative to the location of the fracture projection onto the observation surface;
- the asperity or roughness of faces of real fractures, which causes the fractures to be partially closed, attenuates the seismic response by approximately the percentage of closed area of a fracture to its total area; and
- the characteristics of scattered waves are observed to be dependent on the fracture-filling material—fluid or gas. When filled with fluid, compressional and converted diffractions undergo a polarity reversal over the fracture on the *Z*-component; when filled with gas, this occurs on the *X*-component.

Specific characteristics of wave responses from mesofracture zones

A bed (or a bed-bounded zone) of near-vertical fluid-filled mesofractures, a few meters in height, is a common feature of fractured reservoirs. When a compressional wave encounters such a reservoir, a response of scattered compressional and scattered converted waves is created. These waves are formed as a result of the interference of, respectively, compressional and converted diffractions from mesofractures near the boundaries of the fractured bed. According to kinematic features, these waves correspond to quasi-reflected compressional and converted waves from the boundary coincident with the top of the mesofracture bed and, when the bed is rather thick, from the boundary coincident with its base as well (for the compressional wave).

The relatively high amplitude strength of the scattered waves is caused by (1) the formation of the response from fluid-filled mesofractures due to their deviation from the vertical, (2) a high degree of overlap of interfering diffractions when a scattered wavefront is formed, and (3) the absence of any polarity reversal on the *X*-component of the converted diffraction from each single mesofracture.

The scattered compressional waves exhibit unique amplitude versus offset (AVO curve) behavior. For reservoirs with a zero or minor impedance difference at the top boundary, the scattered compressional waves, recorded at a wide range of CMP offsets, have a zero reflection coefficient at the normal incidence on the fractured bed; this coefficient (when positive) increases rapidly with the offset. In the case of a positive impedance difference at the bed top, a fluid-filled fractured bed is shown to behave in a similar manner.

For reflections from boundaries without fluid-filled fracturing, the AVO curves are significantly different. When the impedance contrast is positive and the contrast in the *Vp/Vs* ratio is small, the curves are at maximum at the normal incidence and have a negative gradient. In the more general case when the impedance contrast has any sign, boundaries in tight rocks at depths typical of oil exploration have very small differences in both *Vp/Vs* ratio and impedance. This results in minimal dependence of the reflection coefficient on the angle of incidence of the wave and, therefore, in a gentle character of the AVO curves of the reflected waves with very low values of their gradients, regardless of the sign.

On the *X*-component, the scattered converted wave from these fractured features is broadly comparable with the *X*-component of compressional reflections, with an amplitude strength that exceeds those of the scattered compressional and converted reflected waves at offsets not exceeding the feature's depth. The scattered converted wave has additional characteristics:

- a polarity reversal over the source location on the *X*-component;
- considerable asymmetry of the wavefields (different wave structure) on opposite sides of a receiver spread;
- a horizontal shift of the zero value of the AVO curve, even if the fracture inclination is minor; and
- a one- or two-pulse signal of the scattered converted wave on the *X*-component which is formed at the top fracture tips only, due to high absorption of shear waves propagating upward from the bed base.

The discovered opaque properties of the mesofractured beds are due to abnormally high absorption of shear wave energy, which may prevent the detection of fractured zones if a target bed lies below such a non-transparent interval.

Specific characteristics of seismic responses from clusters of macrofractures

The response from a macrofractured bed differs from that of a mesofractured bed. First, the amplitude strength of a macrofractured bed is much lower (by a factor of 1.5 to 2). Second, in this case, a wave scattered at the lower tips of the macrofractures is observed, regardless of the wave type.

It has been established that the information about the position and structure of a system (cluster) of fluid-filled, unidirectional, near-vertical macrofractures of dimensions comparable to the seismic wavelength and typical of hydrocarbon reservoirs in tight rocks is contained in the scattered (diffracted) compressional waves and scattered (diffracted) converted waves. The compressional scattered waves are recorded primarily at long offsets, while the converted scattered waves are recorded at intermediate offsets, i.e., at those that are shorter than the cluster depth. Both are formed by the interference of diffractions coming from the upper and lower tips of the macrofractures.

The kinematic characteristics of compressional and converted waves scattered at a macrofracture system are shown to correspond to the hyperbolas of compressional and converted reflections from the hypothetical boundaries that run through the tips of macrofractures. Here, the converted scattered wave, as well as the converted reflection, has a polarity reversal at the point source location.

The scattered converted wavefront is noted to be stable on the *X*-component within a wide range of fracture spacings and stable to variations in fracture inclination and fracture spacing when the degree of variation is moderate (up to 30% to 40% of the average parameter value).

Modeling with the GCM method has confirmed that it is possible to compute an areal scattered converted wave response from a 3D cluster of macrofractures by applying the GCM method to a realistic 3D geologic model comparable to those encountered in actual 3D surveys.

Methods to detect and map fractured reservoirs and local zones of meso- and macrofracturing on CMP seismic sections.—As demonstrated, the amplitude profile of the scattered compressional wave from a fluid-filled fractured bed has a special character in the form of a considerable and positive AVO curve gradient, which allows us to propose two methods by which reservoirs of this type may be recognized.

The first method exploits the difference in the AVO gradient values between scattered and reflected waves. The method can be applied in diverse environments to search for and map fractured reservoirs located, generally, in tight rocks at depths of typical interest to the oil industry. In these environments, the reflecting boundaries are characterized by negligible contrasts in the *Vp/Vs* ratios and acoustic impedances, which result in minimal dependence of the reflection coefficient on the angle of incidence and, therefore, in a very low value of AVO gradient regardless of its sign. In contrast to the specular reflections, as noted above, the AVO gradient of the scattered wave is considerable.

The second method, which uses the principle of opposite gradient signs of reflected and scattered waves, is applicable only to geologic intervals with positive impedance contrasts and small differences in the *Vp/Vs* ratio at the tops of reflecting beds. This method requires far- and near-offset partial CMP stacks. For waves scattered by fluid-filled fractures, when the impedance contrast is either zero or positive, the stack amplitudes at far offsets are always larger than those at near offsets. Their ratio is either much greater than unity (when the impedance contrast is zero) or just greater than unity (when the impedance contrast is positive).

These methods for recognizing compressional waves scattered at long meso- and macrofractured fluid-filled beds are equally applicable to identifying local bed-bounded zones of these types of fracturing.

As noted, the high relative amplitude strength of the converted wave in the seismic response observed at midrange offsets on the *X*-component is typical of both fluid-filled mesofractures and fluid-filled macrofractures. The structure of a scattered converted wavefield is found to be different on opposite sides of a receiver spread owing to the effect of the asymmetry of the converted diffraction on opposite sides of a point source, which is caused by a fracture inclination and polarity reversal. A horizontal shift of the zero value of the AVO curve also has been shown to be caused by fracture inclinations.

The principal indicators that can be used to predict the presence of fluid-filled bed-bounded fracture zones within target intervals from the *X*-component of CMP time sections include:

- pronounced amplitudes of the scattered converted wavefront;
- significant differences in the structure and amplitude strength of a converted wavefield on opposite sides of a receiver spread, even when fracture inclinations are minor; and
- a horizontal shift of the zero value of the AVO curve of the scattered converted wave.

In order to identify fractured zones more effectively, these diagnostic indicators may be complemented by stacking the *X*-component of converted waves from both sides of a receiver spread without a polarity reversal. This approach allows the converted reflections to be reduced almost to zero, while preserving, because of the above asymmetry, the energy of the difference between the wavefronts of waves scattered from a fractured zone and recorded at opposite sides of a receiver spread.

These findings provide a firm basis for direct detection of:

a) meso- and macrofractured beds (zones) when using scattered compressional waves acquired with conventional seismic data acquisition systems involving long offsets, and

b) meso- and macrofractured zones in reservoirs when using scattered converted waves acquired by multicomponent (3C) receivers at short offsets.

Recommendations regarding the identification of fractured reservoirs in a more general formulation.—To this point, the completed investigations have revealed and analyzed an important property of the AVO curve for the case of fluid-saturated fractured targets—the positive gradient. This automatically filters out all waves with negative gradients.

Of all waves from boundaries with positive AVO gradients, we are concerned with those that are characterized by significant AVO gradients of the compressional and converted waves, substantial traveltime asymmetry and amplitude strength of the converted waves, a horizontal shift of the zero value of the AVO curve (when the boundaries are near-horizontal), and azimuthal anisotropy of velocities.

The fractured beds or zones identified on seismic data will need to be correlated with and supported by well data, e.g., scanner data, cross-dipole acoustic data, etc.

Specific characteristics of seismic responses from megafractures (fracture corridors).—The response from a megafracture, the largest near-vertical zone of disturbed rock integrity (often shear in nature), which may reach several meters in thickness, consists of two compressional diffractions and two converted diffractions propagating from the upper and lower tips of the megafracture. The influence of the variation of fracture infill (fluid or gas) manifests as a change of polarity of the X- and Z-components and as a greater wave strength in the case of gas-filled fractures. For most of the various alternative models of internal structure, the converted diffraction from the lower tip of the megafracture dominates in the entire response and on its X-component, and has no polarity reversal.

In seismic 2D surveys conducted over a layered subsurface using a point energy source, these properties of the responses from megafractures are augmented by another feature. The overall strength of the wave response from a megafracture increases substantially as the source moves away from the fracture projection onto the line, at distances comparable to the megafracture depth. In addition to diffractions, the wave response from a megafracture contains compressional and converted duplex (double-bounce, or prismatic) arrivals that first reflect from the underlying boundaries and then from the megafracture itself or vice versa. These duplex waves also are important for identifying the megafractures.

Dynamically pronounced duplex waves from a megafracture in 2D surveys acquired with point sources carry the most complete information about the megafracture when the sources are located sufficiently far from the megafracture projection point, i.e., at distances ranging from 50% to 100% of the fracture depth. In order to identify those waves, and consequently megafractures, in the layered subsurface, it is advisable to apply duplex wave migration.

Cluster nature of wave responses from zones of chaotically oriented fractures and specific characteristics of their wave properties.—We have established that, in zones of chaotic diffuse fracturing (reservoirs in massive rocks), the energy of wave responses increases with increasing irregularity in fracture concentrations, i.e., with strengthening of the cluster structure.

The concept of the cluster nature of the wave response from a zone of chaotic diffuse fracturing is confirmed by the unambiguous quantitative relationship between the degree of irregularity in fracture concentrations and the strength of the response.

The primary feature that influences the response from a boundary-free zone of chaotic fracture concentrations is the change in the ratio of the size of the cluster of macro-heterogeneities that consist of micro-heterogeneities to the seismic wavelength, up to their comparability. Responses from a macrozone of fluid-filled fractures are characterized by higher amplitudes of converted PS-waves relative to those of compressional PP-waves.

References

Aki, K., and P. Richards, 1983, Quantitative seismology: Mir Publishers (Russian translation).

Aldridge, D. F., 1990, The Berlage wavelet: Geophysics, **55**, no. 11, 1508–1511, https://doi.org/10.1190/1.1442799.

Alekseev, A. S., and B. Y. Gelchinsky, 1959, On the ray method of computation of wavefields for inhomogeneous media with curved interfaces, *in* Petrashen, G.I. ed., Problems of the dynamic theory of propagation of seismic waves, **3**, 107–116, Leningrad University Press (in Russian).

Al-Muftah, A., K. Kumar, C. R. K. Murty, T. Lemaux, A. Ranjan, G. Dejoussineau, C. Bertrano, and H. Behira, 2009, Integrated reservoir connectivity study of Ahmadi fractured reservoir in Bahrain field: Reservoir Characterization and Simulation Conference, SPE/EAGE, Extended Abstracts, SPE125400.

Antonenko, M. N., I. Y. Antonova, and V. B. Leviant, 2004, Study of seismic fields scattered from zones of diffuse cavities and fractures by numerical simulation methods: The Russian Geophysics Journal, **2**, 8–20 (in Russian).

Bagrintseva, K. I., 1999, Formation conditions and properties of carbonate reservoirs of oil and gas: Russian State Geological Prospecting University publication (in Russian).

Bakulin, A., V. Grechka, and I. Tsvankin, 2000, Estimation of fracture parameters from reflection seismic data. Part I: HTI model due to a single fracture set: Geophysics, **65**, no. 6, 1788–1802, https://doi.org/10.1190/1.1444863.

Bakulin, A., V. Grechka, N. Karaev, A. Anisimov, and E. Kozlov, 2004, Physical modeling and theoretical studies of seismic reflections from a fault zone: 74th Annual Meeting, SEG, Expanded Abstracts, 1674–1677, https://doi.org/10.1190/1.1845152.

Bathe, K.-J., 1996, Finite element procedures: Prentice Hall.

Bertotti, G., A. Immenhauser, and J. Kenter, 2003, Joint systems and fault-controlled hydraulic fracturing in outcrops – Impact on precitions of reservoir properties: 1st North African/Mediterranean Petroleum and Geosciences Conference, EAGE, Extended Abstracts, T036.

Berzon, I. S., L. I. Ratnikova, and M. I. Rats-Khizgiya, 1966, Seismic converted waves: Nauka Publishers (in Russian).

Bischoff, R., and R. Bejaoui, 2005, Integrated modeling of the mature Ashtart field: 67th Annual Meeting, EAGE, Expanded Abstracts, SPE 94007.

Bohlen, T., and E. H. Saenger, 2006, Accuracy of heterogeneous staggered-grid finite-difference modeling of Rayleigh waves: Geophysics, **71**, no. 4, T109–T115, https://doi.org/10.1190/1.2213051.

Casciano, C., L. Ruvo, B. Volpi, and F. Masserano, 2004, Well test simulation through discrete fracture network modeling in a fractured carbonate reservoir: Petroleum Geoscience, **10**, no. 4, 331–342, https://doi.org/10.1144/1354-079303-590.

Chen, T., M. Fehler, S. Brown, Y. Zhang, X. Fang, D. Burns, and P. Wang, 2010, Modeling of acoustic wave scattering from a two-dimensional fracture: 80th Annual Meeting, SEG, Expanded Abstracts, SM–37, https://doi.org/10.1190/1.3513474.

Coates, R. T., and M. Schoenberg, 1995, Finite-difference modeling of faults and fractures: Geophysics, **60**, no. 5, 1514–1526, https://doi.org/10.1190/1.1443884.

Dorofeeva, E. V., 1986, Tectonic fracturing of rocks and conditions of formation of fractured reservoirs of oil and gas: Nedra Publishers (in Russian).

Druskin, V., and L. Knizhnerman, 1988, A spectral semi-discrete method for the numerical solution of 3-D nonstationary problems in electrical prospecting: Izvestiya, Academy of Sciences, USSR. Physics of the Solid Earth, 24, 63–74.

Eringen, A. C., and E. S. Suhubi, 1974, Elastodynamics, Volume 1: Finite Motions, Academic Press.

Eringen, A. C., and E. S. Suhubi, 1975, Elastodynamics, Volume 2: Linear Theory, Academic Press.

Fang, X., M. Fehler, T. Chen, and D. R. Burns, 2010, Sensitivity analysis of fracture scattering: 80th Annual Meeting, SEG, Expanded Abstracts, 2340–2345, https://doi.org/10.1190/1.3513320.

Frehner, M., 2014, Krauklis wave initiation in fluid-filled fractures by seismic body waves: Geophysics, **79**, no. 1, T27–T35, https://doi.org//10.1190/geo2013-0093.1.

Frehner, M., and S. M. Schmalholz, 2010, Finite-element simulations of Stoneley guided-wave reflection and scattering at the tips of fluid-filled fractures: Geophysics, **75**, no. 2, T23–T36, https://doi.org/10.1190/1.3340361.

Frehner, M., S. M. Schmalholz, E. H. Saenger, and H. Steeb, 2008, Comparison of finite difference and finite element methods for simulating two-dimensional scattering of elastic waves: Physics of the Earth and Planetary Interiors, **171**, no. 1–4, 112–121, https://doi.org//10.1016/j.pepi.2008.07.003.

Gardner, G. H. F., and A. Canning, 1998, AVA analysis after velocity-independent DMO and imaging: Geophysics, **63**, no. 2, 686–691, https://doi.org/10.1190/1.1444368.

Gassmann, F., 1951, Elastic waves through a packing of spheres: Geophysics, **16**, no. 4, 673–685, https://doi.org/10.1190/1.1437718.

Gik, L. D., 1997, Physical modeling of seismic wave propagation in porous and fractured media: Geology and Geophysics, **38**, 804–815 (in Russian).

Gogonenkov, G. N., and L. A. Nekrasova, 2012, Prediction of higher productivity zones with 3D seismics in the Bazhenov Formation of West Siberia: The Russian Geophysics Journal, Special Issue (in Russian).

Goldin, S. V., 2005, Introduction to geometric seismics: Novosibirsk State University (in Russian).

Golubev, V. I., I. B. Petrov, and N. I. Khokhlov, 2013, Numerical simulation of seismic activity by the grid-characteristic method: Computational Mathematics and Mathematical Physics, **53**, no. 10, 1523–1533 (in Russian and English).

Gornyak, Z. V., A. S. Kostyukevych, B. Link, N. Y. Marmalevskyi, V. V. Mershchiy, Y. V. Roganov, and I. Y. Khromova, 2008, Study of vertical inhomogeneities using duplex wave migration: Seismic Technologies, **5**, no. 1, 3–14 (in Russian).

Grechka, V., and M. Kachanov, 2006, Effective elasticity of fractured rocks. A snapshot of the work in progress: Geophysics, **71**, no. 6, W45–W58, https://doi.org/10.1190/1.2360212.

Groenenboom, J., and J. T. Fokkema, 1998, Guided waves along hydraulic fractures: 68th Annual Meeting, SEG, Expanded Abstracts, 1632–1635, https://doi.org/10.1190/1.1820234.

Hall, F., and Y. Wang, 2012, Seismic response of fractures by numerical simulation: Geophysical Journal International, **189**, no. 1, 591–601, https://doi.org/10.1111/j.1365-246X.2012.05360.x.

Haskell, N. A., 1953, The dispersion of waves in multilayered media: Bulletin of the Seismological Society of America, **43**, no. 1, 17–34.

Haugen, G. U., and M. A. Schoenberg, 2000, The echo of a fault or fracture: Geophysics, **65**, no. 1, 176–189, https://doi.org/10.1190/1.1444708.

Helbig, K. and L. Thomsen, 2005, 75-plus years of anisotropy in exploration and reservoir seismics: A historical review of concepts and methods: Geophysics, **70**, no. 6, 9ND–23ND, https://doi.org/10.1190/1.2122407.

Hesthaven, J. S., and T. Warburton, 2008, Nodal discontinuous Galerkin methods: Algorithms, analysis, and applications. Texts in applied mathematics: Springer Verlag.

Hofmann, A. P., A. Price, G. Kaffenberger, R. Godderij, and M. Simpson, 2002, Hanze chalk oil field – the chalk pearl in the Dutch North Sea: 64th Annual Meeting, EAGE, Expanded Abstracts, 211.

Hsu, C. J., and M. Schoenberg, 1993, Elastic waves through a simulated fractured medium: Geophysics, **58**, no. 7, 964–977, https://doi.org/10.1190/1.1443487.

Hudson, J. A., 1991, Crack distributions which account for a given seismic anisotropy: Geophysical Journal International, **104**, no. 3, 517–521, https://doi.org/10.1111/j.1365-246X.1991.tb05698.x.

Hudson, J. A., T. Pointer, and E. Liu, 2001, Effective-medium theories for fluid-saturated materials with aligned cracks: Geophysical Prospecting, **49**, no. 5, 509–522, https://doi.org/10.1046/j.1365-2478.2001.00272.x.

Hughes, T. J. R., 1987, The finite element method: Linear static and dynamic finite element analysis: Dover Publications.

Ivanov, V. D., I. B. Petrov, A. G. Tormasov, and A. S. Kholodov, 1999, Grid-characteristic method to compute deformations on regular grids: Mathematical Modelling, **11**, no. 7, 118–127.

Karaev, N. A., E. A. Kozlov, G. N. Karaev, Y. P. Lukashin, O. M. Prokator, and V. M. Semenov, 2008, Physical modeling of fractured-porous objects: Seismic Technologies, **5**, no. 3, 81–88 (in Russian).

Karaev, N. A., V. B. Leviant, I. B. Petrov, G. N. Karaev, and M. V. Muratov, 2015, Detection and characterization of fracture systems from P-to-S scattering. Potentiality checks by physical modeling and simulations: Seismic Technologies, **12**, no. 1, 1–11 (in Russian).

Karaev, N. A., Y. P. Lukashin, G. N. Karaev, and O. M. Prokator, 2010, Physical modeling of fractured-porous objects with an ability to test Seismic Technologies: Seismic Technologies, **7**, no. 3, 96–104 (in Russian).

Karaev, N. A., Y. P. Lukashin, O. M. Prokator, and V. M. Semenov, 2008, Physical modeling of fractured media: Seismic Technologies, **5**, no. 2, 64–73 (in Russian).

Khromova, I. Y., 2008, Duplex wave migration as a method for mapping fracture zones of tectonic genesis: Oil and Gas Geology, **3**, 37–47 (in Russian).

Khromova, I., B. Link, and N. Marmelevskyi, 2011, Comparison of seismic-based methods for fracture permeability prediction: First Break, **29**, no. 1766, 37–44, https://doi.org/10.3997/1365-2397.2011001.

Kondaurov, V. I., and V. E. Fortov, 2002, Fundamentals of thermomechanics of condensed media: Moscow Institute of Physics and Technology (in Russian).

Korneev, V. 2008, Slow waves in fractures filled with viscous fluid: Geophysics, **73**, no. 1, N1–N7, https://doi.org/10.1190/1.2802174.

Korneev, V., 2011, Normal reflection from a vertical stack of fluid-filled fractures: 81st Annual Meeting, SEG, Expanded Abstracts, 2966–2970, https://doi.org/10.1190/1.3627811.

Kosloff, D. D., and E. Baysal, 1982, Forward modeling by a Fourier method: Geophysics, **47**, no. 10, 1402–1412, https://doi.org/10.1190/1.1441288.

Kostin, V. I., V. V. Lisitsa, G. V. Reshetova, and V. A. Tcheverda, 2011, A finite-difference method for the numerical simulation of seismic wave propagation through multiscale media: Numerical Methods and Programming, **12**, 321–329 (in Russian).

Kostyukevych, A., N. Marmalevskyi, Y. Roganov, and V. Tulchinsky, 2008, Anisotropic 2.5D-3C finite-difference modeling: 70th Conference, EAGE, Extended Abstracts, P043.

Kozlov, E. A., 2006, Media models in exploration seismology: GERS (in Russian).

Kozlov, E. A., N. L. Baranskii, V. F. Sementsov, and N. A. Aksenov, 2004, Separate imaging of specular and scattering geological objects using 3D seismic data: Seismic Technologies, **2**, no. 2, 4–16 (in Russian).

Kremlev, A. N., G. N. Erokhin, L. E. Starikov, and M. A. Zverev, 2008, Prediction of fractured and cavernous reservoirs using scattered seismic waves: Seismic Technologies, **5**, no. 3, 36–39 (in Russian).

Kukudzhanov, V. N., 2004, Decomposition method for elastoplastic equations: Izvestiya of the Russian Academy of Sciences: Mechanics of Solids, **39**, 73–80.

Kulikovskiy, A. G., N. V. Pogorelov, and A. Y. Semenov, 2001, Mathematical problems of the numerical solution of hyperbolic equations: Fizmatlit (in Russian).

Kunin, K. N., E. G. Kartseva, D. E. Kashcheev, A. V. Savelyev, N. V. Zolotoy, A. M. Blank, and N. V. Fedina, 2013, Prediction of the productivity of Bazhenov sediments: 3rd Conference, EurAsian Geophysical Society, Expanded Abstracts (in Russian).

Kuznetsov, O. L., Y. A. Kuryanov, I. A. Chirkin, and S. I. Shlenkin, 2004, Side-viewing seismic locator: The Russian Geophysics Journal, Special Issue, 17–22 (in Russian).

Landa, E. I., 2013, Role of wave field diffraction component in seismic imaging: Seismic Technologies, **10**, no. 1, 1–48 (in Russian).

Leviant, V. B., I. B. Petrov, and F. B. Chelnokov, 2005, Nature of the scattered seismic response of zones of diffuse cavities and fractures in massive rocks: The Russian Geophysics Journal, **6**, 5–19 (in Russian).

Leviant, V. B., I. B. Petrov, and I. Kvasov, 2011, Numerical simulation of the wave response from subvertical macrofractures as probable fluid-conducting channels: Seismic Technologies, **8**, no. 4, 41–61 (in Russian).

Leviant, V. B., I. B. Petrov, and M. V. Muratov, 2012, Numerical simulation of wave responses from subvertical macrofractures system: Seismic Technologies, **9**, no. 1, 1–20 (in Russian).

Leviant, V. B., I. B. Petrov, and S. A. Pankratov, 2009, Study of characteristics of compressional and converted waves of the backscattered response from fractured reservoir zones: Seismic Technologies, **6**, no. 2, 3–11 (in Russian).

Leviant, V. B., I. B. Petrov, M. V. Muratov, and S. A. Bykov, 2013, Stability of P-to-S scattering off a system of fractures: Seismic Technologies, **10**, no. 1, 1–25 (in Russian).

Leviant, V. B., I. B. Petrov, V. I. Golubev, and M. V. Muratov, 2014, Numerical 3D modeling of seismic responses from large vertical macrofracture systems: Seismic Technologies, **11**, no. 2, 5–23 (in Russian).

Leviant, V. B., I. Y. Khromova, E. A. Kozlov, I. N. Kerusov, D. E. Kashcheev, V. V. Kolesov, and N. Y. Marmalevskyi, 2010, Methodology recommendations for using seismic data to estimate reserves of hydrocarbons in carbonate rocks with porosity of fracture-cavernous type: Central Geophysical Expedition (in Russian).

Leviant, V. B., V. A. Miryakha, M. V. Muratov, and I. B. Petrov, 2015, Seismic responses of vertical fractures depending on their thickness: Seismic Technologies, **12**, no. 3, 16–30 (in Russian).

Lisitsa, V. V., G. V., Reshetova, V. A. Pozdnyakov, and V. A. Cheverda, 2012, Numerical modeling of interaction between seismic wavefields and cavernous-fractured reservoirs and its practical application, in Supercomputer technologies in science, education and industry: Moscow State University Press.

Lisitsa, V. V., V. A. Pozdnyakov, G. V. Reshetova, V. G. Khaidukov, V. A. Cheverda, and V. V. Shilikov, 2013, Scattered seismic responses: simulation and imaging. Part 1. Two-dimensional media: Seismic Technologies, **10**, no. 1, 1–26 (in Russian).

Liu, E., and A. Martinez, 2012, Seismic fracture characterization – Concepts and practical applications: EAGE.

MacBeth, C., 1995, How can anisotropy be used for reservoir characterization?: First Break, **13**, no. 1, 31–37.

Magomedov, K. M., and A. S. Kholodov, 1988, Grid-characteristic numerical methods: Nauka (in Russian).

Magomedov, K. M., and A. S. Kholodov, 2017, Grid-characteristic numerical methods: 2nd revised and augmented edition, Universities of Russia series, LiTRes online tutorial for BS and MS seakers, authors A.S. Kholodov and K.M. Magomedov (in Russian).

Marmalevskyi, N., B. H. Link, Y. Roganov, A. Kostyukevych, and Z. Gornyak, 2007, Duplex wave migration – a practical and effective tool for imaging vertical boundaries: 69th Annual Meeting, EAGE, Expanded Abstracts, 283.

Miryakha, V. A., A. V. Sannikov, and I. B. Petrov, 2015, Discontinuous Galerkin method for numerical simulation of dynamic processes in solids: Mathematical Modelling, **27**, no. 3, 96–108 (in Russian).

Nakagawa, S., K. T. Nihei, and L. R. Myer, 2002, Numerical simulation of 3D elastic wave scattering off a layer containing parallel periodic fractures: 82nd Annual Meeting, SEG, Expanded Abstracts, 1967–1970, https://doi.org/10.1190/1.1817080.

Nekrasov, A. S., 2006, Geological and geophysical studies of carbonate reservoirs of oil fields: Perm State University (in Russian).

Newmark, N. M., 1959, A method of computation for structural dynamics: Journal of the Engineering Mechanics Division, **85**, 67–94.

Petrov, I. B., A. V. Favorskaya, A. V. Sannikov, and I. Kvasov, 2013, Grid-characteristic method using high-order interpolation on tetrahedral hierarchical meshes with a multiple time step: Mathematical Modelling, **25**, no. 2, 42–52 (in Russian).

Petrov, I. B., and A. S. Kholodov, 1984, Numerical investigation of certain dynamical problems of the mechanics of a deformable solid body by the grid-characteristic method: U.S.S.R. Computational Mathematics and Mathematical Physics, **24**, 722–739, https://doi.org/10.1016/0041-5553(84)90044-2 (in Russian and English).

Petrov, I. B., A. G. Tormasov, and A. S. Kholodov, 1989, On numerical study of nonstationary processes in multilayered structure of deformable media: Izvestia of USSR Academy of Sciences, MTT Series, no. 4, 89–95 (in Russian).

Petrov, I. B., A. G. Tormasov, and A. S. Kholodov, 1990, On using hybrated grid-characteristic schemes for numerical solution of 3D tasks for deformable solids: Computational Mathematics and Mathematical Physics, **30**, 1237–1244 (in Russian and English).

Pozdniakov, V. A., 2005, Strength of scattered waves: A new seismic attribute for predicting filtration and capacity properties of oil-saturated reservoir: Proceedings of Russian Academy of Sciences: Geophysics Series, **404**, 34–30 (in Russian).

Pozdnyakov, V. A., D. V. Safonov, and V. V. Shilikov, 2009, Prediction of distribution of fracture zones using 3D seismic data within the Yurubcheno-Tokhomskaya zone: Seismic Technologies, **6**, no. 1, 85–90 (in Russian).

Priolo, E., J. M. Carcione, and G. Seriani, 1994, Numerical simulation of interface waves by high-order spectral modeling techniques: The Journal of the Acoustical Society of America, **95**, no. 2, 681–693, https://doi.org/10.1121/1.408428.

Raymer, L. L., E. R. Hunt, and J. S. Gardner, 1980, An improved sonic transit time-to-porosity transform, Paper P, 1–12: Transactions of the 21st SPWLA Annual Logging Symposium.

Revenko, A. V., M. A. Bordyug, and V. D. Nemova, 2013, Seismic modeling of changes in the Bazhenov Formation sediments based on the results of core analysis: 3rd Conference, EurAsian Geophysical Society, Extended Abstracts (in Russian).

Roganov, Y. V., and V. Y. Roganov, 2016, Propagation of waves in periodic fluid-solid layered media: Geophysical Journal (Ukraine), **38**, no. 6, 101–117 (in Russian).

Roganov, V. Y., and Y. V. Roganov, 2017, Low-frequency scattering in a half-space filled by periodical fluid-solid medium with inclined layers: Geophysical Journal (Ukraine), **39**, no. 4, 55–74 (in Russian).

Rüger, A., 1998, Variation of P-wave reflectivity with offset and azimuth in anisotropic media: Geophysics, **63**, no. 3, 935–947, https://doi.org/10.1190/1.1444405.

Saenger, E., and S. Shapiro, 2002, Effective velocities in fractured media: a numerical study using the rotated staggered finite-difference grid: Geophysical Prospecting, **50**, no. 2, 183–194, https://doi.org/10.1046/j.1365-2478.2002.00309.x.

Saenger, E. H., O. S. Krüger, and S. A. Shapiro, 2004, Effective elastic properties of randomly fractured soils. 3D numerical experiments: Geophysical Prospecting, **52**, no. 3, 183–195, https://doi.org/10.1111/j.1365-2478.2004.00407.x.

Sayers, C. M., 2009, Seismic characterization of reservoirs containing multiple fracture sets: Geophysical Prospecting, **57**, no. 2, 187–192, https://doi.org/10.1111/j.1365-2478.2008.00766.x.

Schoenberg, M., 1980, Elastic wave behavior across linear slip interfaces: The Journal of the Acoustical Society of America, **68**, no. 5, 1516–1521, https://doi.org/10.1121/1.385077.

Schoenberg, M., 1984, Wave propagation in alternating solid and fluid layers: Wave Motion, **6**, no. 3, 303–320, https://doi.org/10.1016/0165-2125(84)90033-7.

Seriani, G., and E. Priolo, 1994, Spectral element method for acoustic wave simulation in heterogeneous media: Finite Elements in Analysis and Design, **16**, no. 3-4, 337–348, https://doi.org/10.1016/0168-874X(94)90076-0.

Shevchenko, A. A., 2007, Seismic surveys in wells: Moscow State University (in Russian).

Shlenkin, S. I., G. V. Kashirin, A. V. Masyukov, and V. V. Kharakhinov, 2000, New seismic technologies to study complex oil and gas reservoirs: Proceedings of the international conference dedicated to the 300th anniversary of the mining and geological service of Russia, 493–494 (in Russian).

Sim, L. A., and A. V. Mikhailov, 2008, Faults in the sedimentary cover of platforms and methods for their study: Problems in Tectonophysics, 141–148, Earth Physics Institute (in Russian).

Thomsen, L., 1986, Weak elastic anisotropy: Geophysics, **51**, no. 10, 1954–1966, https://doi.org/10.1190/1.1442051.

Thomson, W. T., 1950, Transmission of classic waves through a stratified solid material: Journal of Applied Physics, **21**, no. 2, 89–93, https://doi.org/10.1063/1.1699629.

Timurziev, A. I., 2010, Development of Sylvester's 'flower models' concept using new kinematic model of tectonic faults: The Russian Geophysics Journal, **2**, 24–31 (in Russian).

Toro, E. F., 1999, Solvers and numerical methods for fluid dynamics: A practical introduction, 2nd ed.: Springer Verlag.

Ursenbach, C. P., 2002, Generalized Gardner relations: 72nd Annual International Meeting, SEG, Expanded Abstracts, 1885–1888, https://doi.org/10.1190/1.1817057.

Usmanov, I. Sh., E. N. Trofimova, A. V. Dyakina, and A. V. Karlov, 2006, Oil-saturated limestones of the Bazhenov Formation. The ways of realization of the oil and gas potential of Khanty-Mansi Autonomous Okrug: Conference Proceedings (in Russian).

Vershinin, A. V., and D. I. Sabitov, 2011, Spectral element modeling of acoustic logging in anisotropic viscoelastic media: 2nd International Scientific and Paractical Conference – Modern Seismic Methods for Oil and Gas Prospecting in Complex Structures (Seismo-2011).

Vlastos, S., E. Liu, I. G. Main, and X.-Y. Li, 2003, Numerical simulation of wave propagation in media with discrete distributions of fractures: effects of fracture sizes and spatial distributions: Geophysical Journal International, **152**, no. 3, 649–668, https://doi.org/10.1046/j.1365-246X.2003.01876.x.

Willis, M., R. Rao, D. Burns, J. Byun, and L. Vetri, 2004, Spatial orientation and distribution of reservoir fractures from scattered seismic energy: 74th Annual Meeting, SEG, Expanded Abstracts, 1535–1538, https://doi.org/10.1190/1.1851133.

Willis, M. E., D. R. Burns, R. Rao, B. Minsley, M. N. Toksöz, and L. Vetri, 2006, Spatial orientation and distribution of reservoir fractures from scattered seismic energy: Geophysics, **71**, no. 5, O43–O51, https://doi.org/10.1190/1.2235977.

Wu, C., J. M. Harris, K. T. Nihei, and S. Nakagawa, 2005, Two-dimensional finite-difference seismic modeling of an open fluid-filled fracture: Comparison of thin-layer and linear-slip models: Geophysics, **70**, no. 4, T57–T62, https://doi.org/10.1190/1.1988187.

Xu, Y., M. Chapman, X.-Y. Li, and I. G. Main, 2010, Effects of fracture spacing on seismic wave propagation – A 3D numerical simulation study on discrete fracture models: 72nd Annual Meeting, EAGE, Expanded Abstracts, P108.

Zheng, Y., X. Fang, M. Fehler, and D. Burns, 2011, Double-beam stacking to infer seismic properties of fractured reservoirs: 81st Annual Meeting, SEG, Expanded Abstracts, 1809–1813, https://doi.org/10.1190/1.3627557.

Zienkiewicz, O. C., and R. L. Taylor, 2000, The finite element method, Fifth edition, Volume 1, The Basis: Butterworth-Heinemann.

Index

A

B

C

D

E

F

G

H

I

K

L

M

N

P

Q

R

S

T

U

W

Z